Premium Calculation in Insurance

NATO ASI Series

Advanced Science Institutes Series

A series presenting the results of activities sponsored by the NATO Science Committee, which aims at the dissemination of advanced scientific and technological knowledge, with a view to strengthening links between scientific communities.

The series is published by an international board of publishers in conjunction with the NATO Scientific Affairs Division

A	Life Sciences	Plenum Publishing Corporation
B	Physics	London and New York
C	Mathematical and Physical Sciences	D. Reidel Publishing Company Dordrecht, Boston and Lancaster
D	Behavioural and Social Sciences	Martinus Nijhoff Publishers
E	Engineering and Materials Sciences	The Hague, Boston and Lancaster
F	Computer and Systems Sciences	Springer-Verlag
G	Ecological Sciences	Berlin, Heidelberg, New York and Tokyo

Series C: Mathematical and Physical Sciences Vol. 121

Premium Calculation in Insurance

edited by

F. de Vylder

and

M. Goovaerts

Catholic University of Louvain, Belgium

and

J. Haezendonck

University of Antwerp, Belgium

D. Reidel Publishing Company

Dordrecht / Boston / Lancaster

Published in cooperation with NATO Scientific Affairs Division

Proceedings of the NATO Advanced Study Institute on
Insurance Premiums
Louvain, Belgium
July 18-31, 1983

Library of Congress Cataloging in Publication Data

NATO Advanced Study Institute on Insurance Premiums
(1983: Louvain, Belgium)
Premium calculation in insurance.

(NATO ASI series. Series C, Mathematical and physical sciences; vol. 121)
"Published in cooperation with NATO Scientific Affairs Division."
Includes index.
1. Insurance-Rates and tables. 2. Insurance-Mathematics. I. Vylder, Florent de. II. Goovaerts, M. J. III. Haezendonck, J., 1940- IV. North Atlantic Treaty Organization. Scientific Affairs Division. V. Title. VI. Series: NATO ASI series.
Series C, Mathematical and physical sciences; Vol. 121.
HG8065.N37 1983 368'.001 84-2011
ISBN-13: 978-94-009-6356-6 e-ISBN-13: 978-94-009-6354-2
DOI: 10.1007/978-94-009-6354-2

Published by D. Reidel Publishing Company
P.O. Box 17, 3300 AA Dordrecht, Holland

Sold and distributed in the U.S.A. and Canada
by Kluwer Academic Publishers,
190 Old Derby Street, Hingham, MA 02043, U.S.A.

In all other countries, sold and distributed
by Kluwer Academic Publishers Group,
P.O. Box 322, 3300 AH Dordrecht, Holland

D. Reidel Publishing Company is a member of the Kluwer Academic Publishers Group

TABLE OF CONTENTS

Scientific committee

J. Beekman (U.S.A.), P. Boyle (Canada), R. De Groot (Belgium), F. Delbaen (Belgium), G.W. de Wit (The Netherlands), H. Gerber (Switzerland), W.R. Heilman (F.R. Germany), J. Janssen (Belgium) W.S. Jewell (U.S.A.), S. Klugman (U.S.A.) E. Kremer (F.R. Germany), C. Angela-Mormino (Italy), A.F. Shapiro (U.S.A.), J. Teugels (Belgium)

Co-sponsors

B.V.V.O.
(Professional association of Belgian Insurance Companies)
N.F.W.O.
(Belgian National Science Foundation)
I.A.W.
(Institute of Actuarial Sciences of Leuven)
Ministerie van Nationale Opvoeding

SCIENTIFIC CONTENTS OF THE COURSE

The Directors

As explained in Gerber's text, a premium calculation principle is a functional that assigns a usually loaded premium to any distribution of claims (see Gerber, An Introduction to Mathematical Risk Theory, (eds.), Huebner Foundation Monograph, (1979), pp. 1-164).
Since the last decade a lot of research, both from the practical as well as from the theoretical point of view has been carried out in the framework of the calculation of insurance and reinsurance premiums. As is often the case in other brances of sciences a new field of applications of mathematics arises. In this new field of applied mathematics a distinction has to be made between premiums calculated according to some premium calculation principles and statistical rules derived in the framework of experience rating.
A principle of premium calculation is a rule, say κ, that assigns a premium π, a real number, to any given risk S, a random variable, given by its distribution function F_S. A review of these premium calculation principles as well as a collection of desirable properties (implied by practical considerations) are given. Implementation of these properties makes it possible to derive some new characterizations of the most important premium principles, a unifying theory is developed. It is also shown how the practical insurance business has made application of these strong results. The systematic study of it will necessarily lead us to the rigorous treatment of reinsurance premiums. Another fundamental subject, both from the theoretical as well as from the practical point of view, which has to be examined in detail is the ordering among risks.

F. de Vylder et al. (eds.), Premium Calculation in Insurance, ix–x.

As far as experience rating is concerned we will limit ourselves to the development of modern credibility theory. This is a topic very well known among actuaries. The last decade it has also drawn the attention of pure statisticians. In the lectures emphasis lies on practical models of credibility theory, i.e. those already currently used in some countries and those, most promising for future applications. The models are explained in full detail, including the statistical problem of structural parameters estimation. For completeness the more academic models are briefly reviewed.

The current state of the art, both of premium principles, as well as experience rating is reported on.
Moreover there is a connection between the subject dealt with in the summer school and some important topics of the mathematics and economics of insurance theory, such as ruin theory, bonus-malus systems, risk exchange, pooling etc.. The relationship with these subjects is also be reported on.

ACKNOWLEDGEMENT

We gratefully acknowledge the award of a grant from the Scientific Affairs Division of NATO to run the Institute, especially the assistance of Dr. C. Sinclair during the tenure of the award, and also the cooperation of the University of Leuven.

OPENING SESSION

INVITED ADDRESS

Prof. Dr. R. VANDEPUTTE
Former Minister of Finance
Chairman of the "Commissie van Verzekeringen - Commission des assurances"

As we know, insurance supposes a grouping of persons or goods, all subject to one and the same sort of risks. The insurer collects a premium from each member of the group and pays the amount covered to the one on whom the risk falls. He takes care of the equilibrium between the accepted contracts and the premiums received.

Insurance is a recent phenomenon. Its application started in England at the end of the 18th century. Expecially during the 20th century it has experienced a large expansion. In many cases insurance companies have become large and powerful financial institutions which, either based on statistical data or in a purely pragmatic way, still take on more new risks with adapted insurance policies. Large insurance companies have dozens of different sorts of contracts but nevertheless it is evident that some products or insurance lines have enlarged more than others.

Insurance is from a social viewpoint a very useful, or even more, a necessary institution. We can hardly imagine our contemporary society without the covering of fire dangers, the covering of a number of professional liabilities, the covering of car accidents, and the advantages of life insurance or group insurance.
Due to the importance of insurance for the population, it seems to be normal that there is a certain control by the government.

The welfare of the country is highly connected with the solvency of the insurers and with the contents of the contracts. Still the intervention of public administration in the activities of insurance companies seems to be different from country to country.

F. de Vylder et al. (eds.), Premium Calculation in Insurance, 3–4.

This diversity finds explanation in the historical evolution of insurance that shows divergent aspects, but especially as a result of the prevailing consenus concerning the relationship between government and private affairs. While in some countries the ideas of a liberal economy are advocated, in other places a larger task is entrusted to the government.

Because the political and economic doctrine, which is the basis of the organisation of the state, does not have the same content everywhere, there are considerable difficulties when an international standardization of insurance is pursued. The agencies of the European Economic Community have been trying for years to obtain a greater standardization. Until now, the results have not fulfilled the trouble they took. This should be recognized because it is the truth. In the field of insurance, as in many other fields, the loyalty to national tradition remains very important.

Nevertheless, international study sessions such as this one, organized here in Leuven, are most helpful. They allow comparisons between existing insurance systems, so everyone can gain from the experience of his colleagues.

I wish you success and I hope that the lecturers will have concern for simplicity and clarity. They should realise that not only is language a difficulty for some participants, but that the situation in their home-countries often does not exist in other countries. So the participants are each obliged to make an effort to understand how things are done elsewhere. Good luck !

SOME MAJOR ISSUES IN ECONOMICS AND INSURANCE DEVELOPMENTS
(Insurance Premiums: Economic Factors for Insurance Pricing Policies)

Orio GIARINI
Secretary General
The Geneva Association

This paper introduces four major issues relating to economics and insurance:

1. The Post Industrial Economic Pattern: the Importance of Insurance.
2. Technology and the Demand for Insurance.
3. Financial Aspects of Insurance.
4. The Limits to Government and the Expansion of Insurance.

1. THE POST-INDUSTRIAL ECONOMIC PATTERN: THE IMPORTANCE OF INSURANCE

It is important to consider in the first place the transformations of industrial societies, as they have developed in the last two centuries. For insurance, the economic environment varies enormously if this is a traditional industrial one or if it is a post-industrial one.

The traditional industrial society can be described as a situation in which the priviledged and by far most important way to produce wealth and welfare, is by the manufacturing process, whereby raw materials are transformed into final usable products. This process can be described by graph number 1.

F. de Vylder et al. (eds.), Premium Calculation in Insurance, 5–16.

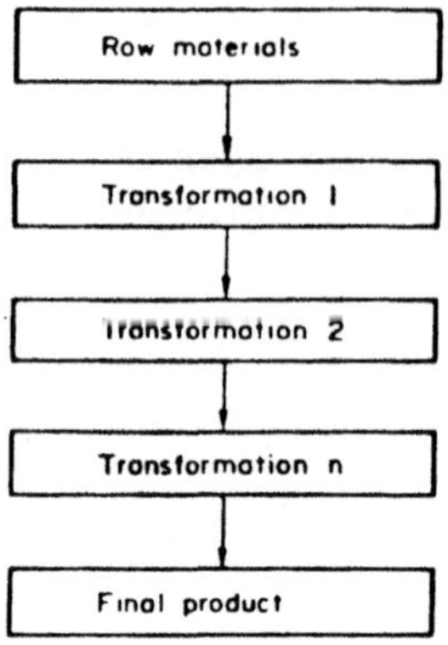

Graph 1: The Building up of the Final Product

In this situation, where the main preoccupation is to produce goods, insurance and other services, although recognized as important, are nonetheless secondary. In other words, they are not as important as production.

An effect of this situation in economics, is the law of Engel: according to this law, the market of insurance expands only when the primary needs of the consumers are satisfied. In other words, people buy insurance only when they have enough food, shelter, clothing etc.. Insurance, in this case, is something almost superfluous, or, at best, something which can only grow and develop after a certain level of wealth has been achieved.

But if we now look at what has happened to insurance premium up to 1979, these hypothesis can be considered as doubtful.

We can observe in Table 1, that insurance premium, at the world level, has grown more rapidly since 1950 than GNP. What is most revealing, is that the real growth of premium income from 1971 to 1979 has also been very high and close to the average for the whole period of thirty years.

It is clear that if the Engel's law would have worked during the period of the 70's, insurance growth should have declined at least as much as the average figure of GNP. It is often said that costs of insurance are of such a nature that when the economic situation is declining, there is always a tendency to postpone buying more insurance.

This has obviously not been the case for the period taken into consideration even if the economic crisis of the 70's had some negative impacts on the insurance development.

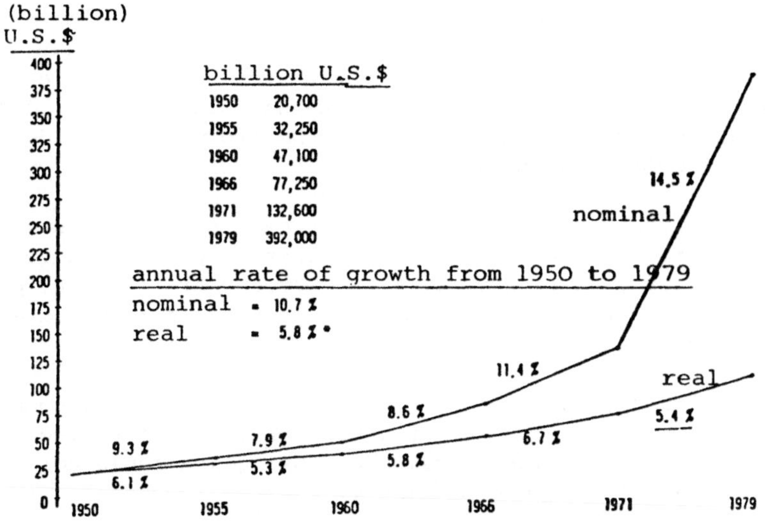

* OECD - Price Index of consumption by IMF

Source : SIGMA [1981]

Table 1: World Premium from 1950 to 1979 (direct business)

We propose to look for the reasons explaining the increasing relative importance of insurance around the world, as compared to the development of the economy, in another direction.

In our view, this is fundamentally due to the fact that the industrial structure has fundamentally changed and that the economic machinery producing wealth and welfare now is clearly post-industriel, worldwide. This means that the transformation and delivery of goods and services in the present economy, instead of being described by Graph 1, can today better be described by Graph 2.

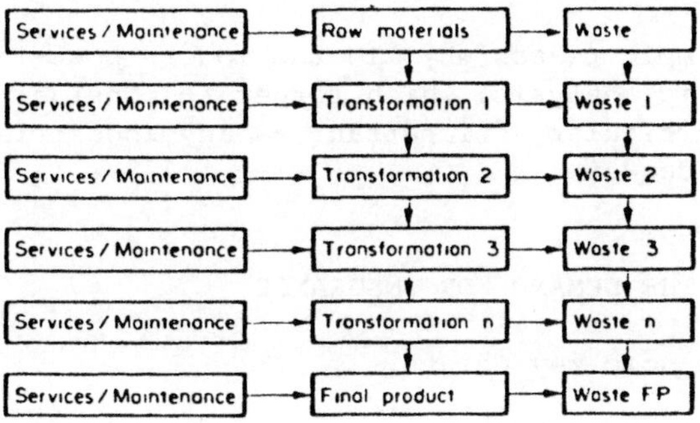

Graph 2: The Structure of the Post-Industrial Society

Graph 2 puts in evidence what now is acknowledged by most people active in any sector of the economic activity, e.g. that services of any sort (maintenance, repair, research, financial services, insurance, management services, waste handling, etc..), represent more and more the essential part of the production and delivery system of goods and services.

It is important to recognize that for each product we buy, be it an automobile or a carpet, the pure cost of production is very seldom higher than twenty to thirty percent of the final prize of these products. More than seventy percent is represented by the cost of making the complex service and delivery system work.

One can also say today, that unless services do work, no product can ever be made available. In other words, services are today as essential as production systems for making available the wealth produced by the economic machinery.

If it is always true that production is essential for developing wealth and welfare, it is now also true that production is useless, unless there is adequate service making this production usable and available. In fact, in the present situation of modern technological development, the pure production system even if it is always essential, is, in economic terms, a minor part of the whole economic system.

This also means that the post-industrial economy is a particularly complex one: the production and delivery systems are very complex and are also, often, very vulnerable. This vulnerability is the reference pattern for the need of an increasing insurance business.

We are then now in an economic situation in which it is essential, in order to have the production system working in a usable way, to get an increasing and more and more qualified insurance supply. Coping with insurable risks is a key gear in the contemporary economic system.

This is the very simple reason why, in the modern post-industrial economy, any economic analysis, which forget to consider services and insurance in particular as important as any industrial activity, is simply inadequate.

2. TECHNOLOGY AND THE DEMAND FOR INSURANCE

I shall sketch here some key points.

2.1. The Demand for Insurance in the Future: a Challenge for the Distribution and Marketing Systems

Starting from the present situation, the insurance industry seems to be pushed today in two fundamentally different directions:

a) the first one concerns insurance products which can be or will be more and more mass marketed. This does not depend only on demand characteristics (such as revenue, age structure, type of activities, etc..) but even more probably on new distribution systems, some of which already exist and some technologically new.

By the next decade, it is quite likely that a considerable amount of business will be offered and bought through:

- home computerized communication systems: these systems should with time even increase the amount of qualified specifications and transparency attached to a policy, taking into account the easy comparison with alternative offers. The payment of premium and the underwriting itself - through these means - will start at that time to be of detectable importance.

The answering of questions such as: how much, when, where, who, is very much a problem of intensive marketing and market research, which can be done with acceptable accuracy.

Of course, the same system will serve other functions, including financial, so that general, global supermarkets will be available at home. How far, again, this will integrate banking, insurance and other related services, is a question which can easily be monitored. In any case, this is primarily a distribution systems' problem: if, today, a supermarket sells wine as well as shoes, it does not mean it has integrated the manufacture of these products, or that it should economically do so.

- even before arriving at the "home available supermarket", it is probable that many insurance products today might be more and more marketed and sold through a variety of distribution channels, and with package deals including other products (again not necessarily combined at the "production" level). One important question here is: how far even the "simple" insurance policies still need a salesman who gets personally in touch with the customer. The answer can be narrowed down by appropriate market and marketing research, incorporating the many experiences in this field promoted by large distribution chains and other organisations in the United States and Europe.

Such developments raise the important and delicate question of the quantity and quality of employment in insurance in the years to come.

By the next decade, the total sales of insurance will continue to increase faster than the world GNP, but it is most likely that employment, in all the functions which are easily done in a mechanical and/or computerized way, will decrease in absolute terms. The general trend up to the seventies which consisted of increasing job opportunities in the service sector and in particular in insurance, is probably definitely being reversed. A specific economic research, to be conducted in several parts of the world simultaneously, would be needed to clarify this point in quantitative and qualitative terms and provide a specific key as to how to handle the matter.

b) In any case, an important contribution to the employment problem might come from the other large sector of insurance expansion, the one having to do with risks which demand a professional technical understanding of their nature, origin, level of vulnerability. This is the high level, high quality type of risks which demand a qualified ability, knowledge and skill to be adequately controlled and covered.

The problems involved are summarized, in economic terms, in the following Table 2:

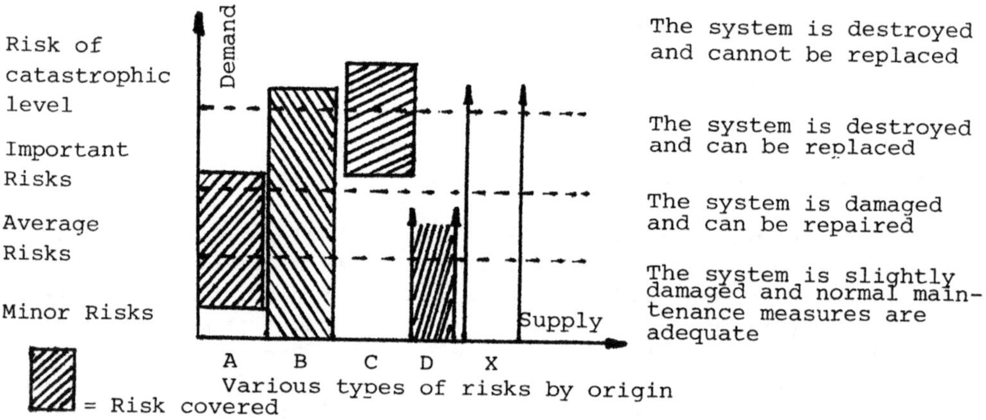

Table 2: Insurance Supply and Demand for Risk Cover

One of the major problems shown by the Table 2 is that the economic demand to cover a risk is in terms of the level of vulnerability: whatever the source or the origin of the risk and damage, the demand side is interested in adequately managing or covering its level of vulnerability.

From the insurance supply side, risks are covered by homogeneous classes, defined by homogeneous origin.

c) It is very important today to re-examine in depth what the notion of homogeneity of risk classes may mean when:

- an industrial risk might today simultaneously refer to an insurance cover, which might be offered from different risk classes. A production in a chemical plant might stop and produce consequential losses; this stop of production might be linked to the recognition of a product which is likely to produce damages when used. In this case, it falls into product liability. Furtheron, this might oblige industry to spend money to recall these products. It is obvious that in a situation in which one single accident produces opportunities of losses in so different directions, the insurance management of such a problem should strive to integrate as far as possible various risk classes (as in this case consequential losses, recall practices and product liability together). This problem has been evidenced very clearly by a recent study of the Geneva Association on "Recall Practices" 1).

- Technology means very often specialization and today, when one wants to cover for instance these risks associated with space activities, one finds that every and each space programme might have a different vulnerability configuration 2).

- A third, very important aspect, is a question of the time of reference related to various classes of risk and in particular those with catastrophic connotations; in other words, the problem that in order to reconstruct a usable universe of risks, it is more and more necessary to take into consideration as time unit not simply the one year of activity as it is normal practice in the economic system, but very often periods of five, ten, twenty years or more. This also is very often the effect of specialization and increase of technological vulnerability. This very important issue has of course a lot to do with fiscal policies. The key economic question here is: which is the correct period of time (how many years) during which an economic activity should be taken into consideration? What is the time unit of reference?

d) It is clear that one of the greatest challenges for the insurance industry today is to upgrade insurance supply, which also means considering the employment problem in this industry.

Up to now when insurable risks were not an essential tool in managing the overall risks of an enterprise, of an individual or even of society, a client could accept to adapt to the classification proposed on the supply side. Once the insurable risks become higher or more important for an industrial company, the profession of the risk managers stepped up, which essentially represents an effort to define the demand side, not in terms of one type of risk but in terms of one level of vulnerability.

This is parallel to a movement where more and more risks, in the industrial sector, enter the upper categories (important and catastrophic risks) and when it becomes even more of a problem to understand, estimate and underwrite risks related to new technologies. Moreover: once an industrialist analyses his risk control demand, in terms of protection and quality control of his product from the production line to the consumer, he becomes more and more conscious of the problem of how to integrate a fire policy, with a consequential loss, with a product liability, with a recall protection etc.. At this stage, the insurance supplier needs greater qualifications to understand the system at risk to be insured at its various vulnerability points and integrate adequately the specific ("vertical") insurance possibilities. In a purely supply oriented situation, the actuary is more than adequate to define rates within homogeneous category. Then, when real demand is defined, the engineer and the economist have to step in.

The key issue here is not mass distribution, but to upgrade the quality of the insurance supply by adequately integrating at least the engineering and the economist type of analysis. In terms of employment, this means that by upgrading its formation in the above directions, insurance might help to counterattack the trend in decreasing global employment, and at the same time improve the rating capacity and expand its markets. In the new technologocal sectors, some of the studies of the "Geneva Association" have shown that in this way many more risks could be underwritten (and therefore use financial capacity) and at the same time avoid accepting underrated risks due to lack of adequate technological judgement 3).

3. FINANCIAL ASPECTS OF INSURANCE

3.1. Interest Rates and "Cash Flow Underwriting"

High interest rates, essentially high real interest rates, are the results of the supply and demand interplay for money. Once Mozart was told by the Austrian Emperor, Joseph II, after the performance of one of his operas "There are too many notes in your composition". "Not one more nor less than needed", answered Mozart. So it is with the level of interest rates: they are as high or as low as the supply and demand for money let them be. There are many factors influencing this supply and demand, in particular the role played by the dollar at the international level and the impact of the American economy on it, but we do not have the time to go into this problem.

A fundamental hypothesis is that the "normal" economic real rate of interest should tend to something in the vicinity of zero 4). If now, to take an example, the U.S. budget deficit this year goes beyond 200 billion dollars and if all this money competes for funds in the market, interest rates might shoot upwards again. And there are many other occasions in the world to increase this pressure. This will not happen if the market for money is preserved from pressures of this size.

In the long run, in any case, interest rates should smooth out and my guess is that up to 1990 they will not be more than one to two points, in real net terms, from the inflation rate. A prerequisite for such stability is of course the avoidance of a major world financial crisis, which we still prefer to place among the great unforeseeable events.

For instance, this means, in my view, that this business has first to maintain or even improve its specific management and financial praxis, paying careful attention to risks and underwriting results first.

3.2. The Limits of Portfolio Economic Theory and of Cash Flow Underwriting

There are many economists these days, who have specialized in finance and are interested in insurance, who have the tendency to overstress the identification of the insurance industry with all other financial sectors and in particular with banking. Extending the "portfolio theory" for instance, they argue that the only difference between a bank and an insurance is that, whereas a deposit in a bank is a kind of claim to withdraw on demand, the premium paid to insurance produces a conditional claim. The word "conditional" means the occurence of the event for which a policy has been underwritten and the premium paid.

In my view this definition is very partial and, if taken as a basis for an economic analysis of insurance is essentially misleading for a variety of reasons:

- first of all, a large part of the insurance business, when it pays its claims, is largely conditioned not by the nominal value at the moment the premium is paid, but by the real cost of the claim. On other words, insurance is often indexed - and in some cases even over-indexed - to inflation by the nature of the business itself. In periods of relatively high interest rates this situation might be overlooked and creates a false impression of wealth and profit.

- after all the "portfolio theory" could be extended to any economic activity, by the mere fact that every and each economic actor tries to make money. This would mean - ad absurdum - that an industrial company too, would have to optimize their industrial investments against other more profitable financial activities outside its normal field of activity. It is not because all the economic activities are expressed in money, that they are all strictly financial. They are also financial.

In our view, even if the financial weight of insurance companies is probably greater than those of an industry, there is a danger that the "portfolio theory" approach introduces more confusion than understanding of the specific functions of each economic activity.

- at first sight then, "cash flow underwriting" and "portfolio theory" seem to go hand in hand, but even admitting the importance of increasing the financial management ability of insurance, the additional danger is to submit insurance (and for that, life insurance in particular) to the logic of a world financial market and operations where the long term has been abolished for the benefit of the short and very short term.

So, if "portfolio theory" at the theoretical level and "cash flow underwriting" at the practical level, are used as a technique or a situation offering the opportunity to better follow and even to make some profit from the economic environment, there is nothing to object to. But the essence of insurance is elsewhere; it is in the ability to identify, define, cover and pay for insurable risks: and it is this qualification which, whatever the world situation, will be the core of the insurance industry way through the year 2000. It is in this direction that professionalism has to be developed and if "cash flow underwriting" is taken as a strategic basis for insurance, it might also have negative effects on the priorities to be given to the underwriting professionalism.

4. LIMITS TO GOVERNMENT AND THE EXPANSION OF INSURANCE

a) If economic growth at a real rate of 6 % per year is not possible for a sustained period, states and governments, whatever their political ideology, provided they remain democratic, will need to discharge more and more of their economic tasks to the privately organised sector.

The states will have to take care that a minimum level of social justice be maintained, either by direct intervention through specific social schemes or by fiscal means such as negative income tax, or rather a blend of all this. But precisely in order to be able to do this, they will have to step out from all the layers of activity where, the excess of security the state has tried to provide in economic terms has become more and more the source of general economic insecurity and instability.

In other words, a new kind of social contract between the state and private organisations will develop, with much more space for the private side to develop life/pension schemes, health protection schemes etc.. It is here, at the interface between state and insurance, that some of the key developments will take place and they will extend in many sectors, including foreign trade. A decisive factor in this evolution will be the capacity to analyse these problems in objective, verifiable ways so as not to let the political demagogy block all possibilities and make things worse for everybody. The conflicting political attitudes can probably be contained in an overall, dispassionate analysis of how to better organise in society the security schemes of tomorrow. A trend which has already started.

b) Although the average economic growth, particularly in pro capita terms, will not be, in the world, of a great order of magnitude, there is still enough to do economically in a world of about 6 billion people at the end of the century.

The fact, that the North American countries went from 64 % of total world premium in 1970 to 46 % in 1980 announces a process of re-equilibration at the world level in terms of insurance activity, which has registered the important developments of Japan and Europe. The rest of the world is bound to follow this trend in a general, even if only moderate and long movement of world redistribution. Limits of governments, worldwide diffusion at various levels of industrialisation, organisation of security do provide insurance a future "with a future". Reinsurers will of course be a key element in this global world.

References:

1) See the "Geneva Papers on Risk and Insurance" Nr. 22, January 1982.

2) This problem is being studied by EUROSAT in Geneva, for a research sponsored by the Geneva Association.

3) Nature and Importance of Economic Losses due to the Utilisation of Computerized Systems in Europe in 1988, by André George (Diebold S.A., Paris), "Geneva Papers on Risk and Insurance" Nr. 3, October 1976.

Nature and Importance of Economic Losses due to the Utilisation of Containers in the Transportation Systems, by Jacques P. Pezier, Donald C. Dreswell and Stanley J. Davenport (SRI - Stanford Research Institute), "Geneva Papers on Risk and Insurance" Nr. 13, October 1979.

Risk and Indirect Losses due to the Utilisation of Computerized Systems, by André George (Pactel SA, Paris), "Geneva Papers on Risk and Insurance" Nr. 13, October 1979 (updated Geneva Papers Nr. 3).

4) The reasons are explained in "The Diminishing Returns of Technology", Pergamon Press, Oxford, 1979.

MAIN CONTRIBUTIONS

RISK CONVOLUTION CALCULATIONS

John A. Beekman

Ball State University, Muncie, Indiana 47306, U.S.A.

Abstract. Convolutions of probability distributions have long been used in collective risk theory to determine the distribution of aggregate claims. Two examples of such a distribution are given. The paper then presents results for the distribution of aggregate claims for the family of claim distributions in which claim values are equi-spaced, and equi-probable. The distribution of claims may be either the Poisson or negative binomial law. Examples and tables are included. A convolution type series for the infinite time ruin function is examined, including approximations and their error analyses.

I. INTRODUCTION

Assume that $X_1, X_2, \ldots$ are independent, identically distributed random variables with common distribution $P(x) = P[X \leq x]$. Assume that $\{N(t), t \geq 0\}$ is a non-negative integer valued stochastic process, independent of $\{X_i\}$. For each $t \geq 0$, let $S(t) = X_1 + X_2 + \ldots + X_{N(t)}$. The distribution of aggregate claims is $P[S(t) \leq x]$, commonly denoted by $F(x,t)$. A long-standing result about $F(x,t)$ is contained in page 20 of Cramér (5).

Theorem. If $E(X) < \infty$, $F(x,t) = \sum_{n=0}^{\infty} P\{N(t) = n\} P^{n*}(x)$ where

$$P^{0*}(x) = \begin{cases} 0, & x < 0 \\ 1, & x \geq 0 \end{cases} \qquad P^{1*}(x) = P(x)$$

$$P^{n*}(x) = \int_{-\infty}^{\infty} P^{(n-1)*}(x-t)\,dP(t), \ n > 1.$$

F. de Vylder et al. (eds.), Premium Calculation in Insurance, 19–30.

The following two examples occur in pages 42 and 43 of Beekman (3).

Example 1. Assume that N(t) has a Poisson distribution, i.e.

$$P[N(t) = n] = \frac{e^{-t}t^n}{n!}, \; n = 0, 1, 2, \ldots$$

and that all of the claims are for one monetary unit, i.e.,

$$P(x) = \begin{cases} 0, & x < 1 \\ 1, & x \geq 1. \end{cases}$$

In this case, the convolutions are easy to calculate.

$$P^{2*}(x) = \int_0^\infty P^{1*}(x-t)dP(t) = P^{1*}(x-1) = \begin{cases} 0, & x < 2 \\ 1, & x \geq 2 \end{cases}.$$

For n = 3, 4, ...

$$P^{n*}(x) = \begin{cases} 0, & x < n \\ 1, & x \geq n \end{cases}.$$

$$F(x,t) = \sum_{n=0}^{\infty} \frac{e^{-t}t^n}{n!} P^{n*}(x) = \sum_{n=0}^{[x]} \frac{e^{-t}t^n}{n!},$$

where [x] is the greatest integer in x.

Example 2. Assume the same distribution for N(t) and that

$$P(x) = \begin{cases} 0, & x < 0 \\ 1 - e^{-x}, & x \geq 0. \end{cases}$$

The convolutions can be determined explicitly.

$$P^{2*}(x) = \int_0^\infty P^{1*}(x-t)dP(t) = \int_0^x [1-e^{-(x-t)}]e^{-t}dt$$

$$= 1 - \sum_{n=0}^{1} \frac{e^{-x}x^n}{n!} = \int_0^x \frac{e^{-t}t}{1!}\,dt\,.$$

By mathematical induction,

$$P^{n*}(x) = \int_0^x \frac{e^{-t}t^{n-1}}{(n-1)!}\,dt$$

the incomplete gamma function.

Thus

$$F(x,t) = \sum_{n=0}^{\infty} \frac{e^{-t} t^n}{n!} P^{n*}(x) .$$

II. EQUI-SPACED, EQUI-PROBABLE CLAIMS

Assume that each X_i assumes the values 1, 2, ..., a with equal probabilities of 1/a. Frequently we will assume that a = 5. The generating function for X_i is

$$G(s) = E\,[s^X] = (s + s^2 + \ldots + s^a)/a = \frac{s}{a}\left(\frac{1 - s^a}{1 - s}\right),$$

for $|s| < 1$. Let $S_n = X_1 + X_2 + \ldots + X_n$, and H(s) = generating function of S_n. By the independence of $\{X_i\}$,

$$H(s) = [G(s)]^n = \left\{\frac{s(1-s^a)}{a(1-s)}\right\}^n , \quad |s| < 1.$$

By definition, the coefficient of s^j in H(s) equals $P[S_n = j]$.
NOTE: $P[S_n = j] = P^{n*}(j)$.

RESULT

$$P[S_n = j] = \sum_{k=0}^{\infty} \frac{(-1)^k}{a^n} \binom{n}{k} \binom{j - ak - 1}{n - 1} .$$

PROOF

$$H(s) = \frac{s^n}{a^n} \sum_{k=0}^{n} \binom{n}{k} (s^a)^k (-1)^k \sum_{i=0}^{\infty} (-1)^i \binom{-n}{i} s^i$$

where

$$\binom{-n}{i} = \frac{(-n)(-n-1)(-n-2)\ldots(-n-i+1)}{i\,(i-1)(i-2)\ldots 1}$$

$$= \frac{1}{a^n} \sum_{k=0}^{n} \sum_{i=0}^{\infty} (-1)^{k+i} \binom{n}{k} \binom{-n}{i} s^{n+i+ak}$$

Let $j = n+i+ak$. Then $i = j-n-ak$ and there is only one term in the "i" series where that occurs. Since $S_n \geq n$, consider a $j \geq n$.

$$P[S_n = j] = \frac{1}{a^n} \sum_{k=0}^{\infty} (-1)^{k+j-n-ak} \binom{n}{k} \binom{-n}{j-n-ak}$$

where

$$\binom{n}{k} = 0 \quad \text{for } k > n.$$

Problem 3, page 63, of Feller (6) implies that

$$\binom{-n}{j-n-ak} = (-1)^{j-n-ak} \binom{j-ak-1}{j-n-ak} .$$

Since $\binom{n}{r} = \binom{n}{n-r}$,

$$\binom{j-ak-1}{j-ak-1-(n-1)} = \binom{j-ak-1}{n-1} .$$

That completes the proof.

The following Theorems, and Corollaries appear in Beekman and Fuelling (1).

Theorem 1. Assume that $\{X_i\}$ are independent random variables, each assuming the values 1, 2, ..., a with probability 1/a. Assume that $\{N(t), t \geqslant 0\}$ is a non-negative integer valued stochastic process independent of $\{X_i\}$. Let $S(t) = X_1 + X_2 + \ldots + X_{N(t)}$, $t \geqslant 0$.

Then

$$P[S(t) = j] = \sum_{n=0}^{j} P[N(t) = n] \sum_{k=0}^{\infty} \frac{(-1)^k}{a^n} \binom{n}{k} \binom{j-ak-1}{n-1}$$

where

$$\binom{x}{k} = 0 \quad \text{for } k > x.$$

Proof of Theorem 1. By using the additivity property of measures, the product rule for probabilities, the definition of conditional probability, and the independence of $\{X_i\}$ and $N(t)$,

$$P[S(t) = j] = \sum_{n=0}^{\infty} P[S(t) = j \text{ and } N(t) = n]$$

$$= \sum_{n=0}^{\infty} P[S(t) = j | N(t) = n] \, P[N(t) = n]$$

$$= \sum_{n=0}^{\infty} \frac{P[S(t) = j, N(t) = n]}{P[N(t) = n]} P[N(t) = N]$$

$$= \sum_{n=0}^{\infty} P[S_n = j]\, P[N(t) = n]\ .$$

At this point, the distribution for the $\{X_i\}$ implies that $S_n \geqslant n$ and $P[S_n = j] = 0$ for $n > j$. This provides the conclusion that

$$P[S(t) = j] = \sum_{n=0}^{j} P[S_n = j] P[N(t) = n].$$

Then apply the previous Result.

Remark. For ease of writing, denote the series for $P[S_n = j]$ by $g(n,j)$.

Corollary 1. Assume the hypotheses of Theorem 1, and also that $N(t)$ has a Poisson distribution, i.e., $P[N(t)=n]=e^{-\lambda t}(\lambda t)^n/n!$ for $n = 0, 1, 2, \ldots$ and $\lambda>0$. Then $P[S(t)=0]=e^{-\lambda t}$, and for $j=1, 2, \ldots$

$$P[S(t) = j] = \sum_{n=1}^{j} \frac{e^{-\lambda t}(\lambda t)^n}{n!}\, g(n,j).$$

Corollary 2. Assume the hypotheses of Theorem 1, and also that $N(t)$ has a negative binomial distribution, i.e.

$$P[N(t) = n] = \frac{\Gamma(n+b)}{n!\Gamma(b)} \left(\frac{c}{c+t}\right)^b \left(\frac{t}{c+t}\right)^n \quad \text{for } c>0,\ b>0.$$

Then

$$P[S(t) = 0] = \left(\frac{c}{c+t}\right)^b \text{, and for } j = 1, 2, \ldots$$

$$P[S(t) = j] = \sum_{n=1}^{j} \frac{\Gamma(n+b)}{n!\Gamma(b)} \left(\frac{c}{c+t}\right)^b \left(\frac{t}{c+t}\right)^n g(n,j).$$

Remark. If one assumes that the parameter λ in a Poisson law has a gamma distribution with parameters $c>0$, $b>0$, then

$$P[N(t) = n] = \int_0^\infty \frac{e^{-\lambda t}(\lambda t)^n}{n!} \frac{c^b}{\Gamma(b)} \lambda^{b-1} e^{-c\lambda} d\lambda.$$

Upon integration, this becomes

$$P[N(t) = n] = \frac{\Gamma(n+b)}{n!\Gamma(b)} \left(\frac{c}{c+t}\right)^b \left(\frac{t}{c+t}\right)^n .$$

As sample calculations, we begin with a fixed integer n = 10 and a = 5.

Then

$$P[S_{10} = 15] = \sum_{k=0}^{\infty} \frac{(-1)^k}{5^{10}} \binom{10}{k} \binom{15-5k-1}{9}$$

$$= \sum_{k=0}^{1} \frac{(-1)^k}{5^{10}} \binom{10}{k} \binom{14-5k}{9}$$

$$= \frac{1}{5^{10}} \left[\binom{14}{9} - \binom{10}{1}\binom{9}{9}\right] = \frac{1{,}992}{5^{10}} .$$

Also

$$P[S_{10} = 20] = \sum_{k=0}^{\infty} \frac{(-1)^k}{5^{10}} \binom{10}{k} \binom{20-5k-1}{9}$$

$$= \sum_{k=0}^{2} \frac{(-1)^k}{5^{10}} \binom{10}{k} \binom{19-5k}{9} = \frac{72{,}403}{5^{10}} .$$

Six lengthy tables for F(x,t) are contained in Beekman-Fuelling (1). Extracts from those tables are now presented.

Table 1. Poisson (λ = 1.0, a = 5)

x	F(x, 10)	x	F(x, 10)
0.	0.000045	40	0.841886
10	0.017958	50	0.965154
20	0.185975	60	0.994877
30	0.544684	70	0.999472

Table 2. Poisson (λ = 1.0, a = 5)

x	F(x, 20)	x	F(x, 20)
0	0.000000	60	0.531692
10	0.000014	70	0.767203
20	0.001017	80	0.910920
30	0.015065	90	0.973522
40	0.087732	100	0.993794
50	0.270411	110	0.998834

Table 3. Poisson (λ = 0.5, a = 5)

x	F(x, 60)	x	F(x, 60)
0	0.000000	80	0.310480
10	0.000000	90	0.525902
20	0.000001	100	0.726422
30	0.000065	110	0.868559
40	0.001118	120	0.947387
50	0.009323	130	0.982362
60	0.044836	140	0.995011
70	0.140035	150	0.998801

Table 4. Negative binomial (b = 10, c = 10)

x	F(x, 20)	x	F(x, 20)
0	0.000017	80	0.813082
10	0.002358	90	0.890585
20	0.024083	100	0.939384
30	0.092067	110	0.968005
40	0.217367	120	0.983818
50	0.381682	130	0.992119
60	0.551883	140	0.996289
70	0.700100	150	0.998305

Table 5. (b = 5; c = 10)

x	F(x, 50)	x	F(x, 50)
0	0.000129	90	0.705238
10	0.005855	100	0.779128
20	0.031795	110	0.837834
30	0.088247	120	0.883068
40	0.174570	130	0.917028
50	0.281766	140	0.941966
60	0.397785	150	0.959929
70	0.511672	160	0.972659
80	0.615604	170	0.981598

The choices for b and c were motivated by previous uses of the negative binomial distribution in Beekman (4) and Weber (8).

III. CONVOLUTION TYPE SERIES FOR $\psi(u)$.

The following Theorem was proved in 1968 (see Beekman (2)), but seemed to have little utility. It now appears to offer potential for practical calculations.

<u>Theorem</u> 2. Assume that P(x) is such that

$$\lim_{x\to\infty} x[1-P(x)] = 0 .$$

Assume that $\{X_i\}$ are independent random variables, each with distribution P(x). Assume that

$$P[N(t) = n] = \frac{e^{-t}t^n}{n!} ,$$

n = 0, 1, 2, ... and that $\{N(t), t\geq 0\}$ is independent of $\{X_i\}$. Assume $\theta>0$, and $u\geq 0$. Let $p_1 = E(X)$, and

$$\psi(u) = P\{\underset{0\leq t<\infty}{\text{supremum}} [S(t)-tp_1(1+\theta)]>u\} .$$

Let

$$H_0^*(x) = 1 \text{ if } x\geq 0 \text{ and } 0 \text{ if } x<0.$$

Let

$$H^*(x) = \frac{1}{p_1}\int_0^x [1-P(y)]dy \text{ for } x\geq 0, \text{ and } 0 \text{ for } x<0;$$

$$H_1^*(x) = H^*(x);\ H_n^*(x) = \int_0^x H_{n-1}^*(x-t)dH^*(t) \text{ for } x\geq 0 \text{ and } 0$$

for $x<0$, $n\geq 1$.

Then

$$\psi(u) = 1-\frac{\theta}{1+\theta}\sum_{n=0}^{\infty}\left(\frac{1}{1+\theta}\right)^n H_n^*(u).$$

<u>Theorem</u> 3. Assume the same hypotheses as in Theorem 2.

For any positive integer N,

$$\psi(u) \doteq 1-\frac{\theta}{1+\theta}\sum_{n=0}^{N}\left(\frac{1}{1+\theta}\right)^n H_n^*(u),$$

with

$$|\text{Error}| \leq (1+\theta)^{-(N+1)} .$$

Proof. $H_n{}^*(u) \le 1$ for all n and u. Hence the |Truncation Error| of

$$\frac{\theta}{1+\theta} \sum_{n=N+1}^{\infty} \left(\frac{1}{1+\theta}\right)^n H_n{}^*(u) \le \frac{\theta}{1+\theta} \sum_{n=N+1}^{\infty} (1+\theta)^{-n} = (1+\theta)^{-(N+1)}.$$

<u>Theorem</u> 4. Assume the same hypotheses as in Theorem 2. Also, assume there exists a positive constant A such that

$$P[X_i \le A] = 1, \; i = 1, 2, \ldots$$

Then, if N is a positive integer and $u \ge N \cdot A$,

$$\psi(u) \doteq 1 - \frac{\theta}{1+\theta} \sum_{n=0}^{N} (1+\theta)^{-n}$$

with

$$|\text{error}| \le (1+\theta)^{-(N+1)} .$$

Proof. For $x \ge A$, $H_1{}^*(x) = 1$ since

$$H_1{}^*(x) = \frac{1}{p_1} \int_0^x [1-P(y)]dy$$

$$= \frac{y}{p_1} [1-P(y)] \Big|_0^x + \frac{1}{p_1} \int_0^x y dP(y) = 1.$$

Since

$$H_n{}^*(u) = P[X_1 + X_2 + \ldots + X_n \le u]$$

where each X_i has distribution $H_1{}^*(x)$,

$$H_n{}^*(u) = 1 \text{ for } n = 0, 1, 2, \ldots, N .$$

<u>Example</u>. Let $a = 5$. Assume that $N(t)$ has a Poisson distribution with mean 10. A prudent actuary could require

$$u \ge 2\, E\{S(t)\} + 2 \text{ Std. Dev. } \{S(t)\}.$$

Thus,

$$u \ge 20p_1 + 2(20p_2)^{1/2} \doteq 90 .$$

Assume $\theta = 0.3$.

$$H_n^*(90) = 1 \text{ for } n = 0, 1, 2, \ldots, 18 .$$

$$\psi(90) \doteq 1 - \frac{0.3}{1.3} \sum_{n=0}^{18} (1.3)^{-n}$$

$$= (1.3)^{-19} \doteq 0.007$$

with error ≤ 0.007 .

If

$$u = 20p_1 + 3(20p_2)^{1/2} \doteq 105,$$

$$\psi(u) \doteq (1.3)^{-22} \doteq 0.003,$$

with error ≤ 0.003.

IV. FUTURE RESEARCH

In many cases, Theorem 4 requires u to be much too high. Let us reconsider Examples 1 and 2 on pages 52 and 53 of Beekman (3). The first claim distribution is

$$P_1(y) = \begin{cases} 0, & y < 2 \\ 0.3, & 2 \leq y < 5 \\ 0.5, & 5 \leq y < 10 \\ 0.8, & 10 \leq y < 20 \\ 1.0, & y \geq 20 \end{cases}$$

Here $p_1 = 8.6$ units. We will assume that $\theta = .3$. For $\psi(u)$ to equal 0.01 (approximately),

$$1 - \frac{0.3}{1.3} \sum_{n=0}^{N} (1.3)^{-n} = (1.3)^{-(N+1)} \doteq 0.01, \ N = 17.$$

But that says $u = 17(20) = N \cdot A = 340$ units. By contrast, the Lundberg asymptotic approximation to $\psi(u)$ has $u \doteq 128$ units, for $\psi(u) = 0.01$.

The second claim distribution is

$$P_2(y) = \begin{cases} 0, & y < 2 \\ 0.3, & 2 \leq y < 5 \\ 0.5, & 5 \leq y < 10 \\ 0.8, & 10 \leq y < 20 \\ 0.85, & 20 \leq y < 30 \\ 0.90, & 30 \leq y < 40 \\ 0.95, & 40 \leq y < 50 \\ 1.00, & y \geq 50 \end{cases}$$

Here p_1 = 11.6 units. Assume $\theta = .3$. Now $u = N \cdot A = 17(50) = 850$ units. Yet the Lundberg approximation implies that $u = 253$ units for $\psi(u) = 0.01$.

The purpose of Examples 1 and 2 was to show the increase needed in u if the retention limit is raised from 20 to 50 units.

The problem with Theorem 4 is that one needs the $H_n{}^*(u)$ multipliers to reduce the values of $(1 + \theta)^{-n}$.

Remark. If

$$1- \frac{\theta}{1+\theta} \sum_{n=0}^{N} (1+\theta)^{-n} H_n{}^*(u) = \varepsilon ,$$

then $\psi(u) \leq \varepsilon$, and $|\text{Error}| \leq \varepsilon$.

Proof. As a function of N, 1- Above Sum is decreasing in N. Furthermore,

$$\lim_{N\to\infty} [1\text{-Above Sum}] = \psi(u).$$

The authors of (1) intend to use MACSYMA to compute $H_n{}^*(x)$ values. That computer language was developed at The Massachusetts Institute of Technology to manipulate entire functions, rather than values. See reference (7).

REFERENCES

(1) Beekman, J. A. and C. P. Fuelling, "Risk Convolution Calculations", Scandinavian Actuarial Journal, 1981: pp. 151-164.

(2) Beekman, J. A., "Collective Risk Results", Trans., Society Actuaries, 20 (1968), pp. 182-199.

(3) Beekman, J. A., Two Stochastic Processes, Almqvist and Wiksell International, Stockholm; also Halsted Press (Wiley), New York, 1974.

(4) Beekman, J. A., "A Stochastic Investment Model", Trans., Society Actuaries, 32 (1980), pp. 9-24.

(5) Cramér, H., Collective Risk Theory, Forsakringsaktiebolaget Skandia, Jubilee Volume, 1955.

(6) Feller, W., An Introduction to Probability Theory and Its Applications, Vol. 1, 3rd Ed., J. Wiley, New York, 1968.

(7) MACSYMA Users' Manual, Laboratory for Computer Science, Massachusetts Institute of Technology, Cambridge, Massachusetts.

(8) Weber, Donald C., "A Stochastic Approach to Automobile Compensation", Proc., Casualty Actuarial Society, 57 (1970), pp. 27-61.

RISK SHARING, INCENTIVES AND MORAL HAZARD

Phelim P. Boyle, Ph.D., F.I.A., F.C.I.A.

University of Waterloo
Waterloo, Ontario, Canada
N2L 3G1

Abstract

Many traditional insurance models have ignored important behavioural features. This paper provides an introduction to a recent class of economic models which rectifies this omission. The principal-agent problem is introduced and analyzed in the presence of moral hazard. The paper discusses the trade-off between risk-sharing and incentives. There is also a brief introduction to modern option pricing theory.

F. de Vylder et al. (eds.), Premium Calculation in Insurance, 31–75.

I. INTRODUCTION

I am pleased to participate in this Summer School and look forward to sharing some ideas with you over the next few days.

At the outset I would like to describe the approach I will take in presenting the material.[1] I aim to present the material in a non-rigorous way and hopefully in an intuitive manner. At the same time I will draw attention to some of the major technical problems. It is pitched at someone who is unfamiliar with the area. The results presented here are unfamiliar to actuaries and insurance mathematicians although they are well known in some other fields.

During the next few minutes I will make some preliminary comments. The purpose of these comments is to place the lectures in perspective and motivate the upcoming material. After this I will outline briefly the topics to be covered during the rest of this lecture and in the lectures that will follow.

One of the central themes of these lectures is RISK-SHARING. Risk-sharing is a common response to uncertainty. Such uncertainty can arise from natural phenomena or social causes. One particular form of risk-sharing is the insurance mechanism. I will be dealing with models which have a natural application in the insurance area but they have been applied in other areas as well. In fact some of the paradigms to be discussed have the capacity to provide a unified treatment of problems in diverse fields.

In constructing models in the social and natural sciences it is customary to use simplifying assumptions. The area of insurance mathematics is no different. Thus in modelling insurance premiums some or all of the following assumptions are often made:

(i) There are no transactions costs.

(ii) The claim amount can be ascertained costlessly.

(iii) The existence of insurance and the type of contract does not change the loss distribution. In other words, the loss distribution is purely a random process.

(iv) Competition and market effects are ignored.

(v) There are no serious informational asymmetries between the insurance company and the insured. The parties involved communicate honestly.

Although these troublesome frictions can be ignored at the modelling stage, many of them crop up again at the time the insurance contract is marketed or designed.

A couple of examples may be of interest. Most actuarial students have gone through the experience of calculating a mathematically designed set of premium rates to be told that the published rates will bear little relationship to them because of competition. Another illustration comes from the life insurance market. Recently in North America, insurance companies have been giving much lower rates to non-smokers in their renewable term contracts. Because of a number of factors there is a tendency for the healthier lives to move their insurance around and so a number of insurance companies are experiencing a lot of difficulty because they have been left with the poorer lives and yet can only charge a pooled or blended premium.

In a practical setting an attempt is often made to deal with these frictions. Ways of dealing with them include explicit contract design, inclusion of incentives, monitoring devices, administrative procedures and legislative measures. Thus while the quantitative models concentrated on the risk as a random process, the contract design and administrative procedures reflected a keen appreciation of the non-random aspects of the risks insured and the potential problems caused by asymmetric information. One of the most obvious illustrations of this is the so-called principle of indemnity which is found in most insurance legislation. This states that an insured is not permitted to recover more than the value of the insured asset under an insurance contract. If the premium for full insurance were P and the asset was worth A then if the risk were purely random there should be no objection to the insured paying say 2P and recovering 2A in the event of an accident. However in most jurisdictions A is the maximum amount that can be insured. The very existence of insurance can alter the claim distribution and this is an example of MORAL HAZARD. Moral Hazard occurs when the insured by virtue of his action (or inaction) can affect the outcome. There are many examples of this phenomenon in the insurance field and in other areas as well.

Whereas moral hazard arises because the actions of a particular insured are not costlessly observable, adverse selection occurs when a standard contract is offered to a group whose probabilities of loss differ. If the individual members of the group have superior knowledge of their own risk characteristics then even if the contract is fairly priced on the basis of the group as a whole only the high risk individuals will take up the contract and the premiums collected will be inadequate. Insurance companies have developed techniques for dealing with adverse selection. For example in group insurance plans which cover a well defined group such as the employees of a particular corporation, membership of the group insurance plan may be mandatory.

Adverse selection has a number of applications beyond the insurance market. Akerlof (1970) has examined an interesting illustration of this phenomenon in the used car market. Buyers of used cars know only the average quality of cars in the market whereas the seller who has owned the car has more accurate information on the car's quality. Those sellers with above average cars will tend to leave the market and as a result the average quality of the remaining cars will decline. Ultimately this process will lead to the collapse of the market. Akerlof develops a simple model of this phenomenon and discusses some of the remedies that could be used to prevent market failure.

In the last few years economic theorists have made considerable progress in constructing models that include moral hazard and adverse selection. In the present lecture series the emphasis will be on moral hazard. Much of this analysis has been concerned with the so-called PRINCIPAL AGENT PROBLEM. Two individuals operating in an uncertain environment wish to engage in risk-sharing. One party, the principal, owns certain resources and engages the second party, the agent, to perform certain tasks. The output is both a function of the effort exerted by the agent and the random state of nature that ensues. The principal and the agent strive to maximize their own welfare. The principal's problem is to design a risk-sharing rule or contract that is efficient both from a risk-sharing viewpoint and a motivational viewpoint. If the agent's effort is unobservable the problem becomes both quite interesting and, indeed, complex. Recent results in this area have provided valuable insights into the interplay between risk-sharing and incentives. Holmstrom (1979) presents a concise treatment of the problem and discusses some applications and extensions.

One of the main aims of the present lecture series is to introduce these developments. Hopefully this will be a useful service since insurance mathematicians and actuaries in general have had little exposure to some of these ideas.

To lay the foundation for the subsequent discussion Section 2 starts with a brief discussion of the concept of utility. Some useful utility functions will be mentioned. Pure risk-sharing will be discussed as well as the concept of Pareto optimality. A key result due to Karl Borch will be discussed and a brief example will be given.

In Section 3 a model will be set up where the output of the process depends both on the agent's action and the random state of nature. Initially it is assumed that the agent's action is observable and that the parties act cooperatively. This is an instructive example in its own right as well as being important for the next stage.

In Section 4 it is assumed that the agent's action is unobservable introducing moral hazard. Now the principal's problem is to design a sharing rule that will not only provide risk-sharing but will also help to motivate the agent. In its most general setting this problem involves mathematical problems which are both subtle and profound.

In one of the extensions I shall look at the role of ancillary information in contract design.

Finally in Section 5 I will discuss some unrelated material that has to do with risk-sharing in financial markets and talk about modern developments in option pricing.

2. UTILITY, PARETO OPTIMALITY AND PURE RISK SHARING

The concept of utility provides a convenient way of modelling preferences. Each individual is assumed to possess a utility function which induces a cardinal preference ordering among different alternatives. Suppose for example that the relevant criterion is money (wealth) then a rational individual will calculate the utility of each alternative and select the one with the highest utility value. He will maximize,

$$U(W_A) \text{ over all } A,$$

where W_A is the wealth associated with A, U(W) is the utility function. It is assumed here that the alternatives are non-stochastic.

It is assumed that U' is > 0 so that the individual prefers more to less. The second derivative can be used to distinguish three distinct types:

$U'' < 0$	Risk averse
$U'' = 0$	Risk neutral
$U'' > 0$	Risk lover (gambler)

In general we assume throughout that individuals are risk averse so that their utility functions are concave. We also assume that the utility functions are of class C^2.

If the alternatives involve uncertain outcomes then the individual will maximize the expected value of his utility function over the various alternatives.

A useful measure of an individual's aversion to risk is given by,

$$RA(W) = -\frac{U''(W)}{U'(W)} \qquad (1)$$

Note that this is a local measure of risk aversion. It is invariant under linear transforms. Furthermore it is closely linked to the maximum risk premium an individual will pay to get rid of risks with a small variance.

Two other related risk measures are useful.

$$\text{Risk Tolerance } RT(W) = \frac{1}{RA(W)} = \frac{U'(W)}{U''(W)} \qquad (2)$$

$$\text{Risk Cautiousness } RT'(W) \qquad (3)$$

RT is only defined for $U'' < 0$.

A special class of utility functions will be of interest to us. These are characterized by having Linear Risk Tolerance. They are called HARA utility functions. HARA stands for Hyperbolic Absolute Risk Aversion because of the form of RA(W). They can be classified as follows:

TYPE	UTILITY FUNCTION U(X)	POSSIBLE OUTCOMES	RISK AVERSION	RISK TOLERANCE
Exponential	$\frac{\exp(-\gamma X)}{\gamma}$	$-\infty < X < \infty$	γ	$\frac{1}{\gamma}$
Logarithmic	$\log(K + X)$	$-K < X < \infty$	$\frac{1}{(K+X)}$	$K + X$
Power D	$(k+X)^{\Theta}$ $\Theta < 1$ $\Theta \neq 0$	$-K < X < \infty$	$(\frac{1-\Theta}{K+X})$	$(\frac{K+X}{1-\Theta})$
Power I	$-(K-X)^{\Theta}$ $\Theta > 1$	$-\infty < X < K$	$(\frac{\Theta-1}{K-X})$	$(\frac{K-X}{\Theta-1})$

In subsequent developments we will use these but sometimes with different parameters.

Pure Risk Sharing

Suppose we have two risk averse individuals with utility functions U_1 and U_2 who wish to engage in pure risk-sharing. The outcome depends on some purely random process and can be costlessly observed. Both individuals share the same probability beliefs. It is assumed that there is no action involved and that the only choice variable involves the selection of the risk-sharing contract which stipulates the relative share of each party under the different possible outcomes.

Assume that x, a random variable, denotes the total output measured in money terms. Individual 1's share is $x - S(x)$ where $S(x)$ is the share of individual 2. If x is positive we have a risky asset whereas if x is negative we are dealing with an insurance type situation. For the moment assume $x > 0$. Depending on the risk-sharing arrangement selected, the expected utilities of the two parties will differ. The different possible combinations are drawn below:

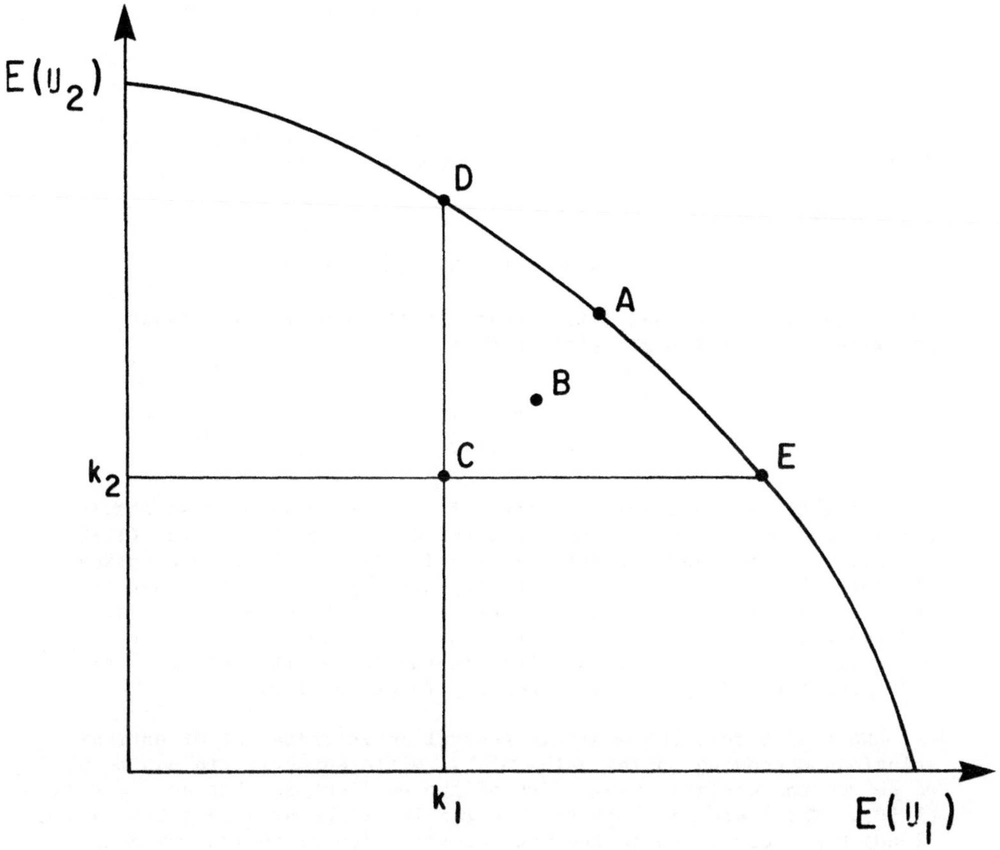

Expected Utility Levels

Point A is preferable to point B because at A both parties have higher expected utility. In fact we see that the optimal solution will be somewhere in the north east boundary - precisely where will depend on the bargaining weights.

It may be that individual i is guaranteed a certain utility level k_i even without the contract. This means that they will only be interested in points within the segment DCE. Points on the north east boundary can be characterised as Pareto optimal soutions. A Pareto optimal solution is one in which one individual cannot be made better off in expected utility terms without making the other worse off. Geometrically the Pareto optimal frontier can be constructed by maximizing,

$$E(U_1) + \lambda\ E(U_2) \qquad \lambda > 0$$

subject to any relevant constraints. For different values of λ different points are obtained on the Pareto frontier.

Assume that,

$$L = \int U_1(x - s(x))\ f(x)dx + \lambda \int U_2(s(x))\ f(x)dx \qquad (4)$$

where f(x) is the probability density function of x. Pointwise optimisation of the Lagrangian leads to,

$$\frac{U_1'(x-s(x))}{U_2'(s(x))} = \lambda \qquad (5)$$

This condition is not only necessary for Pareto optimal risk sharing, but it is also sufficient and was first derived by Karl Borch (1962). It gives a simple and elegant characterization of the Pareto optimal sharing rules in the pure risk-sharing setting. It states that the sharing rule is selected so that the ratio of the marginal utilities for the two individuals is constant for all outcomes. It can be generalized to n individuals. Note in particular that equation (5) is independent of the probability density function of x.

Note also that there are in general an infinite set of optimal solutions depending on the value of λ. Different criteria can be used to select the optimal point. One of the earliest is that due to Nash (1950). There are many others. In the insurance area Hans Buhlmann (1980) has recently suggested another selection criterion which has attractive properties. Buhlmann's result relies on a theorem by Gale and Sobel (1982).

In passing note that bringing in one further condition can result in a unique point. For example if we insist that individual (2) is constrained to obtain an expected utility = k_2 this identifies a unique point on the Pareto frontier.

It should be also noted that the analysis presented here solves for the optimal sharing rule which is really to say that the optimal contract is obtained. This is an enlargement of the concept of insurance premium to the entire contract and proceeds in the directions suggested at the beginning of Section 1.

Equation (5) gives a characterization of the optimal sharing rules. Linear sharing rules are of considerable interest. They are simple and correspond to proportional insurance. There is a relationship between linear sharing rules and the HARA utility functions introduced previously: Wilson (1968) Mossin (1973, p. 113).

Linear Sharing Rules

If the 2 individuals are risk-averse and have the same beliefs then the Pareto optimal pure risk-sharing arrangement will be linear in the total outcome if and only if the utility functions are both of the same HARA class with identical cautiousness.

In financial economics linear sharing rules are extremely important. The implication of this assumption is that individuals pool risky assets and buy a share of the market or total portfolio. The individual investors get rid of all the diversifiable risk in this way.

Note also that $\frac{ds}{dx}$ can be obtained by differentiating (5) with respect to x and then using (5) to elminate λ. We find,

$$\frac{ds}{dx} = \frac{\dfrac{U_1''}{U_1''}}{\dfrac{U_1''}{U_1''} + \dfrac{U_2''}{U_2'}} \qquad (6)$$

Equation (6) illustrates the close connection between the marginal change in the respective shares and the risk tolerances of the parties involved.

<u>Example 1</u> Pure Risk-Sharing with Exponential Utilities

$$U_1(W) = \frac{1-e^{-\gamma_1 W}}{\gamma_1}$$

$$U_2(W) = \frac{1-e^{-\gamma_2 W}}{\gamma_2}$$

Suppose 1 and 2 are insurance companies with intitial wealth W_1 and W_2. We know that the Pareto optimal solution is to pool total claims. Let x denote aggregate claims. From equation (5) s(x) is the solution of,

$$\frac{e^{-\gamma_1 [W_1 - (x-s(x)]}}{e^{-\gamma_2 [W_2 - s(x)]}} = \lambda \qquad (7)$$

which leads to

$$s(x) = \left(\frac{\gamma_1}{\gamma_1+\gamma_2}\right) x + \left(\frac{\gamma_2 W_2 - \gamma_1 W_1}{\gamma_1 + \gamma_2}\right) - \Theta \qquad (8)$$

where $\Theta = \log(\lambda)$

Equation (8) has the predicted linear form. The first term indicates that the share of each company is proportional to its relative risk tolerance. Recall that,

$$\frac{\frac{1}{\gamma_2}}{\frac{1}{\gamma_1} + \frac{1}{\gamma_2}} = \frac{\gamma_1}{\gamma_1 + \gamma_2} \qquad (9)$$

where $\frac{1}{\gamma_i}$ is the risk tolerance of Company i. The second term in (8) is a combination of wealth and risk aversion and the third term Θ is related to the bargaining weight λ.

3. THE COOPERATIVE MODEL WITH NON-PECUNIARY COMPONENT

In the previous development the pure risk sharing arrangement was analyzed. Our next task is to examine a situation where the outcome depends in part on the action taken by one of the parties. However even when the action is selected the actual outcome will still be a random variable because of the random state of nature that occurs. It is convenient to parametrize the agent's action by "a" and think of "a" as effort. The output has a density function conditional upon "a". Furthermore the impact of an increase in "a" is to increase the density function f(x,a) in the sense of first order stochastic dominance. Higher effort in general leads to higher output. Note that for an insurance application the movement is in the opposite direction. Here the action may be the level of care expended or the precautionary actions taken to avoid or reduce loss. In this case the amount of the claim will decrease as the insured expends more effort.

However most of the time we will deal with cases where the outcome corresponds to positive monetary values since this is the dominant model in the literature. Two of the earliest papers in this area are by Spence and Zeckhauser (1971) and Mirless (1971. These two papers used different approaches to model the uncertainty in the output. We will find it convenient to use the approach suggested by Mirless.

The particular assumptions employed to set up the model are detailed below:

(i) Two economic agents share identical beliefs. They wish to share the output which although random depends on the action or the effort of one of the parties.

(ii) The principal with utility only for wealth sets the contract terms. His utility depends only on wealth. He knows not only his own utility function but also that of the agent. Furthermore he knows the distribution of the output given a certain action by the agent. Clearly these are strong assumptions. The principal's utility function is

$U_1(W)$ where $U_1' > 0$

$U_1'' < 0$

(iii) The agent controls the action. He can in some models select an action different from that previously agreed. For the moment we assume that the agent's action "a" is OBSERVABLE by the principal. This will lead to a cooperative sharing rule.

(iv) The action selected in conjunction with the state of nature determines the outcome through a payoff function,

$$x = W(\Phi, a)$$

where Φ denotes the state of nature. We can think of x as being a random variable conditional on "a" with distribution f(x, a). It is assumed that $F(x,a)_a \leq 0$. It is also assumed that f(x,a) is twice differentiable with respect to x and a and that its likelihood function $\frac{f_a(x,a)}{f(x,a)}$ is monotone increasing in x and that it has fixed support.

(v) The agent's utility depends both on his wealth W and the effort expended "a". His utility function H(W,a) is given by

$$H(W,a) = U_2(W) - V(a) \qquad (10)$$

where $V' > 0$ $\qquad U' > 0$
$V'' > 0$ $\qquad U'' < 0$.

More general utility forms are possible but this one is simple and convenient. The utility function here is additive and separable. Note that increasing "a" lowers the agent's utility if no adjustment is made to his wealth.

(vi) In this problem we will assume that the agent's action is OBSERVABLE to the principal. It turns out that this implies that the two parties act cooperatively. Because of the separability of the agent's utility function the Pareto optimal sharing rules are the same as in the pure risk-sharing case.

(vii) It is also assumed that the agent requires a reservation level of utility H_o. To obtain a unique solution we will sometimes set his expected utility equal to H_o.

With this notation the Lagrangian for the problem is given by,

$$\max_{s(x),\, a} L \text{ where}$$

$$L = \int U_1[x - s(x)]\, f(x,a)\, dx + \lambda\left[\int U_2[s(x)]\, f(x,a)dx - V(a)\right] \qquad (11)$$

Pointwise optimization of the Lagrangian yields

$$\frac{U_1'[x - s(x)]}{U_2'[s(x)]} = \lambda \qquad (12)$$

which is similar to equation (5).

Alternatively it can be formulated as

$$\max_{s(x),\, a} \int U_1[x - s(x)]\, f(x,a)\, dx \qquad (13)$$

$$\text{subject to} \int U_2[s(x)]\, f(x,a)\, dx - V(a) \geq H_o$$

The two formulations are equivalent as long as the Pareto frontier is concave. There is a one-one relationship between λ and H_o in this case. Lambert (1981) spells out the conditions which ensure this.

In this case where the agent's action is costlessly observable to the principal we could think of it as the situation where the agent is perfectly honest and tells the principal exactly what action he took. Later on we consider cases where we cannot put quite as much faith in the agent's truthfulness. However for the next example we continue to deal with the cooperative solution.

Example 2 Risk Sharing with Non-pecuniary Returns

It is assumed that the principal and the agent have exponential utility functions for wealth. Output has an exponential distribution with expected value equal to the agent's action.

The assumptions made here are for analytic simplicity and to provide a comparison with a later example.

For the purposes of example (2)

$$U_1(W) = \frac{1 - e^{-\gamma_1 W}}{\gamma_1} \qquad \gamma_1 > 0$$

$$U_2(W) = \frac{1 - e^{-\gamma_2 W}}{\gamma_2} \qquad \gamma_2 > 0$$

$$V(a) = \rho a^2 \qquad \rho \text{ a constant} > 0$$

$$f(x,a)\, dx = e^{-\frac{x}{a}} \frac{dx}{a}$$

Furthermore from the solution to (12) we find,

$$s(x) = \frac{\gamma_1 x + \log(\lambda)}{\gamma_1 + \gamma_2} \qquad (14)$$

Since the Lagrangian L is maximized over all actions "a" by the agent the first order conditions for a maximum at "a" are:

$$\int U_1[x - s(x)]\, f_a(x,a)\, dx + \lambda\, [\int U_2[s(x)[f_a(x,a)\, dx - V_a'(a)] = 0 \quad (15)$$

If we substitute for U_1, s, U_2 and V in this equation and note that,

$$f_a(x,a)\, dx = \frac{(x-a)}{a^3}\, e^{-\frac{x}{a}}\, dx$$

we find that this first order condition becomes

$$\int\{\ [\exp(-\gamma_3 x)]\, \frac{\lambda^{\gamma_{11}}}{\gamma_1} + \frac{\lambda}{\gamma_2}\, [\exp(-\gamma_3 x)]\, \lambda^{-\gamma_{22}}]\}\, \frac{(x-a)}{a^3}\, e^{-\frac{x}{a}}\, dx + 2\lambda\rho a = 0$$

where $\gamma_3 = \frac{\gamma_1\gamma_2}{\gamma_1+\gamma_2}$

$$\gamma_{11} = \frac{\gamma_1}{\gamma_1+\gamma_1}$$

$$\gamma_{22} = \frac{\gamma_2}{\gamma_1+\gamma_2}$$

Let $\alpha = [\frac{1}{a} + \gamma_3]^{-1}$. The above equation becomes

$$(2\lambda^{\gamma_{22}})\rho a(1 + a\gamma_3)^2 = 1 \quad (16)$$

For a given value of λ this can be solved to obtain the optimal value of "a" that will give the agent optimal action.

Another approach is to specify that the agent receives his reservation level of utility H_o. In this case we have,

$$\frac{1}{\gamma_2}\{1 - \int \exp(-\gamma_3 x)\, \lambda^{-\gamma_{22}}\, f(x,a)\, dx\} - \rho a^2 = H_o \quad (17)$$

which can be simplified to

$$\frac{1}{\gamma_2} [1 - \frac{\alpha}{a} \lambda^{-\gamma_{22}}] - \rho a^2 = H_o$$

This can be solved for λ to give

$$\lambda = \{\frac{a}{\alpha} [1 - (H_o + \rho a^2)\gamma_2\}^{-\frac{1}{\gamma_{22}}} \tag{18}$$

In order to obtain numerical solutions let

$\gamma_1 = \gamma_2 = 1 \qquad \rho = .01$

then $\gamma_3 = \gamma_{11} = \gamma_{22} = \frac{1}{2}$ and equations (16) and (18) become,

$$\lambda [(.005)(2+a)^2 \ a]^2 = 1 \tag{16a}$$

$$\lambda [(1+ \frac{a}{2})][1-(H_o + .01a^2)]^2 = 1 \tag{18a}$$

With these numerical values the expected utilities are as follows:

$$\text{Principal} \ [1 - \frac{2}{2+a} \lambda^{.5}] \tag{19}$$

$$\text{Agent} \quad [1 - \frac{2}{2+a} \lambda^{-.5}] -(.01)a^2 \tag{20}$$

<u>Optimal Solution</u> for Example 2

When $\lambda = .5$ the first order condition (16a) implies that the optimal value of a is equal to 5.30310304. In this case the expected utilities are as follows:

Principal	$E(U_1)$	.80635443
Agent	$E(H)$	.33147979
	$E(U_1 + \lambda H)$	.97209432

Comparative Statics

If the agent were to select an action which is not the optimal one the impact on the parties' expected utilities would be as follows: For values of "a" above a* the agent's expected utility is decreased because the entire effort does not compensate for the expected additional monetary benefit. For the principal increases in "a" above a* lead to greater output and hence greater expected utility. Note that the group expected utility function

$$E(U_1 + \lambda H) \text{ is maximized when } a = a^*$$

For this type of cooperative sharing rule the conditions are enough to ensure that the first order conditions give a unique value of "a" and a global maximum as well as a local maximum. In more complex problems this point is often troublesome. Table 1 illustrates numerically that the group utility function is maximized at $a = a^*$.

Table 1 Variation in "a" about the Optimal Action a*

$\gamma_1 = \gamma_2 = 1$ $\quad a^* = 5.30310304$

$\rho = .01$ $\quad \lambda = .5$

Agent's Action	$E(U_1)$	$E(H)$	$E(U_1) + \lambda E(H)$
5.20310302	0.80366606	0.33660927	0.97197070
5.30310304	0.80635443	0.33147979	0.97209432
5.40310307	0.80897016	0.32600505	0.97197269

Variation in Principal's Risk Aversion

If the principal becomes more risk averse then γ_1 increases. The principal's risk-tolerance $\frac{1}{\gamma_1}$ decreases since his share can be written in the form

$$x - s(x) = \frac{\frac{x}{\gamma_1}}{\frac{1}{\gamma_1} + \frac{1}{\gamma_2}} - \frac{\log \lambda}{\frac{1}{\gamma_1} + \frac{1}{\gamma_2}} \tag{21}$$

Increases in γ_1 give the principal lower expected utility and a relatively higher expected utility to the agent. As γ_1 increases the size of the optimal "a" declines and total welfare $E(U_1 + \lambda H)$ also declines. This is illustrated numerically in Table 2.

Table 2 Variation in γ_1. Impact on optimal solution.

$\gamma_2 = 1$

$\lambda = .5$

$\rho = .01$

γ_1	a*	$E(U_1)$	E(H)	$E(U_1) + \lambda E(H)$
.5	7.05931	1.52658797	0.02824840	1.54071217
1.0	5.30310	0.80635443	0.33147979	0.97209432
1.5	4.62688	0.55018861	0.43648565	0.76843144

Variation in Agent's Risk Aversion (γ_2)

As γ_2 increases the agent's risk tolerance declines and he receives a lower share of the output. Total output declines and the principal's expected utility increases.

Table 3 Variation in γ_2

$\gamma_1 = 1$

$\lambda = .5$

$\rho = .01$

γ_2	a*	$E(U_1)$	E(H)	$E(U_1) + \lambda H$
.500	6.40699	.79909821	.78589728	1.19204685
1.000	5.30310	0.80635443	0.33147979	0.97209432
1.500	4.89276	0.80743784	0.17052619	0.89270094

Variation in Agent's Work Aversion

As the agent becomes more work averse total output declines. The expected utility of both parties declines.

Table 4 Variation in Agent's Work Aversion

$\gamma_1 = 1$ $\lambda = .5$

$\gamma_2 = 1$

ρ	a*	$E(U_1)$	E(H)	$E(U_1) + \lambda(H)$
.010	5.30310	0.80635443	0.33147979	0.97209432
.020	3.96912	0.76307840	0.21107842	0.86861760
.030	3.32498	0.73441876	0.13717362	0.80300557
.040	2.92053	0.71258909	0.08399867	0.75458843

4. NON-COOPERATIVE SITUATION - MORAL HAZARD

In many realistic situations the principal may not be able to observe the agent's action. This leads to the moral hazard situation discussed earlier. In mathematical terms the consequence of this assumption is to add an additional constraint to the earlier problem. This constraint reflects the fact that the agent now acts in his own self-interest. As we shall see this leads to a welfare loss.

The cooperative solution is sometimes called the first best solution. When there is moral hazard the sharing rule trades off risk-sharing for incentives and the solution is sometimes called the second best.

In this case when the principal offers a contract to the agent he is aware that the agent is work averse. To encourage the agent to put in extra effort he will tend to structure the sharing rule so that the agent is rewarded for favourable outcomes and penalized for unfavourable outcomes. This change in the sharing rule interferes with the risk-sharing that is also taking place so that both parties are worse off than in the pure risk-sharing situation.

There are a number of important technical assumptions needed here to ensure that a solution exists and is unique. We comment on some of these later. Some of them have been referred to already. We shall see that this case involving moral hazard is a much more difficult problem than the cooperative case.

Notation The notation corresponds with that used earlier

$U_1(\cdot)$	-	Principal's utility
$U_2(\cdot)$	-	Agent's utility for wealth
$V(\cdot)$	-	Agent's disutility for effort
x	-	Output in money terms
a	-	Action (effort) of agent
s(x)	-	Agent's share of output
f(x,a)	-	Probability density of output for a given a

$U_i > 0$ - $i = 1, 2$

$U_{1} \leq 0$ -

$U_2 < 0$ - [otherwise more hazard can be avoided]

V - is separable, convex and twice differentiable

f(x,a) - twice differentiable with respect to both its arguments and it has fixed support

s(x) - Share going to the agent when output is x. It is assumed that x does not change with a, ie. x has fixed support

It is assumed that only x, the output, is observable. Since we can only contract on observables, the sharing rule s is a function of x alone. Note the literature is somewhat confusing here since the principal can and does use other paramters in the sharing rule. More on this later.

The optimal rule is given by

$$\max_{s(x),a} \quad E\left\{U_1[x-s(x)]\right\} \tag{22}$$

$$\text{subject to} \quad E\left\{H(s(x),a)\right\} \geq H_0 \tag{23}$$

$$a \in \underset{a' \in \Omega}{\operatorname{argmax}} \; E\left\{H(s(x),a')\right\} \tag{24}$$

where Ω is the set of all possible actions and the notation "argmax" denotes the set of arguments that maximize the function that follows.

The inclusion of the equation (24) distinguishes this from the cooperative problem. In the problems we deal with there will be a unique interior maximum and the sharing rule will be differentiable. Note this is not true in general, Holmstrom (1979), Clarke & Darrough (1980) and Grossman & Hart (1982).

Equations (22), (23), (24) give rise to the following first order conditions. Note: They need not be sufficient and in some cases will not even be necessary.[2]

$$\max_{s(x),a} \int U_1[(x - s(x)] f(x, a) dx \qquad (25)$$

subject to $\int\{U_2[s(x)] - V(a)\} \quad f(x,a) \geq H_o \qquad (26)$

and $\int\{U_2[s(x)] f_a (x, a)\} dx = V'(a) \qquad (27)$

As Holmstrom points out it may be necessary to put bounds on s(x) to avoid non-existence of the solution. Note that here the optimization is taking place over a class of functions - s(x) and the properties of these functions may have important mathematical repercussions. For a discussion on this see Clarke & Darrough (1980).

The equations (25), (26), (27) may be written in terms of a Lagrangian.

$$L = E(U_1) + \lambda [E(H)-H_o] + \mu [\frac{\partial}{\partial a} E[H(s)] \qquad (28)$$

Pointwise optimisation leads to:

$$\frac{U_1'[x - s(x)]}{U_2[s(x)]} = \lambda + \mu \frac{f_a(x,a)}{f(x,a)} \qquad (29)$$

for all x.

Equation (29) gives the conditions satisfied by s(x) for all x at the optimum value of a.

<u>Note 1</u> If $\mu = 0$ equation (27) is non-binding so we recover the pure risk-sharing case (equation (12)).

<u>Note 2</u> The second term on the right-hand side of equation (28)

$$\mu \frac{f_a (x,a)}{f(x,a)}$$

indicates that the functional form of s(x) depends on the distribuiton of x and its functional relation to a. This is in sharp contrast to the pure risk-sharing situation.

Note 3 The term $\mu \frac{f_a(x,a)}{f(x,a)}$ measures the departure from pure risk-sharing. It arises from the incentive aspect of the problem. Observe that $\frac{f_a(x,a)}{f(x,a)}$ is the derivative of the maximum liklihood function log f when a is the unknown paramter. The ratio $\frac{f_a}{f}$ measures how strongly one can infer from x that the agent did not take the optimal action. The agent is rewarded or penalized according to the extent of this deviation.

The equation for μ can be found by obtaining the first order condition for the agent's action by finding $\frac{\partial L}{\partial a}$ and recognizing that equation (27) must also hold. This gives

$$\int U_1[x-s(x)]\, f_a(x,a)dx + \mu \left\{\int U_2(s(x))\, f_{aa}(x,a)dx - V''(a)\right\} = 0 \quad (30)$$

The necessary conditions are given by equations

(26)→ Agents $E(H) \geq H_o$

(27)→ Agent's action yields an optimum utility level for himself

& (30)→ the action maximizes the principal's $E(U_1)$ subject to both constraints

where the sharing rule is given by equation (29) at the optimal value of the parameters.

In general this is a complicated set of equations to solve even if we ensure that a unique solution exists.

However before proceeding to discuss a specific solution some comments on the economic sequence of events are helpful.

Step 1 The principal solves the optimization problem for a given level of H_o. This leads to an optimum triplet (λ^* μ^*, A) where A will be the optimal action that the agent will take.

Step 2 The principal offers the agent the following sharing rule

$$s^*(x) = s^*(x, P^*) \qquad (31)$$

where

$$P^* = [\lambda^* + \mu^* \frac{f_a(x, a)}{f(x, a)}] \text{ evaluated at } a = A$$

Note that s* is set by the principal and that A in particular is a parameter selected by the principal.

Step 3 The agent computes his expected utility for the sharing rule given by (31). Because of the way s*(x) was computed the agent's expected utility is maximized if and only if he takes an action a = A.

Note When the optimal sharing rule has been obtained it is a function of A through (31). However when we are examining the impact of the agent's action on the agent's expected utility after A has been selected the only place where this variation impacts is through the probability density function f(x,a). This observation helps explain why the derivative of the agent's expected utility with respect to the agent's action for the optimal sharing rule is

$$\int U_2[s(x)]\ f_a(x, a)\ dx - V'(a) = 0 \qquad (32)$$

even though the optimal sharing rule is a function of A. We can think of A as being under the principal's control so that it is invariant under variations in the agent's action "a".

The equations (26), (27), (30) subject to equation (29) are in general difficult to solve. At the time of writing I am aware of only 2 exact solutions and both of these assume a risk-neutral principal. It turns out that an explicit solution for equation (29) for s(x) can be obtained if both the principal and the agent have HARA utility functions with identical cautiousness. This means that U_1 and U_2 belong to the HARA class with,

$$RT'(U_1) = \frac{-d}{dW}\left(\frac{U_1'}{U_1''}\right)$$

$$= \frac{-d}{dW}\left(\frac{U_2'}{U_2''}\right) = RT'(U_2) \qquad (33)$$

For the power utility class let

$$U_i = [\alpha W_i + B_i]^{1-\frac{1}{\alpha}} \qquad \alpha \neq 0,\ \alpha \neq 1$$

then equation (29) can be solved to yield

$$s(x) = \frac{P^{\alpha}\,[\alpha x + B_1] - B_2}{\alpha(1 + P^{\alpha})} \qquad (34)$$

$$\text{where } P = \lambda + \mu \frac{f_a(x,a)}{f(x,a)} \qquad (35)$$

The comparison with the first best case is easily made by putting $\mu = 0$ in which case $P = \lambda$.

$$\text{When } U_i(W) = \frac{1-\exp[-\gamma_i(W)]}{\gamma_1}$$

the solution to (29) becomes

$$s(x) = \frac{\gamma_1 x + \log(P)}{\gamma_1 + \gamma_2} \qquad (36)$$

where P is again given by equation (35).

It is interesting to note that the assumption of HARA utility with identical risk cautiousness leads to an explicit functional form for s(x). This observation has been exploited by Boyle and Butterworth (1982) in developing numerical methods for the derivation of the optimal solution. We now examine a specific case which leads to a numerical solution.

Example 3 Risk Sharing with Moral Hazard

This is the counterpart of Example 2 except that "a" is no longer observable.

$$U_i(W) = 1 - e^{-\gamma_i(W)} \qquad \gamma_i > 0$$

$$V(a) = \rho a^2 \qquad \rho \text{ is a constant} > 0$$

$$f(x, a)dx = e^{-\frac{x}{a}} \frac{dx}{a}$$

With these assumptions the expected utilities are as follows:

Principal

$$\frac{1}{\gamma_1} \{1 - \int \exp(-\gamma_3 x)\ P^{\gamma_{11}}\ f(x, a)\ dx\} \tag{37}$$

Agent

$$\frac{1}{\gamma_2} \{1 - \int \exp(-\gamma_3 x)\ P^{-\gamma_{22}}\ f(x, a)\ dx\} - V(a) \tag{38}$$

where the form of the sharing rule (36) has been used to derive these equations and γ_3, γ_{11} and γ_{22} are as in Example 2.

The three equations (26), (27) and (30) can be written out under these assumptions in the form,

Agent's reservation level

$$\frac{1}{\gamma_2} \{1 - \int \exp(-\gamma_3 x)\, P^{-\gamma_{22}} f(x, a)dx\} - V(a) \geq Ho \qquad (39)$$

Nash Condition

$$\frac{1}{\gamma_2} \{\int \exp(-\gamma_3 x)\, P^{-\gamma_{22}} f_a(x, a)dx\} + V'(a) = 0 \qquad (40)$$

Adjoint Equation

$$\frac{1}{\gamma_1} \{\int \exp(-\gamma_3 x)\, P^{\gamma_{11}} f_a(x, a)dx\} + \frac{\mu}{\gamma_2} \{\int \exp(-\gamma_3 x)\, P^{-\gamma_{22}}$$

$$f_{aa}(x, a)dx + V_a\} = 0 \qquad (41)$$

where f(x, a) is the probability density function for an exponential variate with mean a.

Equations (39), (40) and (41) can be transformed into equations involving the incomplete gamma function. Equation (39) becomes

$$\frac{1}{\gamma_2} \{\frac{1}{\gamma_2} - \frac{a}{\mu} Z^{\gamma_{11}} B\, \Gamma(\gamma_{11}, \Theta)\} - V(a) \geq H_o \qquad (42)$$

where

$$\frac{1}{\alpha} = \frac{1}{a} + \gamma_3$$

$$Z = \frac{\mu\alpha}{a^2}$$

$$\Theta = \frac{a}{\alpha}\left[\frac{a\lambda}{\mu} - 1\right]$$

$$B = \exp(\Theta)$$

$$\Gamma(y, \Theta) = \int_{\Theta}^{\infty} e^{-t}\, t^{y-1}\, dy$$

This integral is defined for $\Theta > 0$ which implies $\frac{a\lambda}{\mu} > 1$. To see the economic origin of this inequality note that,

$$P = \lambda + \mu\,\frac{f_{,a}}{f} > 0$$

where $\frac{f_{,a}}{f} = \frac{x - a}{a^2}$ for exponential output. Hence

$$P = \lambda + \frac{\mu(x - a)}{a^2} > 0 \qquad \text{for all } x$$

Since the minimum value of x is zero this implies that

$$\lambda - \frac{\mu}{a} > 0 \text{ or } \frac{a\lambda}{\mu} > 1$$

The Nash condition becomes

$$-\frac{1}{\gamma_2}\left\{\frac{\gamma}{a\mu} B\, [Z^{\gamma_{11}}\, \Gamma([1 + \gamma_{11}, \Theta) - \lambda Z^{-\gamma_{22}}\, \Gamma(\gamma_{11}, \Theta)\right\} - V'(a) = 0 \qquad (45)$$

and finally the adjoint equation gives

$$\frac{\alpha B}{\gamma_1 \mu a}\left\{Z^{(\gamma_{11}+1)}\, \Gamma(\gamma_{11} + 2, \Theta) - \lambda\, Z^{\gamma_{11}}\, \Gamma(\gamma_{11} + 1, \Theta)\right\}$$

$$+ \frac{\alpha B}{\gamma_2 \mu a} \; z^{-\gamma_{22}} \{ z^2 \Gamma(\gamma_{11} + 2,\Theta) - 2\lambda z \Gamma(\gamma_{11} + 1,\Theta) + \lambda^2 \Gamma(\gamma_{11},\Theta)$$

$$- \frac{2\mu}{a} [z \Gamma(\gamma_{11} + 1,\Theta) - \lambda \Gamma(\gamma_{11},\Theta)]$$

$$- \frac{\mu^2}{a^2} \Gamma(\gamma_{11}, \Theta)\} + \mu V''(a) = 0 \qquad (44)$$

Furthermore the principal's expected utility under these assumptions becomes

$$\frac{1}{\gamma_1} \{1 - \frac{a}{\mu} \; z^{(1 + \gamma_{11})} \; B \Gamma(1 + \gamma_{11},\Theta)\} \qquad (45)$$

At this stage we have three non-linear equations (42), (43) and (44) for three unknowns - the optimal triplet (λ, μ, a). These can be solved using standard numerical methods. Boyle and Butterworth (1982) have given further details and in fact have checked the solution using two different algorithms on two different computers.

For the purpose of a numerical solution different parameter values were used but the following are most revealing for current purposes:

γ_1 = 1

γ_2 = 1

ρ = .01

H_o = .30748102

The reason for the unusual value of H_o is to make comparisons with the pure risk-sharing case easier. With these parameter values the optimal triplet is given by:

λ^* = .5

μ^* = .65582671

A = 4.22933049

From equation (36) the optimal sharing rule is:

$$s^*(x) = \frac{x}{2} + \frac{1}{2} \log \left[.5 + \mu \frac{(x-A)}{A^2} \right] \tag{46}$$

This gives us immediately an interesting comparison with the first best or pure risk-sharing result where:

$$s(x) = \frac{x}{2} + \frac{1}{2} \log (.5) \tag{47}$$

Both solutions have the same value of λ and for some purposes this facilitates comparisons. Note that if x lies below A then the agent gets less under the second best case whereas if x exceeds A the agent does better under the second best arrangement. The incentive effects of s* are reflected in this arrangement since x is really used as signal concerning the agent's action.

Still keeping $\lambda = .5$ the agent's share under both situations is given below.

Table 5 Agent's share under first best and second best solution

$\gamma_1 = \gamma_2 = 1$ $\rho = .01a^2$ $\lambda = .50$

Output x	First Best	Second Best
1.01	-.342	-.527
.51	-.092	-.251
1.01	0.158	0.024
1.51	0.408	0.297
2.01	0.658	0.570
3.01	1.158	1.112
3.51	1.408	1.381
4.01	1.658	1.650
4.23* = A	1.768	1.768
4.51	1.908	1.919
5.01	2.158	2.186
10.01	4.658	4.835

Another meaningful way to compare the two solutions is to hold H_0 constant in both cases. Table 6 gives the comparison for both cases.

Table 6 Comparison of First Best and Second Best Solutions

Parameter or Function	First Best	Second Best
H_o = .30748102	.30748102	.30748102
Optimal a	5.40558576	4.22933049
Optimal λ	.45513126	.5
Optimal μ	0.0	.65582671
$E(U_1)$	.81780397	.79865085
E(H)	.30748102	.30748102
$E(U_1) + \lambda E(H)$	.95774819	.95239136

Table 6 indicates that the total output declines as one moves from the first best to the second best situation. Furthermore the welfare loss is absorbed entirely by the principal in this case because we have set up the comparison with the agent receiving the same expected utility under both solutions.

Table 7 illustrates how the agent's expected utility changes as a is varied about the optimal A for the second best solution. As predicted his expected utility is maximised for a = A.

Table 7 Agent's Expected Utility for Small Perturbations around A

λ = .50

$\gamma_1 = \gamma_2 = 1$ $\rho = .01a^2$ $f \sim e^{-\frac{x}{a}}$

a	Agent's Expected Utility
4.20933	.307471
4.21933	.307479
4.22933	.307481
4.23933	.307479
4.24933	.307471

The impact of changing γ_1 and γ_2 on the optimal solution is similar in both the first best and the second best situations. Furthermore if the agent becomes more work averse the total expected output declines and also the total welfare as well. These results are also predictable and in line with the pure risk-sharing situation.

The numerical approach can and has been applied to other HARA type utilities and to different output densities. It appears to offer certain advantages for work in this area.

Advantages of HARA assumptions:

(1) It is simple to solve for the optimal sharing rule. This enables some general sharing rule properties to be identified.

(2) The numerical solution of examples is greatly facilitated.

(3) The usual fundamental technical difficulties are minimized or eliminated.

(4) Since the first-best solution for this entire class always gives a linear sharing rule, comparison between the first-best and second-best solutions is made easier.

(5) Extensions to more complex problems involving
multiagents
multiperiods
multiactions by agent
may be facilitated.

The disadvantage is of course that the restriction on the form of the utility function limits the generality of the solutions.

<u>Role of Additional Information on Optimal Sharing Rule</u>

In analyzing the second-best solution it was assumed that the only information available on the agent's action was indireclty through the total output. It would seem reasonable that additional information on the agent's action might lead to an improvement in overall welfare. This corresponds well with practice where one finds additional information brought into risk-sharing agreement. Sometimes monitoring is used or there is an investigation if the output is below a certain level. In the insurance area if the insured does not live up to the terms of the contract the benefits may be forfeited. Often it may be costly to acquire this additional information.

In the present section it is assumed that the additional information is a signal denoted by y. We follow Holmstrom (1979) here. Assume y is observed by both parties and hence it can be used in constructing a sharing rule. Let F(x, y, a) denote the joint distribution of x and y for a given "a" and let f(x, y, a) be the density function. Both f, a and f, aa are assumed to exist. The sharing rule in this case is s(x, y).

Using the same notation as before the optimal sharing rule satisfies

$$\frac{U_1'[x - s(x,y)]}{U_2'[s(x,y)]} = \lambda + \mu \frac{f_a(x, y, a)}{f(x, y, a)} \qquad (48)$$

where λ μ and a satisfy the generalization of (26), (27) and (30) to include the presence of y. These are the usual conditions that the agent's expected utility be greater or equal to his reservation value, the Nash condition and the first order condition for a constrained maximum.

As Holmstrom (1979) states the term $\frac{f,a}{f}$ has the same interpretation as before. However for a given x this term can vary with y. Thus for the same value of x but different y values the agent may receive less remuneration. Note that if a value of y indicates that less can be inferred about a from x, the deviation from optimal risk-sharing should be smaller. At one extreme the signal y might indicate that $f_a(x, y, a) = 0$ and so optimal risk-sharing should be employed.

It is of interest to investigate conditions under which a signal is valuable. A signal is said to be <u>valuable</u> if both the principal and the agent can be made strictly better off with a contract of the form s(x, y) than they are with a contract of the form s(x).

Holmstrom (1979) proves that a signal is valuable if and only if it is false that

$$\frac{f_a(x, y, a)}{f(x, y, a)} = h(x, a) \qquad (49)$$

When (49) holds a contract of the form s(x) will satisfy (48). When it is false then the optimal contract has the form s(x, y).

Equation (49) can be solved to give,

$$f(x, y, a) = g(x, y)\, h(x, a) \qquad (50)$$

Note that this is the condition for a sufficient statistic if one views a as a random parameter, de Groot (1970). When (50) holds x is a sufficient statistic for the pair (x, y) with respect to "a" which means that x carries all the information about a and y adds no new information.

The material presented in these lectures has provided an introduction to the principal agent problem. Currently there is considerable research activity in this field and a number of extensions and refinements to the models discussed here have been made. The applications of these models encompass a very wide range covering problems in microeconomics, financial economics, labour markets, accounting and insurance. It is hoped that the material presented here will make it easier to access this literature.

5. AN INTRODUCTION TO OPTION PRICING

Options are very important in financial economics for a number of reasons. In the last decade a number of new options markets have opened and there has been a tremendous volume of trading in these markets. Options on common stocks, bonds, other financial assets and commodities are now traded in these markets. In the 1970's there has been significant academic research in the pricing of these options and some of this research is incorporated in models used by practitioners. Since an option is a security whose pay-off is related in a contractual way to some other underlying asset, very many corporate liabilities can be expressed as options or combinations of options. Indeed financial managers have shown considerable ingenuity in creating new financial instruments with option features to enlarge the range of choices available to investors.

The purpose of this section is to provide some background on options and introduce some of the basic results and ideas in modern option pricing. In particular they key valuation result due to Black and Scholes (1973) will be derived and analyzed.

First we start with some definitions. An option is a security which confers the right to buy or sell a given asset within a given time subject to certain conditions. The asset will generally be described as the underlying asset. This could be a common share, an ounce of gold, a piece of real estate or a case of claret.

An option to purchase the underlying asset is known as a call option whereas an option to sell the asset is called a put option.

When the asset is bought or sold under the terms of the option agreement the option is said to be exercised. The last date on which an option can be exercised is called the maturity date. The amount exchanged when the option is exercised is called the exercise price, or strike price or striking price.

If an option can only be exercised at maturity it is known as a European option. If the option can be exercised at any time during its lifetime it is known as an American option. The options traded on the organized option exchanges tend to be American options. For expositional purposes we will find it more convenient to deal with European call options.

As an example, consider the following call option where the underlying asset is a common share of Canadian Pacific Limited. This is a very large diversified company which is traded on the Toronto and other stock exchanges. On July 13, 1983 the closing price for one share of CPL was $45.50. At the same time the November $45 call option had a closing price of $3.625. This means that an investor could acquire the right to purchase one share, of CPL on or before the third Friday in November by paying $3.625 on July 13, plus an additional amount of $45 at the time of exercise. If the stock price rose between July and November the option could be quite valuable. However if the stock price dropped to $45 and stayed there the option would have no value. As another example consider the October DFl 360 call option to purchase one share of ABN stock quoted on the Amsterdam Stock Exchange. On July 22 this option traded for DFl 26.50 while the stock price on the same day closed at DFl 384.50.

In theory an infinite number of options could be written on a particular stock. On the organised options exchanges only a limited number of contracts standardised by maturity date and exercise price are permitted. The current option price reflects the variation on these parameters and some clear patterns emerge. For a call option the value of the option increases as the exercise price decreases. As time to maturity increases the call value increases. If the underlying asset increases in price so will the value of the associated call.

The problem of determining a valuation formula for an option in terms of the various relevant state variables has a distinguished history in financial economics. Bachelier (1900) made a very important contribution in this area. The problem was attacked again by Samuelson (1965) and Merton (1971). Thorpe and Kassouf (1967) also made a valuable contribution. However modern option pricing started with the seminal paper by Black and Scholes (1973). These authors derived a formula for valuing a European call option on a non-dividend paying stock. This development has been described as the most important single advance in the theory of financial economics during the 1970's.

It is worth summarizing the main features of the Black-Scholes approach. To derive their model the authors postulated certain ideal assumptions. It turns out that the model is actually quite robust with respect to these assumptions and that each of them has been relaxed in subsequent research. Since the returns on the option and the returns on the underlying stock are intimately related it is possible to exploit the relationship by forming a portfolio of buying one call option and selling an appropriate number of stock shares short in such a way that the portfolio has no risk. Drawing on a basic principle of financial economics this implies the portfolio must earn the riskless interest rate. In order to find an expression for the portfolio return a theorem from stochastic calculus known as Ito's lemma is invoked. This lemma shows how to differentiate functions of stochastic variables. At this stage an equation for the option price is constructed. Since we know the value of the option with certainty at the maturity date this boundary condition can be used to solve the equation.

It is now convenient to list the assumptions.

(i) The underlying asset is a share of common share which is assumed to pay no dividends. The option is a European call option.

(ii) There are no frictions such as taxes or transaction costs.

(iii) Markets are open continuously for trading.

(iv) Investors can borrow or lend at a known constant riskless interst rate R.

(v) The stock price dynamics are such that the ratio of successive stock prices follows a log-normal distribution. This means that the log of successive stock prices follows a normal distribution. It turns out that the local price dynamics of the stock price under this assumption can be written in the form:

$$\frac{dS}{S} = \mu \; dt + \sigma \; d\tilde{Z} \tag{51}$$

where

$S(t)$ - Stock price at time t

μ - Constant

σ - Constant

$d\tilde{Z}$ - A Gauss-Weiner process such that $E(d\tilde{Z}) = 0$

$$E(d\tilde{Z})^2 = dt$$

Equation (51) implies that the stock price return over a small time interval dt can be decomposed into two parts: a deterministic drift term = μ dt; and a stochastic term of $\sigma\, d\tilde{Z}$.

(vi) It is assumed that the option price W is a function of t and S(t) and that W is twice continuously differentiable with respect to S and once continuously differentiable with respect to t. These mathematical requirements are needed to use Ito's lemma.

Construction of Hedged Portfolio

Assume an investor purchases one call option W and simultaneously sells short λ shares of stock. His net investment in the portfolio P is

$$P = W - \lambda S \qquad (52)$$

For the moment λ is not specified.

In a small time interval dt the portfolio P changes to dP where

$$dP = dW - \lambda\, dS$$

to obtain dS we note from (51) that

$$dS = (S\mu dt + S\sigma d\tilde{Z})$$

Since W is a function of t and S (a stochastic variable) we need to use stochastic calculus to obtain an expression for dW. It turns out that,

$$dW = \frac{\partial W}{\partial t} dt + \frac{\partial W}{\partial S} [dS] + \frac{1}{2} \frac{\partial^2 W}{\partial S^2} [dS]^2 \qquad (53)$$

By substituting for [dS] from (51) and retaining only terms up to order dt we note that dW can be rewritten.

$$dW = \frac{\partial W}{S} [\ dt + \ dZ] + \frac{\partial W}{t} dt + \frac{1}{2} \quad S \ \frac{\partial^2 W}{S} dt \qquad (54)$$

Note that equation (53) corresponds to the terms in the Taylor series expansion of W(S, t) as a function of the two variables S and t.

By using equations (51) and (54) the expression for dP can be rewritten.

$$dP = \alpha dt + [\frac{\partial W}{\partial S} - \lambda]S\sigma d\tilde{Z} \qquad (55)$$

where

$$\alpha = [\mu S \frac{\partial W}{\partial S} - \lambda\mu S + \frac{\partial W}{\partial t} + \frac{1}{2}\sigma^2 S^2 \frac{\partial^2 W}{\partial S^2}]$$

Equation (55) indicates that the change in value of P; dP, can be written as the sum of two terms. The first term, αdt, is a deterministic term while the second term is a stochastic term governed by the same stochastic process which produced the local variation in the stock price.

Recall that λ was not specified. We can ensure that the stochastic component of dP vanishes if we set

$$\lambda = \frac{\partial W}{\partial S} \qquad (56)$$

Note that this has to be done continuously as time changes. A rigorous derivation and deeper discussion is given by Ingersol (1982). If condition (56) is used the expression for α becomes:

$$\alpha = \frac{\partial W}{\partial t} + \frac{1}{2}\sigma^2 S^2 \frac{\partial^2 W}{\partial S^2} \qquad (57)$$

Note in particular that the term involving μ drops out. This has an important economic sequel.

The process of eliminating the stochastic component of the return is known as hedging. The concept of hedging has become extremely important both in the theoretical and practical aspects of modern financial economics. It lies at the very heart of the derivation of the Black-Scholes equation.

The No-Arbitrage Principle

A colourful statement of this principle is that there are no free lunches in the capital markets. It states that an investor cannot earn a riskless profit with a zero net investment. This implies that if two investments have identical payoffs in all states of the world then they must sell for the same price. If they did not an investor could exploit the difference and earn arbitrage profits. In a free and competitive market this profit opportunity would be competed away.

Construction of Black-Scholes Equation

In the present case we note that portfolio P is riskless and hence must earn the riskless rate. In the time interval dt the interst earned on the portfolio is

$$RPdt$$

where R is the continuously compounded riskless rate. Hence we have the following equation:

$$dP = [\frac{\partial W}{\partial t} + \frac{1}{2}\sigma^2 S^2 \frac{\partial^2 W}{\partial S^2}]\ dt = PRdt$$

Divide across by dt and substitute for P to obtain

$$\frac{\partial W}{\partial t} + \frac{1}{2}\sigma^2 S^2 \frac{\partial^2 W}{\partial S^2} = R[W - S\frac{\partial W}{\partial S}]$$

This can be rearranged to give:

$$\frac{1}{2}\sigma^2 S^2 \frac{\partial^2 W}{\partial S^2} + RS\frac{\partial W}{\partial S} - RW + \frac{\partial W}{\partial t} = 0 \qquad (58)$$

It is convenient to change the time variable t to T = t* - t where t* is the maturity date of the option. With this substitution (58) becomes:

$$\frac{1}{2}\sigma^2 S^2 \frac{\partial^2 W}{\partial S^2} + RS\frac{\partial W}{\partial S} - \frac{\partial W}{\partial T} - RW = 0 \qquad (59)$$

Equation (59) is a second order parabolic differential equation satisfied by the option price. This is the famous Black-Scholes differential equation.

To solve the equation boundary conditions are required. Note that when $T = 0$, the option is equal to the greater of $S - E$ or 0 where E is the exercise price of the option.

$$W(S, 0) = \text{Max}[0, (S-E)] \tag{60}$$

Also note that if the stock price is zero the option is worthless so that:

$$W(0, T) = 0 \tag{61}$$

Different methods are available for the solution of this equation. Here we write out the solution and show that it satisfies the partial differential equation and the associated boundary conditions.

Solution of Black-Scholes Equation

The solution to equation (59) subject to (60) and (61) is:

$$W(S, T) = SN(d_1) - E\exp(-RT)\, N(d_2) \tag{62}$$

$$\text{where } d_1 = [\log \frac{S}{E} + (R + \frac{1}{2}\ {}^2)T]/\sigma \sqrt{T}$$

$$d_2 = d_1 - \sigma \sqrt{T}$$

$$N(y) = \frac{1}{\sqrt{2\Pi}} \int_{-\infty}^{y} e^{-\frac{u^2}{2}} du$$

Note that $N(\cdot)$ is the univariate standard cumulative normal density function. Note also that the option price

= Current Stock Price (Probability function)

- Discounted Exercise Price (another probability function)

It is straightforward to check that the option price (12) satisfies the boundary conditions. In order to show that W satisfies (59) the various partial derivatives are required.

Partial Derivatives of W

We list here the partial derivatives of W with respect to the variables of interest.

$$\frac{\partial W}{\partial S} = N(d_1)$$

$$\frac{\partial^2 W}{\partial S^2} = \frac{Z(d_1)}{S\sigma\sqrt{T}}$$

$$\frac{\partial W}{\partial T} = E\ \exp(-RT)\ [\frac{Z(d_2)\sigma}{2\sqrt{T}} + R]$$

$$\frac{\partial W}{\partial R} = TE\ \exp(-RT)\ N(d_2)$$

$$\frac{\partial W}{\partial E} = -\ \exp(-RT)\ N(d_2)$$

$$\frac{\partial W}{\partial \sigma^2} = \frac{SZ(d_1)}{2\sigma}\ \sqrt{T}$$

The comparative statics of the option price are summarized in the following table.

Increase in Variable	Change in Option Price
S	Increase
T	Increase
E	Decrease
R	Increase
σ^2	Increase

Note in particular that an increase in the variance or risk of the underlying stock leads to an increase in the option price.

Four of the five variables S, T, E and R are readily observable in an empirical situation. Only the fifth variable σ^2 needs to be estimated. Considerable work has been done in connection with estimating σ^2. However the estimation of the variance remains a difficult problem.

A number of empirical tests of the Black-Scholes formula and its extensions have been conducted with reasonably good agreement between the theoretical predictions and the actual empirical observations.

An Alternative Valuation Expression and Preference-Free Valuation

It has been already noted that the parameter μ drops out of the final valuation equation (59) and does not occur in the solution - equation (62). At first sight this is a surprising result. This result can be traced back to the hedging procedure where the close relationship between the option's return and that of the stock allowed us to construct a riskless hedge whose return was independent of the stock price drift.

Cox and Ross (1976) and Harrison and Kreps (1979) have developed this point to arrive at another valuation formula which provides useful additional insight and is very convenient for analytical work. Their valuation procedure requires that the actual lognormal density of the stock price be replaced by an associated lognormal density. This associated density has an expected return equal to the riskless rate and the same local variance as the actual stock price return. If the original density is

$$L(\mu, \sigma^2, T)$$

the associated density is

$$\bar{L}((r - \frac{1}{2}\sigma^2), \sigma^2, T)$$

Note that if the current stock price is S(t) the expected stock price, assuming the associated density is,

$$S(t) \exp(rT)$$

so that in this case the stock earns the riskless rate. In practice the stock will <u>not</u> earn the riskless rate but this transformation produces an algorithm which gives the correct answer.

Consider the following integral over the terminal stock price S* conditional on the current stock price S(t),

$$\exp(-RT) \int_{S^*>E} (S^*-E)\, \bar{L}\,(r-\frac{1}{2}\sigma^2),\ \sigma^2, T)dS^* \qquad (63)$$

the value of this integral is

$$SN(d_1) - E\exp(-RT)\, N(d_2) \qquad (64)$$

i.e., precisely W(S, T) the option price.

This algorithm enables us to value the call option by discounting the terminal option value integrated over an associated terminal stock price value. The associated terminal stock price distribution is obtained by requiring that the return on the stock be equal to the riskless rate.

Equation (63) closely resembles the expression for the net-stop-loss premium in insurance mathematics. Some observations in this connection have been made previously by Boyle (1978).

Extensions of the Basic Model

The Black-Scholes model has been extended in a number of different directions. It has proved robust with respect to the various assumptions.

The model has been extended to value dividend paying stocks, American options, put options and a whole series of complicated options.

This approach has also been used to price various corporate liabilities such as bonds, common stock, preferred stock, warrants and commodity options.

The basic hedging argument has been used to derive models of the term structure of interest rates. Boyle (1980) and the references therein discusses some of these recent developments.

This is a fast developing field and it is impossible to give an adequate review here. A good early survey is given by Smith (1976). Ingersol (1982) presents a good discussion of option pricing and the important aspects of modern finance in a concise manner. More recent contributions to the literature appear in the North American financial economics journals such as The Journal of Finance, The Journal of Financial Economics and the Journal of Financial and Quantitative Analysis.

Risk Reducing Properties of Options

An investor or portfolio manager can use options to reduce the risk associated with common stock ownership. For example, he can sell calls against some of the common stocks in the portfolio. This essentially truncates the probability distribution of the stock price distribution and reduces the variance of this distribution. Options can be used to change the distribution of returns on financial assets in much the same way that reinsurance can alter the distribution of an insurance company's liabilities.

Footnotes

1 The author's approach to this field has been stimulated and enriched by many discussions with Dr. John Butterworth. Section 4 of these lecture notes is based on joint work with Dr. Butterworth. However the present author bears responsibility for any errors.

2 The paper by Grossman and Hart gives a specific example illustrating this paradoxical point.

References

Akerlof, G. (1970), "The Market for Lemons: Qualitative Uncertainty and the Market Mechanism," Quarterly Journal of Economics, 84, pp. 488-500.

Bachelier, L. (1900), "Theory of Speculation" (English translation in: P. Cootner, ed., 1964, The Random Character of Stock Market Prices, MIT Press, Cambridge, Massachusetts) pp. 17-78.

Black, F. and M. Scholes (1973), "The Pricing of Options and Corporate Liabilities," Journal of Political Economy, 81, pp. 637-659.

Borch, K. (1962), "Equilibrium in a Reinsurance Market," Econometrica, Vol. 30, No. 3, pp. 424-444.

Boyle, P.P. (1978), "The Poisson-Exponential Model and the Non-Central Chi-Squared Distribution," Scandinavian Actuarial Journal, 108-111.

Boyle, P.P. (1980), "Recent Models of the Term Structure of Interest Rates with Actuarial Applications," Transactions of the 21st International Congress of Actuaries, Switzerland, June 19-26, 1980. Vol. 4, pp. 95-105.

Boyle, P.P., and J.E. Butterworth (1982), "The Principal Agent Problem: Numerical Solutions," Working Paper, Faculty of Commerce, University of British Columbia, Vancouver, B.C., Canada, V6T 1Y8.

Buhlmann, H. (1980), "An Economic Premium Principle," Astin Bulletin, 11, pp. 52-60.

Clarke F. and M. Darrough (1980), "Optimal Incentive Schemes, Existence and Characterization," Economic Letters, 5, pp. 305-310.

Cox, J.C. and Ross, S.A. (1976), "The Valuation of Options for Alternative Stochastic Processes," Journal of Financial Economics, 3, 145-166.

DeGroot, M. (1970), "Optimal Statistical Decisions," New York: Mcgraw Hill Book Company.

Gale, D. and J. Sobel (1982), Journal of Mathematical Economics, 9, pp. 51-59.

Grossman, S. and O. Hart (1983), "An Analysis of the Principal Agent Model," Econometrica, Vol. 51, No. 1, pp. 7-46.

Harrison, J.M. and Kreps, D.M. (1979), "Martingales and Arbitrage in Multiperiod Securities Markets," Journal of Economic Theory, 20, pp. 381-408.

Holmstrom, B. (1979), "Moral Hazard and Observability," Bell Journal of Economics, Vol. 10, No. 1, pp. 74-91.

Ingersol, J. (1982), "Notes on the Theory of Financial Decisions: Class Lecture Notes," Working Paper 26, Graduate School of Business, University of Chicago, 1101 East 58th Street, Chicago, IL 60637, U.S.A.

Lambert, R.A. (1981), "Managerial Incentives in Multiperiod Agency Relationships," Unpublished Ph.D. Thesis, Stanford University.

Merton, R.C. (1971), "Theory of Rational Option Pricing," Sloan School of Management, Working Paper, 574-71, (MIT Cambridge, Massachusetts).

Mirless, J.A. (1971), "An Exploration into the Theory of Income Taxation," Review of Economic Studies.

Mossin, J. (1973), "Theory of Financial Markets," Prentice-Hall, Inc., Englewood Cliffs, New Jersey.

Nash, J.F. (1950), "The Bargaining Problem," Econometrica, 28, pp. 122-137.

Samuelson, P.A. (1965), "Rational Theory of Warrant Pricing," Industrial Management Review 6, pp. 13-31.

Smith, C.W. (1976), "Option Pricing: A Review," Journal of Financial Economics, 3, pp. 3-51.

Spence, M. and R. Zeckhauser, "Insurance, Information and Individual Action," American Economic Review, 63. pp. 134-139.

Thorpe, E.O. and S.T. Kassouf, (1967), "Beat the Market," Random House, New York.

STATE-DEPENDENT UTILITY, THE DEMAND FOR INSURANCE AND THE VALUE OF SAFETY

Pierre Dehez

and

Jacques H. Drèze

Abstract

How much should a society spend for the safety of its members? The paper treats that question within the framework of the theory of public expenditures. The methodological foundations of the subjective approach, which underlies the concept of marginal willingness-to-pay for safety, are reviewed. The paper also includes a complete study of the individual demand for life insurance and for (private) safety.

F. de Vylder et al. (eds.), Premium Calculation in Insurance, 77–107.

Public projects increasing safety, or entailing risks of human deaths, have inescapably raised the question: "How much should a society spend for the safety of its members?"; alternatively stated: "What is the social value of a human life?" Answers provided by economists, public administrators and other scientists[2] have come to be classified under two headings: The "human capital" approach and the "willingness-to-pay" approach[3]. Broadly speaking, the first approach seeks an objective measure of the value of a person's life in his or her earning ability, whereas the second approach takes as its starting point the person's subjective tastes. Some of the literature is directed towards relating the two approaches and in particular towards formulating assumptions under which a person's human capital provides a lower bound to the value placed by that person on her life.

It seems to be now generally recognized that "willingness-to-pay" is the logical approach to assessing the benefits of public safety. The approach becomes operational when coupled with the modern theory of public expenditure, initiated by Samuelson (1954). Yet, the methodological foundations of the subjective approach - state-dependent preferences - do not seem to be universally understood, leaving room for occasional doubt or error. The present paper starts (section 1) with a review of these foundations, which involve preferences revealed through gambling, insurance and life protection. The economics of individual demand for insurance and safety are the subject of sections 2 and 3 respectively. The "public goods" aspect is taken up briefly in section 4[4].

Decisions affecting survival probabilities should preferably be analyzed in an intertemporal framework. But such an analysis is at once quite complicated, and methodological issues are more conveniently discussed in the simpler static framework used here.

1. REVEALED PREFERENCES AND EXPECTED UTILITY

1.1 Economic choices affecting survival probabilities are of necessity choices with uncertain consequences; they must accordingly be studied within the framework of a theory of decision-making under uncertainty. The behavioral theory of "expected utility" (or "moral expectation"), in the modern treatment due to von Neumann and Morgenstern (1947) and Savage (1954), provides the starting point for developing such a framework.

In its simplest form, it concerns an individual choosing among bets b,b'..., involving sums of money w_i, w_i'..., staked on events i generated by random devices with objectively known probabilities p_i. The other circumstances affecting the person (her health and wealth, the prices at which money buys commodities, her family status and so on) are taken as given and unaffected by the bets. The theory defines axiomatically "consistent behavior": Choices among arbitrary bets should be well-defined and transitive; if two bets stake identical prizes on some events, the choice among these bets is unaffected by identical modifications of the common prizes ("strong independence" axiom or "sure thing principle"). Under a technical continuity assumption, these axioms lead to the "moral expectation theorem":

(i) There exists a real-valued function of money wealth, $u(w)$, called utility, such that a bet b entailing a higher level of expected utility is always preferred to a bet b' entailing a lower level of expected utility; that is "$\sum_i p_i u(w_i) > \sum_i p_i u(w_i')$ if and only if b is preferred to b' ".

(ii) The function u is defined up to positive linear transformations; that is $v(w) = \alpha + \beta u(w)$, $\beta > 0$, is also a utility.

It should be emphasized that no other interpretation is proposed for u besides the fact that it represents correctly the choices among bets, made by a particular person under given circumstances.

The extension to choices involving consequences more diversified than payments of money, staked on events more substantive than the outcomes of random devices, requires some care. Suppose, for example, that a person setting out for a day of sailing is offered two "bets", of which the first promises a pint of beer if the day is warm and a bowl of soup if the day is cold; whereas the second promises a pint of beer if the day is cold and a bowl of soup if the day is warm. No valid analysis of this situation could ignore the obvious point that "beer on a hot day" and "beer on a chilly day" are two different "consequences". Similarly, in the analysis of life insurance, one cannot ignore the obvious point that "an indemnity in case of death" is not the same consequence as "an indemnity in case of life".

We shall not treat here the case where the definition of consequences includes references to events with unknown, subjective probabilities[5]. Assuming that probabilities are objectively known entails a drastic simplification (at a logical, not only technical level), and enables us to concentrate attention on the utility aspects, which are our primary concern.

Consider the simple case of two events, the event of life indexed $i = 1$, and the event of death, indexed $i = 0$. (These two events could be associated with the outcome of the sailing trip mentioned above...) Recognizing that payments of money in case of life, and payments of money in case of death, are different "consequences", we begin with the definition of two utility functions for money wealth, $u_1(w_1)$ for the case of life and $u_0(w_0)$ for the case of death.

Each of these reflects (is constructed from) *conditional bets* involving sums of money staked on occurrences generated by a random device. Thus, $u_1(w_1)$ is constructed from bets conditioned on the event of life: If and only if you ("the person") return safely from the sailing trip, a roulette wheel is to be spun, and according to the outcome you receive prizes w_{1i} with probabilities p_i. It is assumed that "you" make consistent choices among such bets, i.e. (conditional) choices that can be represented by a (conditional) function $u_1(w_1)$,*defined up to a positive linear transformation.*

Similarly, $u_0(w_0)$ is constructed from bets conditioned on the event of death: If and only if you die at sea, a roulette wheel is to be spun, and your heirs receive prizes w_{0i} with probabilities p_i. Again, $u_0(w_0)$ is thereby defined *up to a positive linear transformation.*

So far, we have simply applied the expected utility theory of von Neumann and Morgenstern to conditional bets. We have constructed two utility functions, say $\alpha_1 + \beta_1 u_1(w_1)$ and $\alpha_0 + \beta_0 u_0(w_0)$, with α_1, β_1, α_0 and β_0 still undefined ($\beta_1 > 0$, $\beta_0 > 0$). And we are assuming that the two events (life and death) have known probabilities p_1, p_0, with $p_1 + p_0 = 1$.

As a next step, we can introduce choices among money payments staked upon the events "life" and "death". A simple example is the straight life insurance contract whereby "you" pay (now) a premium y to receive an indemnity 1 (at the end of the day) in case of death[6]. Suppose that, for $y = \bar{y}$, you are *indifferent* between signing and not signing such a contract. We will then *define* your "utility gain" in case of death, $\alpha_0 + \beta_0 u_0(w_0+1-\bar{y}) - \alpha_0 - \beta_0 u_0(w_0) = \beta_0[u_0(w_0+1-\bar{y}) - u_0(w_0)]$, weighted by its probability p_0, as equal to your "utility loss" in case of life, $\beta_1[u_1(w_1) - u_1(w_1-\bar{y})]$, weighted by its probability $p_1 = 1 - p_0$:

$$p_0\beta_0[u_0(w_0+1-\bar{y}) - u_0(w_0)] + (1-p_0)\beta_1[u_1(w_1-\bar{y}) - u_1(w_1)] = 0 \tag{1.1}$$

$$\frac{\beta_0}{\beta_1} = \frac{1-p_0}{p_0}\,\frac{u_1(w_1)-u_1(w_1-\bar{y})}{u_0(w_0+1-\bar{y})-u_0(w_0)} . \tag{1.2}$$

Consistency requires that the ratio $\frac{\beta_0}{\beta_1}$ be independent of the particular insurance contract used to define it, so long as indifference prevails. Denoting the right hand side of (1.2) by r_β, we have that $\beta_0 = r_\beta\beta_1$, so that the utility in case of death is $\alpha_0 + r_\beta\beta_1 u_0(w_0)$[7].

It should be emphasized that no other interpretation is proposed for r_β besides the fact that it represents correctly the choices made among insurance contracts.

The final step consists in offering the person (you) an opportunity to reduce the probability of death, say from p_0 to $p_0-\delta$, by acquiring a special safety equipment at a cost x. Suppose that, for $x = \bar{x}$, you are indifferent between acquiring and not acquiring the equipment. We will then *define* your expected utility gain, through higher safety, $\delta[\alpha_1 + \beta_1 u_1(w_1-\bar{x}) - \alpha_0 - r_\beta \beta_1 u_0(w_0-\bar{x})]$, as equal to your expected utility loss due to the cost $\bar{x}$[8], $p_0 r_\beta \beta_1[u_0(w_0) - u_0(w_0-\bar{x})] + (1-p_0)\,\beta_1[u_1(w_1) - u_1(w_1-x)]$; thus

$$(1.3)\qquad p_0[\alpha_0 + r_\beta \beta_1 u_0(w_0)] + (1-p_0)[\alpha_1 + \beta_1 u_1(w_1)]$$

$$= (p_0-\delta)[\alpha_0+r_\beta\beta_1 u_0(w_0-\bar{x})] + (1-p_0+\delta)[\alpha_1+\beta_1 u_1(w_1-\bar{x})],$$

$$(1.4)\qquad \alpha_0 - \alpha_1 = \beta_1 u_1(w_1-\bar{x}) - r_\beta \beta_1 u_0(w_0-\bar{x})$$

$$+ \frac{1}{\delta}[p_0 r_\beta \beta_1\{u_0(w_0-\bar{x}) - u_0(w_0)\} + (1-p_0)\beta_1\{u_1(w_1-\bar{x}) - u_1(w_1)\}].$$

Consistency requires that the difference $\alpha_0 - \alpha_1$ be independent of the particular safety gain used to define it, so long as indifference prevails. Denoting the right-hand side of (1.4) by $d_\alpha \beta_1$, we have that $\alpha_0 = \alpha_1 + d_\alpha \beta_1$, so that the utility in case of death is $\alpha_1 + \beta_1[d_\alpha + r_\beta u_0(w_0)]$.

It should be emphasized that no other interpretation is proposed for d_α besides the fact that it represents correctly the choices made among safety levels. We return below to this important point.

Putting pieces together, and using again the moral expectation theorem, we have a representation of choices in terms of an expected utility which is state-dependent and defined up to a *single* positive linear transformation, namely:

$$(1.5)\qquad Eu(w_0,w_1;p_0,\alpha_1,\beta_1) = p_0[\alpha_1+\beta_1\{d_\alpha+r_\beta u_0(w_0)\}] + (1-p_0)[\alpha_1+\beta_1 u_1(w_1)]$$

$$= \alpha_1 + \beta_1[p_0\{d_\alpha+r_\beta u_0(w_0)\} + (1-p_0)u_1(w_1)].$$

The choices of origin and unit of scale for u_0 and u_1 have been arbitrary throughout - for instance, $u_0(0) = u_1(0) = 0$; $u_0(1) = u_1(1) = 1$. The choice of α_1 and β_1 is still open, and there is no loss of generality in setting $\alpha_1 = 0$, $\beta_1 = 1$, so that

$$(1.6)\qquad Eu(w_0,w_1;p_0) = p_0[d_\alpha+r_\beta u_0(w_0)] + (1-p_0)u_1(w_1).$$

The important point is that d_α and r_β are *uniquely* defined (given our conventions) by (1.2) and (1.4), and relate unambiguously the origins and units of scale of u_0 and u_1. *No additional degree of freedom is left to rescale or relocate either* u_0 *or* u_1 *separately.* We underscore this feature because it is overlooked in a number of otherwise valuable contributions to the literature[9].

1.2 Formula (1.6) is the more general formulation of the expected utility approach to the valuation of survival probabilities, as introduced in Drèze (1962) and found in Bergström (1978), Cook and Graham (1977) and Jones-Lee (1974, 1976). It is usually presented in the form

$$(1.7)\quad Eu(w_0,w_1;p_0) = p_0u_0(w_0) + (1-p_0)u_1(w_1)$$

without explicit reference to d_α and r_β. It must then be understood that the respective scalings of u_0 and u_1 satisfy (1.2) and (1.4). That is, if $u_1(0) = 0$, $u_1(1) = 1$, then $u_0(0) = d_\alpha$, $u_0(1) = d_\alpha + r_\beta$ as defined there. The only other standard restriction placed on u_0 and u_1 is concavity - i.e. risk aversion[10].

This formulation is consistent with the propositions that a reduction of p_0 by δ increases expected utility by $\delta[u_1(w_1) - u_0(w_0)]$, if w_0 and w_1 are taken as given; and that such a reduction, accompanied with a reduction of both w_0 and w_1 by $\bar{x}$, leaves expected utility unchanged. But these propositions are tautological, to the extent that they merely retreive a step in the construction of (1.7).

In other words, the magnitude of $\bar{x}$ is *in no way* restricted by the abstract formulation (1.7): *A person's willingness-to-pay for increased safety is a matter for that person to decide;* the consistency axioms underlying (1.7) have no implications whatever for that decision. *This is precisely as it should be,* considering the variety of motives which may influence the decision. Thus, a rich bachelor without heirs may still be unwilling to buy additional safety at any positive price, thereby suggesting that he goes through life without enjoying it at all and would just as well face death as a reduction in his lavish (but unrewarding) consumption standard. At the same time, a poor widow working overtime to keep her children at subsistence might decline a well-paid but riskier job, thereby suggesting that she does not want to take chances with her children becoming orphans.

Yet, additional assumptions about either behavior, or market opportunities, or both enable the theorist to draw some implications from the abstract formulation (1.7). Such assumptions, and their implications, are considered in sections 2 and 3.

2. THE DEMAND FOR INSURANCE

2.1 State-dependent utility functions provide the proper tool to study the demand for insurance, when the indemnities are staked on events which are relevant for the definition of final consequences (events which affect preferences). Life insurance is an obvious example - see Drèze (1962). Insurance of irreplaceable commodities, like heirlooms, is another example - see Cook and Graham (1977).

To study the demand for insurance - which has some implications for the demand for safety - we retain and simplify the notation of section 1, in particular of (1.7):

$$\text{(2.1)} \qquad Eu(w_0,w_1;p) = pu_0(w_0) + (1-p)u_1(w_1),$$

where w_0 is "wealth in case of death", w_1 is "wealth in case of life" and $p(= p_0$, subscript omitted) is the probability of death. Denoting non-human wealth by w, we write $w_0 = w + a_0$ and $w_1 = w + a_1$. About u_0 and u_1, we make the following assumption:

Assumption 1

u_0 and u_1 are twice differentiable non-decreasing concave functions; u_1 is strictly increasing.

Differentiability is assumed for convenience, so that we can use derivatives instead of the heavier formalism of subgradients. Concavity is justified by the existence of opportunities to gamble at (nearly) fair odds; see the references in footnote 10. We assume that more wealth is strictly preferred to less wealth in case of life, whereas more wealth is preferred or indifferent to less wealth in case of death. Our comments will be geared to the more typical situation where u_0 is strictly increasing.

Both u_0 and u_1 depend upon other variables which are kept constant, and ignored, in our static analysis. In particular, they depend upon future contingencies, the conditional probabilities of which (given the states of death 0 or life 1) are assumed constant. Thus, u_0 and u_1 may be viewed as well-defined conditional expected utilities (where the expectation is taken over the future contingencies).

We wish to study insurance contracts, involving the definite payment of an initial premium y, giving right to a gross indemnity k(y;p) in case of death. For the holder of such a contract, expected utility is given by

$$\text{(2.2)} \qquad Eu(y) = pu_0(w_0+k(y;p)-y) + (1-p)u_1(w_1-y).$$

An optimal policy corresponds to the value of y for which (2.2) is maximal. Properties of optimal policies stand out more clearly under the following assumptions, motivated by convenience:

Assumption 2

$$\lim_{w \to 0} u_1'(w) = \infty .$$

Assumption 3

$$k(y;p) = \frac{\alpha}{p} y.$$

Assumption 2 guarantees that y maximizing (2.2) is strictly less than w_1, thereby dispensing us with the need to introduce a liquidity constraint explicitly. Assumption 3 introduces the standard stipulation that price is independent of quantity. The parameter α is an index of *fairness*. When $\alpha = 1$, insurance is fair in the sense that $pk(y;p) = y$. The premium is equal to the actuarial value of the policy. For $\alpha > 1$ (resp. < 1), the insurance is more than fair (resp. less than fair). Although the loading factor typically results in less than fair insurance, still tax provisions may restore fairness - see Jones-Lee (1976). Some implications of fairness are exhibited below.

2.2 Necessary and sufficient conditions for an optimal insurance contract, i.e. for a maximum of (2.2), are:

$$(2.3) \qquad (\alpha-p)u_0'(w_0+\frac{\alpha}{p}y-y) - (1-p)u_1'(w_1-y) = 0;$$

$$(2.4) \qquad D =_{def} (\alpha-p)^2u_0''(w_0+\frac{\alpha}{p}y-y) + p(1-p)u_1''(w_1-y) \leqslant 0.$$

Under Assumption 1, the second-order condition is automatically satisfied. If it were stipulated that y must be non-negative, condition (2.3) should be replaced by

$$(2.5) \qquad y[(\alpha-p)u_0' - (1-p)u_1'] = 0, \ (\alpha-p)u_0' - (1-p)u_1' \leqslant 0.$$

The first-order condition implies

Proposition 1

(i) *Under* (2.3), $u_1'-u_0' \gtreqless 0$ *according as* $\alpha \gtreqless 1$.

(ii) *Under* (2.5), $\alpha \geqslant 1$ *implies* $u_0' \leqslant u_1'$.

(iii) *When* $\alpha \leqslant 1$ *(resp* $\alpha=1$*), then* $y > 0$ *only if (resp if and only if)* $u_1'(w_1) < u_0'(w_0)$.

Implicit differentiation of (2.3) leads to the following formulae ($D \neq 0$):

$$(2.6) \qquad \frac{\partial y}{\partial w_0} = -p(\alpha-p)u_0'' \,/\, D, \quad \frac{\partial y}{\partial w_1} = p(1-p)u_1'' \,/\, D;$$

$$(2.7) \qquad \frac{\partial y}{\partial w} = \frac{\partial y}{\partial w_0} + \frac{\partial y}{\partial w_1} = p(1-p)u_1' \left(\frac{u_1''}{u_1'} - \frac{u_0''}{u_0'}\right) / D;$$

$$(2.8) \qquad \frac{\partial y}{\partial p} = [p(u_0'-u_1') + (\alpha-p)\frac{\alpha}{p}\, y\, u_0''] / D = [p\,\frac{1-\alpha}{1-p}\, u_0' + (\alpha-p)\frac{\alpha}{p}\, y\, u_0''] / D;$$

$$(2.9) \qquad \frac{\partial(\frac{\alpha}{p}y - y)}{\partial p} = \left[\frac{\alpha-p}{1-p}(1-\alpha)u_0' - \frac{\alpha}{p}\, y(1-p)u_1''\right] / D$$

$$= \left[1 - \alpha + \frac{\alpha}{p}\, y\, (1-p)\left(-\frac{u_1''}{u_1'}\right)\right]u_1' \,/\, D;$$

$$(2.10) \qquad \frac{\partial y}{\partial \alpha} = -[pu_0' + (\alpha-p)yu_0''] / D;$$

$$(2.11) \qquad \frac{\partial(\frac{\alpha}{p}\, y - y)}{\partial \alpha} = [(1-p)yu_1'' - (\alpha-p)u_0'] / D;$$

$$(2.12) \qquad \left.\frac{\partial y}{\partial p}\right|_{\frac{\alpha}{p} = \text{constant}} = -u_1' \,/\, D.$$

The main implications of these formulae are:

Proposition 2

Under (2.3) *and strict risk aversion* ($D < 0$):

(i) $\frac{\partial y}{\partial w_0} \leq 0$, $0 \leq \frac{\partial y}{\partial w_1} \leq 1$, $\frac{\partial y}{\partial w} \leq 1$, $\frac{\partial y}{\partial w} \gtreqless 0$ according as $-\frac{u_1''}{u_1'} \gtreqless -\frac{u_0''}{u_1'}$;

(ii) $-1 \leq \frac{\partial(\frac{\alpha}{p}\, y - y)}{\partial w_0} \leq 0$; $\alpha \geq 2p$ *implies* $\frac{\partial y}{\partial w_0} \geq -1$;

(iii) $\frac{\partial y}{\partial p} > 0$ *if and only if* $-(w_0 + \frac{\alpha}{p}\, y - y)\, \frac{u_0''}{u_0'} > (1-\alpha)\, \frac{p}{1-p}\, \frac{w_0 + \frac{\alpha}{p} y - y}{\frac{\alpha}{p}\, y - y}$;

$\alpha \geq 1$ *implies* $\frac{\partial y}{\partial p} \geq 0$; $\left.\frac{\partial y}{\partial p}\right|_{\frac{\alpha}{p} = \textit{constant}} > 0$;

(iv) $\frac{\partial(\frac{\alpha}{p}\, y - y)}{\partial p} \leq 0$ *if* $\alpha \leq 1$;

(v) $\frac{\partial y}{\partial \alpha} < 0$ *if and only if* $-(w_0 + \frac{\alpha}{p}\, y - y)\, \frac{u_0''}{u_0'} > \frac{w_0 + \frac{\alpha}{p}\, y - y}{\frac{\alpha}{p}\, y - y} \geq 1$;

$\frac{\partial y}{\partial \alpha} < 0$ *implies* $\frac{\partial y}{\partial p} > 0$;

(vi) $\frac{\partial(\frac{\alpha}{p}\, y - y)}{\partial \alpha} > 0$; $(-\frac{u_0''}{u_0'}) \leq (-\frac{u_1''}{u_1'})$ *implies* $\frac{\partial(\frac{\alpha}{p}\, y)}{\partial \alpha} > 0$.

The first of these propositions is straightforward. The demand for insurance increases with wealth in case of life (say labor income), and decreases with wealth in case of death (say survivors' annuity); an increase in non-human wealth w is never spent entirely on insurance; because such an increase raises w_0, it may even lead to less insurance if (but only if) death strenghtens risk aversion.

Two measures of risk aversion, introduced by Arrow (1971) and Pratt (1964), are commonly used. Absolute risk aversion is measured by $R^A = -\frac{u''}{u'}$ ("twice the risk premium per unit of variance, for infinitesimal risks"). Relative risk aversion is measured by $R^R = -w\frac{u''(w)}{u'(w)}$, a quantity that has the advantage of being dimensionless [11]. Clearly, both measures may differ when evaluated from u_0 or u_1 respectively.

When $R_0^A > R_1^A$, an increase in non-human wealth w reduces the purchase of insurance because, then, the marginal utility of w decreases faster in case of death than in case of life.

Somewhat surprising is the fact that an increase in w_0 could possibly lead to a *larger* reduction of the amount y spent on insurance ($\partial y/\partial w_0 < -1$). However, the second proposition indicates that the reduction in *net insurance benefits* never exceeds the reduction in w_0, and this is the more relevant comparison. When $p > \frac{\alpha}{2}$, ($p > \frac{1}{2}$ in case of fair insurance), then the premium exceeds the net benefit ($y > \frac{\alpha}{p}y - y$). An increase in w_0 matched by an equivalent reduction in net benefits (leaving final wealth in case of death unchanged) would be accompanied by a larger reduction in y.

The effect on y of an increase in the probability of death p is unambiguously positive if one assumes that the terms of the insurance contract $(\frac{\alpha}{p})$ remain unchanged. (An optimally insured person taking up a new risky activity covered under the terms of his policy should increase the amount of coverage). However, when the terms of the policy are adjusted to the increase in p $(\frac{d\frac{\alpha}{p}}{dp} = \frac{-\alpha}{p^2})$, then the sign of $\frac{\partial y}{\partial p}$ remains unambiguously non-negative in case of fair insurance, but depends upon the degree of risk aversion when insurance is less than fair.

Proposition 2 (iii) states that an increase in death probability, fully reflected in the terms of the insurance contract, leads to an increase in the amount spent on insurance if and only if R_0^R, the relative risk aversion measure in case of death, exceeds a certain quantity. For α not too far from unity and p not too far from zero, this quantity is apt to be small, suggesting again an increase in y when p increases. The role of R_0^R is understandable, considering that wealth in case of death, $w_0 + \frac{\alpha}{p} y - y$, would decrease by an amount $\frac{\alpha}{p^2} y\,dp$ if p increased without y being adjusted; how markedly this decrease affects u_0' (u_1' unchanged) matters, and is measured by R_0^A or R_0^R.

Using formula (2.9), Proposition 2 (iv) states that the net insurance benefit $\frac{\alpha}{p} y - y$ is a decreasing function of p whenever insurance is less than fair. In case of fair or more than fair insurance, the effect depends upon R_1^A or R_1^R. This is again understandable since unchanged benefits would require a higher y, hance a lower $w_1 - y$, affecting u_1' (u_0' unchanged) to an extent measured by R_1^A or R_1^R.

Propositions 2 (v) and 2 (vi) consider the effect of the "degree of fairness" α on the premium y and the net benefits $\frac{\alpha}{p} y - y$. It is seen that a reduction of the loading factor always leads to more net coverage. The effect on the premium paid may go in either direction, because the savings due to increased fairness may offset the costs of increased coverage. The premium will invariably be reduced when

$$R_0^R \geq \frac{w_0 + \frac{\alpha}{p} y - y}{\frac{\alpha}{p} y - y},$$

an inequality that may be expected to hold when the net benefits $\frac{\alpha}{p} y - y$ are a major component of wealth in case of death.

2.3 For the sake of completeness, we shall make reference to special circumstances where the premium takes the form of a deduction from w_1 *leaving* w_0 *unchanged*, i.e. a deduction from a_1. This is the case, in particular, for flight insurance bought at an airport with currency. If the plane crashes, it could well happen that whatever cash a passenger was carrying is also lost. In that case, the cost of insurance is zero in case of death. The optimal insurance corresponds to

$$\text{(2.13)} \qquad \max_y \left[p u_0\left(w_0 + \frac{\alpha}{p} y\right) + (1-p) u_1(w_1 - y)\right]$$

with the first-order condition

$$\text{(2.14)} \qquad y[\alpha u_0' - (1-p) u_1'] = 0, \quad \alpha u_0' - (1-p) u_1' \leq 0.$$

By implicit differentiation of this condition when $y > 0$, $D < 0$, we find

$$\text{(2.15)} \qquad \frac{\partial y}{\partial p} = - \frac{u_1' + \frac{\alpha^2 y}{p^2}(-u_0'')}{\frac{\alpha^2}{p} u_0'' + (1-p) u_1''} > 0.$$

Thus, in this case, the sign of $\frac{\partial y}{\partial p}$ is unambiguously positive. Further calculations reveal however that the sign of $\frac{\partial \frac{\alpha}{p} y}{\partial p}$ is now ambiguous for all α and depends upon R_1^R.

As a side comment on this situation, it may be remarked that a passenger for whom $u_0'(w_0) = u_1'(w_1)$ would definitely buy flight insurance at a fair price, or at a less than fair price such that $\alpha > 1-p$. Indeed, for such a passenger, (2.14) is violated at $y = 0$[12].

3. THE DEMAND FOR SAFETY

3.1.1 We now turn to the implications of consistent behavior for an individual's demand for safety. This issue can be approached in two ways. First, one could assume that the probability of death p is a function of the expenses x incurred to "buy safety": $p = p(x)$, $p' < 0$. Interest then centers on a characterization of optimal expenditures on safety and of the impact on these expenditures of marginal changes in w_0, w_1 and p. Alternatively, one could define an individual's marginal willingness-to-pay for safety as the amount of wealth, here denoted ϕ, which would exactly compensate the individual for an infinitesimal increase in death probability, at a constant level of expected utility. Interest then centers on evaluation of the marginal willingness-to-pay for safety, and of the impact on this quantity of marginal changes in w_0, w_1 and p.

We begin with the characterisation of optimal expenditures, and the evaluation of the marginal willingness-to-pay, for safety. We consider first an individual who either carries no life insurance or carries a policy the terms of which (y and $\frac{\alpha}{p}$ y) are kept constant in the face of a change in the safety level p; we turn next to an individual who carries an optimal life insurance policy, satisfying (2.3), the terms of which (y and $\frac{\alpha}{p}$ y) are adjusted to a change in p. The "comparative statics" analysis of changes in w_0, w_1 and p are treated in the (increasingly technical) sections 3.2, 3.3 and 3.4.

Throughout, we shall use the following general expression for expected utility:

$$(3.1) \qquad Eu(x,y) = p(x)u_0(w_0 + \frac{\alpha}{p} y-y-x) + [1-p(x)]u_1(w_1-y-x),$$

where the expenditures on safety, x, are deducted from both w_1 and w_0. Special cases are conveniently obtained by keeping y, $\frac{\alpha}{p}$ y and/or x fixed.

3.1.2 To begin with the simpler problem, assume that y and $\frac{\alpha}{p}$ y are given. Then optimal expenditures on safety are characterized by the first and second order conditions for a maximum of (3.1) with respect to x, namely[13]

(3.2) $$p'(u_0-u_1) - pu_0' - (1-p)u_1' = 0;$$

(3.3) $$\Delta = 2p'(u_1'-u_0') + pu_0'' + (1-p)u_1'' + p''(u_0-u_1) \leq 0.$$

We return below to the second-order conditions. As for (3.2), it is helpful to rewrite it as[14]

(3.4) $$\frac{-1}{p'} = \frac{u_1-u_0}{Eu'}$$

where $Eu' = pu_0' + (1-p)u_1'$. The right hand side of (3.4) is also the general expression for marginal willingness-to-pay for safety:

(3.5) $$\phi = \left.\frac{dw}{dp}\right|_{\overline{Eu}} = -\frac{\partial Eu}{\partial p} \Big/ \frac{\partial Eu}{\partial w} = \frac{u_1-u_0}{Eu'},$$

Hence, at the optimal level of expenditures on safety, the marginal willingness to-pay is equal to the marginal cost of safety, i.e. to the reciprocal of the marginal product $(-p')$ of expenses on safety.

When (3.4) holds, ϕ is positive ($p' < 0$). In general, ϕ is positive whenever suicide is undesirable ($u_1 > u_0$). A sharper conclusion is reached under the minimal assumption that suicide is never desirable when the pecuniary consequences of death are not favorable:

Assumption 4

$u_1(w_1) > u_0(w_0)$ whenever $w_0 \leq w_1$.

The conclusion is stated in proposition 4 below.

When (3.2) is satisfied, $u_1 > u_0$. Therefore, assuming non-increasing returns to safety expenditures ($p'' \leq 0$), a sufficient condition for the second order condition to hold is $u_1' \geq u_0'$ (at the optimal level of expenditures), a condition satisfied when fair or more than fair insurance is available. As is shown in Appendix A.1, Eu is quasi-concave in w and p if

(3.6) $$2(u_1-u_0)(u_0'-u_1')Eu' + (u_1-u_0)^2Eu'' \leq 0$$

where $Eu'' \geq pu'' + (1-p)u_1''$. Quasi-concavity is therefore satisfied whenever $(u_1'-u_0')(u_1-u_0) \geq 0$ everywhere.

3.1.3 When the individual carries a life insurance contract, the terms of which are adjusted to a change in p, then

(3.7) $$\frac{\partial Eu}{\partial x} = p'(u_0-u_1) - pu_0'(1 + \frac{\alpha}{p^2}yp') - (1-p)u_1'.$$

The first-order condition for an optimal insurance coverage is still

(2.3) $(\alpha-p)u_0' - (1-p)u_1' = 0,$

implying $Eu' = \alpha u_0'$. Using this condition, and setting $\frac{\partial Eu}{\partial x} = 0$, we obtain:

(3.8) $$\frac{-1}{p'} = \frac{u_1-u_0}{Eu'} + \frac{y}{p}$$

the right-hand side being equal to ϕ, the marginal willingness-to-pay for safety. It now includes an additional term, reflecting the lower cost of insurance under reduced risks. In Appendix A.2, it is shown that the second-order conditions are satisfied when $\alpha \geqslant 1$. In Appendix A.3, it is shown that when $\alpha = 1$, Eu is quasi-concave in w and p.

Proposition 4

Under Assumption 4, (3.8) *implies*

(3.9) $$\phi \geqslant \frac{\alpha-p}{\alpha(1-p)}(w_1-w_0) + \frac{1-\alpha}{1-p}\frac{y}{p} = \frac{u_1'}{Eu'}(w_1-w_0-\frac{y}{p}) + \frac{y}{p}$$

whereas (3.5) *implies*

(3.10) $$\phi \geqslant \frac{u_1'}{Eu'}(w_1-w_0-\frac{\alpha y}{p}).$$

Proof

By concavity of u_1 (assumption 1):

$$u_1(w_1-y-x) \geqslant u_1(w_0+\frac{\alpha y}{p}-y-x) + (w_1-w_0-\frac{\alpha y}{p})u_1'(w_1-y-x),$$

$$\frac{u_1-u_0}{Eu'} \geqslant \frac{u_1(w_0+\frac{\alpha y}{p}-y-x)-u_0(w_0+\frac{\alpha y}{p}-y-x)}{Eu'} + (w_1-w_0-\frac{\alpha y}{p})\frac{u_1'}{Eu'};$$

using Assumption 4,

$$\frac{u_1-u_0}{Eu'} \geqslant (w_1-w_0-\frac{\alpha y}{p})\frac{u_1'}{Eu'}.$$

This establishes (3.10). When (2.3) holds, $\frac{u_1'}{Eu'} = \frac{\alpha-p}{\alpha(1-p)}$, and (3.8) implies

$$\phi = \frac{u_1-u_0}{Eu'} + \frac{y}{p} \geqslant \frac{\alpha-p}{\alpha(1-p)}(w_1-w_0-\frac{\alpha y}{p}) + \frac{y}{p} = \frac{\alpha-p}{\alpha(1-p)}(w_1-w_0) + \frac{1-\alpha}{1-p}\frac{y}{p}. \quad ||$$

This proposition entails *sufficient conditions under which "a person's life is worth more than her human capital"*[15].

It seems indeed natural to regard $w_1 - w_0$ as a measure of "human capital". *If insurance is available at fair odds,* then Proposition 4 implies $\phi \geqslant w_1 - w_0$ - either through (3.9) with $\alpha = 1$, or through (3.10) with $y = 0$ and $u_1' \geqslant Eu'$ (see Proposition 1).

It may seem surprising that Assumption 4 is not in itself sufficient to establish that a person's life is worth more than her human capital. There are two reasons for this. First, one must reckon with insurance benefits. Of course, one could *define* "human capital" *net of insurance benefits*, i.e. $w_1 - (w_0 + \frac{\alpha}{p} y)$. Proposition 4 implies that $\phi \geqslant w_1 - w_0 - \frac{\alpha}{p} y$ whenever $u_1' \geqslant Eu'$, i.e. whenever $u_1' \geqslant u_0'$. This brings up the second point. Safety is bought by reducing w_0 as well as w_1. When $u_0' > u_1'$, the "opportunity cost" of safety bought from non-human wealth exceeds the opportunity cost in terms of human wealth - explaining why ϕ may fall short of human wealth.

Indeed, assume that safety could be bought from human wealth alone[16]. Then, under a fixed insurance contract, we have:

$$\text{(3.11)} \qquad \left.\frac{dw_1}{dp}\right|_{\overline{Eu}} = \frac{\partial Eu}{\partial p} \Big/ \frac{\partial Eu}{\partial w_1} = \frac{u_1 - u_0}{(1-p)u_1'} \geqslant \frac{w_1 - w_0 - \frac{\alpha y}{p}}{1-p} ,$$

implying

$$\text{(3.12)} \qquad (1-p)\frac{dw_1}{dp} \geqslant w_1 - w_0 - \frac{\alpha y}{p} .$$

When the terms of the insurance contract are adjusted to changes in risk, (3.12) becomes

$$\text{(3.13)} \qquad (1-p)\frac{dw_1}{dp} = w_1 - w_0 - \frac{\alpha - 1}{\alpha - p}\frac{\alpha y}{p} \geqslant w_1 - w_0 - \frac{\alpha y}{p} .$$

The expected value of the amount which an individual is willing to pay in case of life to reduce the probability of death is at least equal to human capital net of insurance benefits, under Assumption 4[17].

3.2 Turning to "comparative statics", we consider first the effect of marginal changes in w_0, w_1 and p on optimal safety expenditures x, for a person who either carries no life insurance or carries a policy the terms of which (y and $\frac{\alpha}{p} y$) are kept constant. (The more general case of simultaneous adjustments in x and y is more complicated, and no definite conclusions emerge.) By a change in p, we mean an exogeneous change in death probability which does not affect p'.

Implicit differentiation of (3.2) leads to the following formulae ($\Delta \neq 0$):

(3.14) $\frac{\partial x}{\partial w_0} = [- p'u_0' + pu_0''] / \Delta, \quad \frac{\partial x}{\partial x_1} = [p'u_1' + (1-p)u_1''] / \Delta;$

(3.15) $\frac{\partial x}{\partial w} = [p'(u_1'-u_0') + pu_0'' + (1-p)u_1''] / \Delta;$

(3.16) $\frac{\partial x}{\partial p} = - (u_1'-u_0') / \Delta.$

The main implications of these formulae are:

Proposition 5

Under (3.2) *and* (3.3) with $\Delta < 0$:

(i) $\frac{\partial x}{\partial w_0} \gtrless 0$ *according as* $\frac{-u_0''}{u_0'} \gtrless \frac{-p'}{p}$;

(ii) $\frac{\partial x}{\partial w_1} > 0$;

(iii) $u_1' - u_0' \geqslant 0$ *implies* $\frac{\partial x}{\partial w} > 0$;

(iv) $u_1' - u_0' \geqslant 0$ *and* $p'' \geqslant 0$ *imply* $\frac{\partial x}{\partial w} \leqslant 1$;

(v) $\frac{\partial x}{\partial p} \gtrless 0$ *according as* $u_1' - u_0' \gtrless 0$.

The sign of $\frac{\partial x}{\partial w_0}$ is ambiguous, because an increase in w_0 has two conflicting effects on the marginal rate of substitution in (3.5). It reduces the *numerator* by a quantity $- u_0'\ dw_0$, thereby reducing the demand for safety (the pecuniary consequences of death are less deterring). But it also reduces the *denominator* by a quantity $pu_0''\ dw_0$ (the expected marginal utility of wealth, which defines the opportunity cose of buying safety, decreases with w_0).

The sign of $\frac{\partial x}{\partial w_1}$ is positive, confirming to intuition: An increase in w_1 increases the numerator *and* reduces the denominator of the right hand side in (3.5). When fair insurance is available, so that $u_1' - u_0'$ is non-negative, an increase in non-human wealth induces additional spending on safety *(iii)*. When safety is acquired at non-decreasing marginal cost ($p'' \geqslant 0$), the marginal propensity to spend on safety is less than one. Note also that $p'' = 0$ would imply

(3.17) $\frac{\partial x}{\partial w} = \frac{p'(u_1'-u_0')+Eu''}{2p'(u_1'-u_0')+Eu''} = \frac{1}{2} + \frac{1}{2}\frac{Eu''}{\Delta} \geqslant \frac{1}{2},$

i.e. a marginal propensity to spend on safety of at least one half. Notice that, for an individual carrying an optimal *fair* insurance policy ($u_0' = u_1'$), the terms of which are not adjusted, $p'' = 0$ would even imply $\frac{\partial x}{\partial w} = 1$: An increment in non-human wealth would be spent entirely on safety! (This is because insurance is bought at constant marginal cost whereas safety is bought at increasing marginal cost, and consumption is subject to diminishing marginal utility.) Casual empiricism thus suggests strongly that if (3.2) holds, $p'' > 0$ and $p''(u_0 - u_1)$ is a non-negligible component of Δ.

When $\frac{\partial x}{\partial p} > 0$, the expenses incurred to buy additional safety would be reduced if the individual experienced an exogenous decrease in death probability and conversely. This plausible reaction is entirely due to the change in the expected marginal utility of wealth (the opportunity cost of buying safety), the sign of which is given by the sign of $u_1' - u_0'$.

3.3 We turn next to the effect of marginal changes in w_0, w_1 and p on the marginal willingness to pay for safety ϕ. We begin with the case of a person who either carries no life insurance or carries a policy the terms of which are kept constant. And we assume that expenditures on safety x are also kept constant.

Differentiating the right-hand side of (3.5), we find:

$$\frac{\partial \phi}{\partial w_0} = -\frac{u_0' + \phi p u_0''}{Eu'}; \tag{3.18}$$

$$\frac{\partial \phi}{\partial w_1} = \frac{u_1' - \phi(1-p)u_1''}{Eu'}; \tag{3.19}$$

$$\frac{\partial \phi}{\partial w} = \frac{u_1' - u_0' - \phi[pu_0'' + (1-p)u_1'']}{Eu'}; \tag{3.20}$$

$$\frac{\partial \phi}{\partial p} = \phi \frac{u_1' - u_0'}{Eu'}. \tag{3.21}$$

These formulae enable us to state the counterpart of proposition 5:

Proposition 6:

Under (3.5):

(i) $\frac{\partial \phi}{\partial w_0} \gtrless 0$ *according as* $\frac{-u_0''}{u_0'} \gtrless \frac{1}{p\phi}$;

(ii) $\frac{\partial \phi}{\partial w_1} > 0$;

(iii) $u_1' - u_0' \geqslant 0$ *implies* $\frac{\partial \phi}{\partial w} \geqslant 0$;

(iv) $u_1' - u_0' \geqslant 0$ *implies* $\frac{\partial \phi}{\partial w_1} \geqslant 1$;

(v) $\frac{\partial \phi}{\partial p} \gtrless 0$ *according as* $u_1' - u_0' \gtrless 0$.

These propositions follow from (3.18)-(3.21), upon remembering that $u_1' \geqslant u_0'$ also implies $u_1' \geqslant Eu'$. These inequalities always hold when insurance is available on fair (or more than fair) terms.

The sign of $\frac{\partial\phi}{\partial w_0}$ is ambiguous, for the same reasons which explained why the sign of $\frac{\partial x}{\partial w_0}$ is ambiguous. For the sake of completeness, we note that the marginal rate of substitution between w_1 and p, namely

$$(3.22) \qquad \phi_1 =_{def} \left.\frac{dw_1}{dp}\right|_{\overline{Eu}} = -\frac{\partial Eu}{\partial p} \Big/ \frac{\partial Eu}{\partial w_1} = \frac{u_1-u_0}{(1-p)u_1'}$$

unambiguously decreases with w_0. Indeed, from (3.22),

$$(3.23) \qquad \frac{\partial\phi_1}{\partial w_0} = \frac{-u_0'}{(1-p)u_1'} < 0.$$

Proposition 6 (iv) may seem startling: When human wealth increases, willingness-to-pay for safety increases *even more*. However, this does not imply a marginal propensity to spend on safety of one or more. Indeed, the marginal propensity is defined by $\frac{\partial x}{\partial w_1}$, with the properties listed in proposition 5.

Proposition 6 (v) states that marginal willingness-to-pay for safety decreases with safety, when fair insurance is available.

3.4 For a person who carries an optimal life-insurance policy, satisfying (2.3), the results of section 3.3 need to be extended. The starting point is the definition of ϕ in (3.8), instead of (3.5). More importantly, differentiation of the right-hand side of (3.8) requires allowance for the adjustments in the insurance premium y, and the indemnity $\frac{\alpha}{p}y$, associated with changes in w_0, w_1 or p. The relevant expressions were given in formulae (2.6)-(2.8) above. Thus, (3.18) now takes the form

$$(3.24) \qquad \frac{\partial\phi}{\partial w_0} = -\frac{u_0'+\phi p u_0''}{Eu'} + \frac{\partial\phi}{\partial y}\frac{\partial y}{\partial w_0},$$

and similarly for (3.19)-(3.21). (Indeed, $\frac{\partial Eu}{\partial y} = 0$ does *not* imply $\frac{\partial\phi}{\partial y} = 0$.) Details of the calculations are given in Appendix A.4, where use is made of the first-order condition (2.3). The resulting formulae are ($D \neq 0$):

$$(3.25) \qquad \frac{\partial\phi}{\partial w_0} = \frac{-1}{\alpha} + \frac{1-\alpha}{\alpha(1-p)}\frac{\partial y}{\partial w_0} - \alpha p(1-p)\frac{u_1-u_0}{(Eu')^2}\frac{u_0''\,u_1''}{D};$$

$$(3.26) \qquad \frac{\partial\phi}{\partial w_1} = \frac{1}{\alpha} + \frac{\alpha-1}{\alpha(1-p)}\left(1-\frac{\partial y}{\partial w_1}\right) - \alpha(\alpha-p)(1-p)\frac{u_1-u_0}{(Eu')^2}\frac{u_0''\,u_1''}{D};$$

$$(3.27) \qquad \frac{\partial\phi}{\partial w} = \frac{\alpha-1}{\alpha(1-p)}\left(1-\frac{\partial y}{\partial w}\right) - \alpha^2(1-p)\frac{u_1-u_0}{(Eu')^2}\frac{u_0''\,u_1''}{D};$$

$$(3.28) \qquad \frac{\partial \phi}{\partial p} = \frac{\alpha - 1}{\alpha(1-p)} \left[\frac{u_1 - u_0}{Eu'} \alpha(\alpha - p) \frac{u''_0}{D} - \frac{\partial y}{\partial p} \right] + \alpha^2 (1-p) \frac{y}{p} \frac{u_1 - u_0}{(Eu')^2} \frac{u''_0 u''_1}{D}$$

These formulae enable us to state a partial counterpart to proposition 6. With optimal life insurance, the sign of $u'_1 - u'_0$ is also the sign of $(\alpha-1)$, since (2.3) implies $u'_1 = \frac{\alpha-p}{1-p} u'_0$. A positive sign thus corresponds to more than fair terms.

The absence of moral hazard requires that

$$(3.29) \qquad u_1(w_1 - x - y) > u_0(w_0 + \frac{\alpha}{p} y - x - y).$$

This does not necessarily follow from assumption 4, when $w_0 + \frac{\alpha}{p} y > w_1$.

Proposition 7

Under (3.8) *and* (3.29), *with strict risk aversion* ($D < 0$):

(i) $\alpha \geq 1$ *implies* $\frac{\partial \phi}{\partial w_1} \geq \frac{1}{\alpha}$;

(ii) $\alpha \geq 1$ *implies* $\frac{\partial \phi}{\partial w} \geq 0$;

(iii) $\alpha = 1$ *implies* $\frac{\partial \phi}{\partial p} \leq 0$.

Propositions 6 (i) and 6 (ii) have no meaningful counterparts here. Propositions 6 (iii) and 6 (iv) are closely matched by 7 (i) and 7 (ii). But 6 (v) is much sharper than 7 (iii). Also, 7 (iii) is somewhat counter-intuitive. Combining 6 (v) and 7 (iii), we remark: For a person who is optimally insured on fair terms, or a person who is not insured due to unfair terms, marginal willingness-to-pay for safety *increases* with safety. Thus, assertions regarding the sign of $\frac{\partial \phi}{\partial p}$ must be carefully justified in terms of underlying assumptions.

4. THE PROVISION OF PUBLIC SAFETY

4.1 Public projects increasing safety, or entailing risks of human deaths, affect simultaneously the survival probabilities of many individuals. In such contexts, public safety has all the properties of a pure public good[18]. An efficient provision of public safety is thus governed by the same principles as an efficient production of public goods. This viewpoint, implicitly adopted by Jones-Lee (1974, 1976) and Mishan (1971) in the guise of cost-benefit analysis, is developed formally in Bergström (1978).

Restating their analysis in the notation of the present paper, let N denote a set of individuals indexed by $i = 1...n$, whose death probabilities p^i depend upon the level z of expenditures on public safety; and let t^i be the tax levied on individual i to finance these expenditures, $\sum_i t^i = z$. Then, omitting the explicit reference to x^i and y^i, we may rewrite (3.1) as

$$(4.1) \qquad Eu^i = p^i(z)u_0^i(w_0^i-t^i) + [1-p^i(z)]\, u_1^i(w_1^i-t^i).$$

A Pareto-efficient program of public expenditures z and taxes t^i, $i = 1\ldots n$, solves

$$(4.2) \qquad \underset{z,\{t^i\}}{\text{Max}} \sum_i \lambda^i\{p^i(z)u_0^i(w_0^i-t^i) + [1-p^i(z)]\, u_1^i(w_1^i-t^i)\} - \lambda(\sum_i t^i - z),$$

where λ is a Lagrange multiplier and where the non-negative multipliers $\{\lambda^i\}$ characterize a particular Pareto optimum[19]. The first-order conditions for a solution of (4.2) are

$$(4.3) \qquad \sum_i \lambda^i[\frac{\partial p^i}{\partial z}(u_0^i - u_1^i)] + \lambda = 0\,;$$

$$(4.4) \qquad -\lambda^i[p^i u_0^{i\prime} + (1-p^i)u_1^{i\prime}] - \lambda = 0, \; i = 1\ldots n.$$

Eliminating the multipliers between (4.3) and (4.4), we obtain

$$(4.5) \qquad \sum_i \frac{\partial p^i}{\partial z}\frac{u_1^i - u_0^i}{Eu^{i\prime}} = -1.$$

Using the definition of ϕ^i in (3.5), thus taking the terms of all insurance contracts as given, we may also write (4.5) as

$$(4.6) \qquad \sum_i \frac{\partial p^i}{\partial z}\phi^i = -1.$$

Let then

$$(4.7) \qquad \sum_i \frac{\partial p^i}{\partial z} = \frac{\partial \sum_i p^i}{\partial z} =_{\text{def}} \frac{-1}{c},$$

so that $c = -\frac{dz}{d\sum_i p^i}$ is the marginal cost of saving one life through public expenditures. Then,

$$(4.8) \qquad c = \sum_i \frac{\partial p^i}{\partial z}\phi^i \Big/ \sum_i \frac{\partial p^i}{\partial z} = \frac{1}{n}\sum_i \phi_i + \sum_i \phi^i\left(\frac{\frac{\partial p^i}{\partial z}}{\sum_i \frac{\partial p^i}{\partial z}} - \frac{1}{n}\right)$$

$$= \frac{1}{n}\sum_i \phi^i - c\,\text{Cov}\,(\phi^i, \frac{\partial p^i}{\partial z}).$$

If the reductions in death probabilities $\frac{\partial p^i}{\partial z}$ are uncorrelated, across individuals, with the measures of willingness-to-pay ϕ^i, then (4.8) reduces to

$$(4.9) \qquad c = \frac{1}{n} \sum_i \phi^i =_{\text{def}} \overline{\phi}.$$

An efficient provision of public safety is obtained when the marginal cost of saving a life is equal to the average willingness-to-pay for safety of the population members. This proposition seems now well understood. Note however the role of the lump-sum tax levies in (4.4).

4.2 Two implications of this analysis seem worth mentioning, in relation to sections 2 and 3.

First, public safety is typically in the nature of a *local* public good[20]. Traffic lights, sidewalks, highway lights or coatings provide obvious examples. The condition (4.9) must then be applied to a local population N, which will be different in each case. It follows that society's marginal expenditure to save a life (society's valuation of life) will also be different in each case. In particular, assume that individual preferences (u_0^i, u_1^i) are identically distributed over individuals in two distinct populations N and N'; the distribution of ϕ^i, $i \in N$, will still differ from the distribution of ϕ^i, $i \in N'$, if the distributions of wealth (w_0^i, w_1^i) or the risk levels p^i differ between the two populations. More specifically, we have seen in propositions 6 and 7 that $\frac{\partial \phi^i}{\partial w_1^i}$ is positive, and greater than or equal to one if life-insurance is available on fair terms. *We may then expect $\overline{\phi}$ to increase with the average value of human wealth in the relevant population.*

It is not surprising that society should be prepared to spend more in order to save a "statistical life" when more "human capital" is at stake.

Propositions 6 and 7 also tell us that $\frac{\partial \phi^i}{\partial w^i}$ is non-negative, when fair insurance is available. *We may then expect $\overline{\phi}$ to increase with the average level of non-human wealth in the relevant population.* This conclusion is less appealing on ethical grounds, since the non-human wealth is not itself in jeopardy. The conclusion follows from the greater willingness of wealthier individuals to "buy safety". This greater willingness would also normally be reflected in higher taxes t^i levied to finance the public expenditures at stake.

As for differences in risk levels, we have seen in section 3 that the sign of $\frac{\partial \phi^i}{\partial p^i}$ is in general ambiguous, so that no definite conclusions emerge on that score.

4.3 Formula (4.9) is based on the assumption that the terms of all insurance contracts $(y^i, \frac{\alpha}{p^i} y^i)$ are given, and unaffected by public safety. An the formula was obtained without taking explicitly into account private expenditures on safety x^i.

If we start from the more general formulation

$$(4.10) \qquad Eu^i = p^i(z,x^i)u_0^i(w_0^i + \frac{\alpha}{p^i} y^i - y^i - x^i - t^i) + [1-p^i(z,x^i)]u_1^i(w_1^i - y^i - x^i - t^i),$$

extend (4.2) accordingly, and use the first-order conditions (2.3) and (3.8), then (4.5) becomes

$$(4.11) \qquad \sum_i \frac{\partial p^i}{\partial z}\left[\frac{u_1^i - u_0^i}{Eu^{i\prime}} + \frac{y^i}{p^i}\right] = -1.$$

This still leads to (4.6)-(4.9), but with ϕ^i now given by (3.8) instead of (3.5). The implications for individual expenditures on insurance and safety follow from (2.8) and (3.16).

In the light of formula (4.11), it is natural to raise the following issue. If the terms of individual insurance contracts are *not* adjusted to improvements in public safety, then individuals will reveal a willingness-to-pay for safety ϕ^i equal to $(u_1^i - u_0^i) / Eu^{i\prime}$, whereas the true benefit to society includes an additional term y^i / p^i, reflecting the savings accruing to life insurance companies. Under fair insurance ($\alpha = 1$), the expected savings of the life insurance companies is precisely equal to $\sum_i \frac{\partial p^i}{\partial z} \frac{y^i}{p^i}$. Given that life-insurance contracts are typically based on mortality tables which are up-dated at irregular intervals, our analysis provides a justification for the participation of insurance companies in the financing of public expenditures on safety.

Footnotes

[1] Paper prepared for the Geneva Conference on *The Value of Life and Safety*, Geneva, March-April 1981. The authors are grateful to Paul Champsaur and Denis Moffet for interesting discussions and helpful suggestions.

[2] The contributors known to us include medical doctors, actuaries, sociologists, civil engineers, statisticians, economists,...

[3] See Jones-Lee (1976) and Linnerooth (1979) for a survey of the literature.

[4] We draw freely on the contents of an earlier paper by one of us - Drèze (1962). That paper, published in the *Revue Française de Recherche Opérationnelle*, apparently contains the first exposition of the "willingness-to-pay" approach. It did influence the work of Michael Jones-Lee (1976), but seems to have otherwise escaped attention, so that some repetition may be in order.

[5] See Drèze (1961), Karni, Schmeidler and Vind (1982) and the references cited there.

[6] Your wealth is thus reduced from w_1 to w_1-y in case of life, but increased from w_0 to w_0+1-y in case of death.

[7] It should be clear from (1.2) that an unknown p_0 raises a logical difficulty; indeed (1.2) is then a relation between the products $p_0\beta_0$ and $p_1\beta_1$, and provides no information on the ratio β_0/β_1 as distinct from the ratio p_0/p_1.

[8] This is an expectation, because it takes on different values under life and death respectively.

[9] See, for instance, Linnerooth (1979): "The utility of the state of death can be arbitrarily set at zero..." (p. 60); "Without any loss in generality the level of consumption at which a person can no longer survive can be set at zero..." (p. 61); "Again without any loss in generality the utility of this point can be arbitrarily set at zero." (p. 64).

[10] For a justification of the concavity assumption based on the existence of opportunities for gambling at nearly fair-odds, see Drèze (1971, section 2.1) or Raiffa (1968, section 4.13).

[11] Reasonable values of R^R suggested in the literature range from 1 (characterizing logarithmic utility functions - see Arrow (1971)) to 2 (rationalizing some observed portfolio choices - see Blume and Friend (1975)) or more (if one wishes to rationalize low deductibles in property or liability insurance - see Drèze (1981)).

[12] The fact that the cash used to pay the premium would have been lost in case of a plane crash explains why our results are at variance with those obtained by Eisner and Strotz (1961).

13 If it were stipulated that x must be non-negative as well as the arguments of u_0 and u_1, then (3.2) should be generalized, in the same manner that (2.3) was.

14 (3.4) is a differential equivalent of (1.4), which could have been written as $\frac{1}{\delta} = \frac{-1}{dp} = \frac{u_1 - u_0}{dEu}$.

15 See Bergström (1978).

16 As when a person boarding a plane rents a parachute against cash...

17 This is precisely the proposition found in Bergström (1978), who uses assumption 4 in the guise of a "thought experiment" involving a rentier cousin.

18 See Samuelson (1954) for the original presentation, and Milleron (1972) for an excellent survey of the relevant theory.

19 That is, the set of Pareto optima is generated by letting the relative weights $\lambda^i / \sum_j \lambda^j$ range over the unit simplex in R^n.

20 On this notion, see e.g. Greenberg (1977), Stiglitz (1977) or Tiebout (1956).

References

Arrow, K.J. (1971), *Essays in the Theory of Risk-Bearing*, Amsterdam, North-Holland.

Bergstrom, Th.C. (1978), "Is a Man's Life Worth More than his Human Capital?" mimeo, C-10, University of Michigan.

Blume M.E. and I. Friend (1975), "The Asset Structure of Individual Portfolios and Some Implications for Utility Functions", *Journal of Finance*, 30, 585-603.

Cook, P.J. and D.A. Graham (1977), "The Demand for Insurance and Protection: The Case of the Irreplaceable Commodity", *Quarterly Journal of Economics*, 9, 1 143-156.

Drèze, J.H. (1961), "Les fondements logiques de l'utilité cardinale et de la probabilité subjective" in *La Décision*, Colloques Internationaux du CNRS (Paris), 73-87.

Drèze, J.H. (1962), "L'utilité sociale d'une vie humaine", *Revue Française de Recherche Opérationnelle*, 22, 139-155.

Drèze, J.H. (1971), "Market Allocation under Uncertainty", *European Economic Review*, 2, 2, 133-165.

Drèze, J.H. (1981), "Inferring Risk Tolerance from Deductibles in Insurance Contracts", *The Geneva Papers on Risk and Insurance*, 20, 48-52.

Eisner R. and R.H. Strotz (1961), "Flight Insurance and the Theory of Choice", *The Journal of Political Economy*, 69, 4, 355-368.

Greenberg, J. (1977), "Pure and Local Public Goods: A Game Theoretic Approach", *Public Finance*, A. Sandmo, ed., Lexington, Massachusetts, Heath and Co.

Jones-Lee, M. (1974), "The Value of Changes in the Probability of Death or Injury", *Journal of Political Economy*, 99, 835-849.

Jones-Lee, M. (1976), *The Value of Life: An Economic Analysis*, London, Martin Robertson.

Karni, E, D. Schmeidler and K. Vind (1982), "On State Dependent Preferences and Subjective Probabilities", Working Paper 88, The Johns Hopkins University, Baltimore.

Linnerooth, J. (1979), "The Value of Human Life: A Review of the Models", *Economic Inquiry*, 17, 52-74.

Milleron, C.(1972), "Theory of Value with Public Goods: A Survey Article", *Journal of Economic Theory*, 5, 419-477.

Mishan, E.J. (1971), "Evaluation of Life and Limb: A Theoretical Approach", *Journal of Political Economy*, 79, 685-705.

Pratt, J. (1964), "Risk-Aversion in the Small and in the Large", *Econometrica*, 32, 122-136.

Raiffa, H. (1968), *Decision Analysis; Introductory Lectures on Choices under Uncertainty*, Reading Mass., Addison-Wesley Publishing Company.

Samuelson, P.A. (1954), "The Pure Theory of Public Expenditures", *Review of Economics and Statistics*, 36, 387-389.

Savage, L.J. (1954), *The Foundations of Statistics*, New York, Wiley.

Stiglitz, J.E. (1977), "The Theory of Local Public Goods" in *The Economics of Public Services*, M. Feldstein and R. Inman (eds.) London, MacMillan

Tiebout, C.M. (1956), "A Pure Theory of Local Expenditures", *Journal of Political Economy*, 64, 416-424.

Von Neumann, J. and O. Morgenstern, (1947), *Theory of Games and Economic Behavior*, Princeton, Princeton University Press.

APPENDIX

A.1 Let us consider the following expected utility function

(A.1) $$Eu = pu_0(w+a_0) + (1-p)u_1(w+a_1).$$

We have

(A.2) $$\frac{\partial Eu}{\partial p} = u_0 - u_1;$$

(A.3) $$\frac{\partial Eu}{\partial w} = pu_0' + (1-p)u_1' = Eu';$$

(A.4) $$\frac{\partial^2 Eu}{\partial p^2} = 0, \quad \frac{\partial^2 Eu}{\partial p \partial w} = u_0' - u_1', \quad \frac{\partial^2 Eu}{\partial w^2} = pu_0'' + (1-p)u_1'' = Eu''.$$

The condition for quasi-concavity of Eu, namely $d^2Eu\Big|_{\overline{Eu}} \leq 0$ takes the form:

(A.5) $$\frac{\partial^2 Eu}{\partial p^2}\left(\frac{\partial E}{\partial w}\right)^2 - 2\frac{\partial^2 Eu}{\partial p \partial w}\frac{\partial Eu}{\partial p}\frac{\partial Eu}{\partial w} + \frac{\partial^2 Eu}{\partial w^2}\left(\frac{\partial E}{\partial p}\right)^2 \leq 0,$$

or

(A.6) $$2(u_0'-u_1')(u_1-u_0)Eu' + (u_1-u_0)^2 Eu'' \leq 0$$

A.2 The first partial derivatives of the expected utility function (3.1) are given by:

(A.7) $$\frac{\partial Eu}{\partial x} = p'(u_0-u_1) - Eu' - \frac{\alpha y}{p} u_0'p';$$

(A.8) $$\frac{\partial Eu}{\partial y} = (\alpha-p)u_0' - (1-p)u_1'.$$

Let (x^*,y^*) denote a point where these derivatives are simultaneously equal to zero. At such a point, the second partial derivatives are given by:

(A.9) $$\frac{\partial^2 Eu}{\partial x^2} = \alpha u_0' \frac{p''}{p'} + (1-p)u_1'' + p\left(1 + \frac{\alpha y}{p^2}p'\right)^2 u_0'' + 2\frac{\alpha-1}{\alpha-p}u_0'p'$$

(A.10) $$\frac{\partial^2 Eu}{\partial y^2} = \frac{(\alpha-p)^2}{p} u_0'' + (1-p)u_1''$$

(A.11) $$\frac{\partial^2 Eu}{\partial x \partial y} = \frac{1-\alpha}{\alpha-p}u_0'p' + (1-p)u_1' - (\alpha-p)\left(1 + \frac{\alpha y}{p^2}p'\right)u_0''.$$

The determinant of the hessian matrix at (x^*,y^*) is

$$\text{(A.12)}\quad \frac{\partial^2 Eu}{\partial x^2}\frac{\partial^2 Eu}{\partial y^2} - \left(\frac{\partial^2 Eu}{\partial x \partial y}\right)^2 = (\alpha \frac{p''}{p'} + 2\,\frac{\alpha-1}{\alpha-p}\,p')u_0'\,\frac{\partial^2 Eu}{\partial y^2} + \alpha\,\frac{1-p}{p}\,(1+y\frac{p'}{p})^2 u_0'' u_1''.$$

Assuming that $p'' \geqslant 0$ and $\alpha \geqslant 1$, we have

$$\text{(A.13)}\quad \frac{\partial^2 Eu}{\partial x^2} \leqslant 0;\quad \frac{\partial^2 Eu}{\partial w^2} \leqslant 0;\quad \frac{\partial^2 Eu}{\partial x^2}\frac{\partial^2 Eu}{\partial y^2} - \frac{\partial^2 Eu}{\partial x \partial y} \geqslant 0$$

i.e. the second-order conditions are verified at (x^*,y^*).

A.3 Let us fix $\alpha = 1$ and consider the following expected utility function

$$\text{(A.14)}\quad Eu = pu_0(w+a_0+\frac{1-p}{p}\,(p,w)) + (1-p)u_1(w+a_1-y(p,w))$$

where $y = y(p,w)$ satisfies the optimality condition (2.3), i.e. $u_0' = u_1'$. Following (2.7) and (2.8), we have

$$\text{(A.15)}\quad \frac{\partial y}{\partial p} = (1-p)y\,u_0''\,/\,pD$$

$$\text{(A.16)}\quad \frac{\partial y}{\partial w} = p(1-p)(u_1''-u_0'')\,/\,D$$

with $D = (1-p)(pu_1'' + (1-p)u_0'')$, $D \neq 0$. The first and second partial derivatives of Eu are given by

$$\text{(A.17)}\quad \frac{\partial Eu}{\partial p} = (u_0-u_1) - yu_1'\,/\,p$$

$$\text{(A.18)}\quad \frac{\partial Eu}{\partial w} = Eu' = u_0'$$

$$\text{(A.19)}\quad \frac{\partial Eu^2}{\partial p^2} = \frac{y}{p}\,u_1''\,\frac{\partial y}{\partial p} = (1-p)y^2 u_0'' u_1''\,/\,p^2 D$$

$$\text{(A.20)}\quad \frac{\partial^2 Eu}{\partial w^2} = u_1''(1-\frac{\partial y}{\partial w}) = (1-p)u_0''u_1''\,/\,D$$

$$\text{(A.21)}\quad \frac{\partial^2 Eu}{\partial p \partial w} = -u_1''\,\frac{\partial y}{\partial p} = -\,(1-p)yu_0''u_1''\,/\,pD\,.$$

The condition for quasi-concavity of Eu is given by (A.5). Here, it takes the form

$$\text{(A.22)}\quad (1-p)u_0''u_1''\,(\frac{\partial Eu}{\partial p} + \frac{y}{p}\,\frac{\partial Eu}{\partial w})^2\,/\,D \leqslant 0$$

a condition which is verified under the Assumption 1.

A.4 For a person carrying an optimal amount of insurance, we know that

$$(2.3) \qquad (\alpha-p)u_0' - (1-p)u_1' = 0;$$

$$(3.12) \qquad \phi = \frac{y}{p} + \frac{u_1-u_0}{Eu'}.$$

The derivatives of ϕ are evaluated as follows:

$$(A.23) \qquad \frac{\partial\phi}{\partial w_0} = \frac{-u_0'Eu'-(u_1-u_0)pu''_0}{(Eu')^2} + \frac{\partial\phi}{\partial y}\frac{\partial\phi}{\partial w_0},$$

where $\frac{\partial y}{\partial w_0}$ is given by (2.6) and

$$(A.24) \qquad \frac{\partial\phi}{\partial y} = \frac{1}{p} + \frac{[-u_1'-u_0'(\frac{\alpha}{p}-1)]Eu'-(u_1-u_0)[(\alpha-p)u_0''-(1-p)u_1'']}{(Eu')^2}.$$

It follows from (2.3) that $Eu' = \alpha u_0' = \frac{\alpha(1-p)}{\alpha-p}u_1'$, so that

$$(A.25) \qquad \frac{\partial\phi}{\partial y} = \frac{1}{p} - \frac{\alpha-p}{\alpha(1-p)} - \frac{\alpha-p}{\alpha p} - \frac{u_1-u_0}{(Eu')^2}[(\alpha-p)u_0'' - (1-p)u_1'']$$

$$= \frac{1-\alpha}{\alpha(1-p)} - \frac{u_1-u_0}{(Eu')^2}[(\alpha-p)u_0'' - (1-p)u_1''].$$

Using (2.6) and (A.25) we obtain

$$(A.26) \qquad \frac{\partial\phi}{\partial w_0} = \frac{-1}{\alpha} + \frac{1-\alpha}{\alpha(1-p)}\frac{\partial y}{\partial w_0} - \frac{u_1-u_0}{(Eu')^2}[pu_0'' + \frac{\partial y}{\partial w_0}\{(\alpha-p)u_0''-(1-p)u_1''\}]$$

$$= \frac{-1}{\alpha} + \frac{1-\alpha}{\alpha(1-p)}\frac{\partial y}{\partial w_0} - \frac{u_1-u_0}{(Eu')^2}\frac{pu_0''}{D}[(\alpha-p)^2u_0''+p(1-p)u_1''-(\alpha-p)^2u_0''$$

$$+ (\alpha-p)(1-p)u_1'']$$

$$= \frac{-1}{\alpha} + \frac{1-\alpha}{\alpha(1-p)}\frac{\partial y}{\partial w_0} - \frac{u_1-u_0}{(Eu')^2}\frac{u_0''\,u_1''}{D}p\alpha(1-p).$$

By the same reasoning, using (2.7),

$$(A.27) \qquad \frac{\partial\phi}{\partial w_1} = \frac{u_1'Eu'-(u_1-u_0)(1-p)u''_1}{(Eu')^2} + \frac{\partial y}{\partial w_1}\left[\frac{1-\alpha}{\alpha(1-p)} - \frac{u_1-u_0}{(Eu')^2}\{(\alpha-p)u_0''-(1-p)u_1''\}\right]$$

$$= \frac{\alpha-p}{\alpha(1-p)} + \frac{1-\alpha}{\alpha(1-p)}\frac{\partial y}{\partial w_1} - \frac{u_1-u_0}{(Eu')^2}\frac{(1-p)u_1''}{D}[(\alpha-p)^2u_0''+p(1-p)u_1''$$

$$+ p(\alpha-p)u_0''-p(1-p)u_1'']$$

$$= \frac{1}{\alpha} + \frac{\alpha-1}{\alpha(1-p)}(1 - \frac{\partial y}{\partial w_1}) - \frac{u_1-u_0}{(Eu')^2}\frac{u_1''\,u_0''}{D}\alpha(\alpha-p)(1-p).$$

Combining (A.26) and (A.27):

$$\frac{\partial\phi}{\partial w_0} + \frac{\partial\phi}{\partial w_1} = \frac{1-\alpha}{\alpha(1-p)}\left(1 - \frac{\partial y}{\partial w_0} - \frac{\partial y}{\partial w_1}\right) - \frac{u_1-u_0}{(Eu')^2}\frac{u''_0\, u''_1}{D}\alpha^2(1-p). \tag{A.28}$$

Still by the same reasoning, using (2.8),

$$\begin{aligned}\frac{\partial\phi}{\partial p} &= \frac{-y}{p^2} + \frac{\frac{\alpha y}{p^2} u'_0 Eu' - (u_1-u_0)(u'_0 - u'_1 - \frac{\alpha y}{p} u''_0)}{(Eu')^2} \\ &\quad + \frac{\partial y}{\partial p}\left[\frac{1-\alpha}{\alpha(1-p)} - \frac{u_1-u_0}{(Eu')^2}\{(\alpha-p)u''_0 - (1-p)u''_1\}\right] \\ &= \frac{1-\alpha}{\alpha(1-p)}\left[-\frac{u_1-u_0}{Eu'} + \frac{\partial y}{\partial p}\right] + \frac{u_1-u_0}{(Eu')^2}\left[\frac{\alpha y}{p}u''_0 - \frac{\partial y}{\partial p}\{(\alpha-p)u''_0 - (1-p)u''_1\}\right] \\ &= \frac{1-\alpha}{\alpha(1-p)}\left[\frac{\partial y}{\partial p} - \frac{u_1-u_0}{Eu'}\{1 + \frac{p}{D}((\alpha-p)u''_0 - (1-p)u''_1)\}\right] \\ &\quad + \frac{u_1-u_0}{(Eu')^2}\frac{\alpha y}{p}u''_0\left[1 - \frac{\alpha-p}{D}\{(\alpha-p)u''_0 - (1-p)u''_1\}\right] \\ &= \frac{1-\alpha}{\alpha(1-p)}\left[\frac{\partial y}{\partial p} - \frac{u_1-u_0}{Eu'}\frac{u''_0}{D}\alpha(\alpha-p)\right] + \frac{u_1-u_0}{(Eu')^2}\frac{\alpha^2 y}{p}(1-p)\frac{u''_0\, u''_1}{D}.\end{aligned} \tag{A.29}$$

SEPARATION OF RISK PARAMETERS

F. DELBAEN

Department of Mathematics
VRIJE UNIVERSITEIT BRUSSEL
Pleinlaan 2
B-1050 Brussels Belgium

§ 1. Introduction

We use the same notation as in the previous paper : "Weighted Markov processes with an application to risk theory". So (Λ, L) is the measurable space of risk parameters. A difficulty in risk theory is the impossibility of observing the risk parameter in a finite time interval. It follows that an insurer has "to guess" what the actual risk is. He can do this using observations up to time t and he can try to predict the actual risk for the future. However the observations he made of a fixed insurance taker are influenced by random effects and if the underwriter had bad luck in the observation period, the insurance company will decide that he has bigger risk and accordingly will charge a higher premium than needed. A theoretical way to overcome this difficulty is to make the observation time large enough. Unfortunately it turns out that the variance for fixed $\lambda \in \Lambda$ may be rather large so that there is a big overlap between small risks with bad luck and bigger risks with good luck. Let us try to make this clear.
Let (Λ, L) be the positive real numbers and for each

F. de Vylder et al. (eds.), Premium Calculation in Insurance, 109–119.

$\lambda \in \Lambda$ let $Z_t : D \to \mathbb{N}$ be a Poisson proces with parameter λ. Let γ be a structure distribution on Λ. One way of fixing pure premiums will be to take into account the number of accidents up to time $t > 0$. From the law of large numbers it follows that for $\mathbb{P}$ allmost all $w \in \Omega$ one has $\frac{Z_t(w)}{t} \to y(w)$.

(there is hope that the observed risk is allmost equal to the actual risk)

However the conditional distribution of Y given Z_t turns out to have density $f_{Z_t,t}(\lambda)$ where $f_{k,t}(\lambda) = \frac{e^{-\lambda t}}{k!}(\lambda t)^k$ and hence there is a big overlap between different $f_{k,t}$, $k=0,\ldots$. Let us now introduce a measure that gives a certain idea how different distributions overlap.

If n functions $f_1 \ldots f_n$ on a measure space (Λ, L, γ) are given then they have disjoint support ($f_k . f_l = 0$ if $k \neq l$) if and only if

$$\max_{k \leq n} |f_k| = \sqrt{\sum_{k \leq n} |f_k|^2} = \sum_{k \leq n} |f_k|$$

In general however one has strict inequalities. In most applications we have that the f_k are positive and either $\int f_k = 1$ for all k or $\sum_{k \leq n} f_k = 1$

In these cases we can introduce

$$\frac{1}{n} \int \max_{k \leq n} f_k \, d\gamma \quad \text{or} \quad \int \max_{k \leq n} f_k \, d\gamma$$

as a measure of non-disjointness. If the parameter is

one then the function f_k have disjoint support. If the parameter is close to one then the f_k are allmost disjoint (this is not an easy exercise).
Let us turn to Poisson case. Here a person with parameter γ will cause k accidents in time interval t with probability $e^{-\lambda t} \frac{(\lambda t)^k}{k!}$. If one should use the exact number $\frac{k}{t}$ as a measure of the risk one arrives at considering the integral

$$\int \max_{k \geq 0} e^{-\lambda t} \frac{(\lambda t)^k}{k!} \, d\gamma(\lambda)$$

$$\text{Now } \max_{k \geq 0} e^{-\lambda t} \frac{(\lambda t)^k}{k!} \approx e^{-\lambda t} \frac{(\lambda t)^{\lambda t}}{(\lambda t)!} \approx \frac{\text{constant}}{\sqrt{\lambda t + 1}}$$

using Stirling's formulae.

It follows that $\int \max_{k \geq 0} f_{k,t}(\lambda) \, d\gamma(\lambda) \to 0$ as $t \to \infty$.
We will interpret this as follows. If $t \to \infty$ then $\frac{N_t}{t}$ will become closer to the exact value (law of large numbers) but the distribution around the exact value has too small concentration (this is precisely the content of the Central Limit Theorem which in case of a Poisson distribution is related to Stirling's formulae). The solution is of course evident:one should not use $\frac{Z_t}{t}$ as a value of the risk parameter but we should divide the real line in different intervals and estimate in which interval the parameter λ will be. This model will be developped in the following paragraph.

§ 2. Description of the model

We introduce a finite number of classes $C=\{1,\dots,K\}$. Each group $j \in C$ corresponds to a fixed subset L_j of Λ Of course $L_1 \dots L_k$ have to form a measurable partition of Λ. An insurance company will try to charge a premium according to the rule: If at time t we estimate a person being in class j then we charge π_j.
Since we can only use observations prior to t we have to construct sets $A_1 \dots A_K$ in F_t such that if the trajectory is in A_j we will charge π_j.
If we charge π_j and the underwriter has $\lambda \notin L_j$ then of course we made a mistake. The aim is to choose $A_1 \dots A_k$ so that the probability of making mistakes is as small as possible. If we compare this new situation with the one in the introduction the resemblance is clear : we want disjoint distributions but this time we require that they should be concentrated in prescribed sets $L_1 \dots L_K$.

§3. A maximum likelihood method

To make a mathematical approach feasible we introduce two assumptions
A1) For all $t \geq 0$ we require that all $\mathbb{P}^\lambda$ are equivalent on F_t (i.e they all have the same null sets)

A2) the mapping $Y:\Omega \to \Lambda$ is measurable (up to null sets) for $\tilde{F}_\infty$. The problem is now to construct $A_1 \ldots A_K$ a F_t measurable partition of D such that $(\star)\ \sum_{k\neq j} \mathbb{P}_\gamma[A_k x L_j]$

$\sum_{k\neq j} \mathbb{P}_\gamma[A_k x L_j]$ is minimal

For assemption A3) and the Random Nikodym Theorem we deduce the existence of a function $h_t : \Omega \to \mathbb{R}$ measurable for G_t such that

$$\forall B \in G_t \qquad \int_B d\mathbb{P}_\gamma = \mathbb{P}_\gamma \ (B) = \int_\Omega h_t(\eta,\lambda)\ d\mathbb{P}^\circ d\gamma$$

where $\mathbb{P}^\circ = \mathbb{P}^{\lambda\circ}$ is a fixed element out of $(\mathbb{P}^\lambda)\lambda \in \Lambda$

Expression $(\star)$ becomes

$$\sum_{k\neq j} \mathbb{P}_\gamma(A_k x L_j) = \sum_k (\mathbb{P}_\gamma(A_k x \Lambda) - \mathbb{P}_\gamma\ (A_k x L_k))$$

$$= 1 - \sum_k \mathbb{P}_\gamma[A_k x L_k]$$

$$= 1 - \sum_k \int_{A_k x L_k} h_t(\eta,\lambda) d\mathbb{P}^\circ d\gamma$$

It turns out we have to maximize

$$\sum_k \int_{A_k x L_k} h_t(\eta,\lambda)\ d\mathbb{P}^\circ d\gamma$$

$$= \sum_k \int_{A_k} d\mathbb{P}^\circ \quad \int_{L_k} h_t(\eta,\lambda) d\gamma(\lambda)$$

$= \sum_k \int_{A_k} d\mathbb{P}^\circ \; p_k^{(t)}(\eta)$ where $p_k^{(t)}(\eta) = \int_{L_k} h_t(\eta,\lambda)d\gamma$

is F_t measurable

It is clear that this expression is always smaller than

$$\int d\mathbb{P}^\circ \max_{k \leq K} p_k^{(t)}(\eta)$$

and it equals this expression only if

$$A_k = \{\eta \mid p_k^{(t)}(\eta) \geq p_s^{(t)}(\eta) \quad \text{for all } s \leq K\}$$

Let us define: $\varphi_t \; D \to C$ as follows

$$\varphi_t(\eta) = k \text{ if for } s < k \quad p_s^{(t)}(\eta) < p_k^{(t)}(\eta)$$

$$\text{for } s > k \quad p_s^{(t)}(\eta) \geq p_k^{(t)}(\eta)$$

Let $A_k^{(t)} = \{w \mid \varphi_t(w)=k\}$

Remark that the sets $A_1^{(t)} \ldots A_k^{(t)}$ are optimal !

<u>Theorem</u> : If $t \to \infty$ then for all $k \leq K$

$$Z^{-1}(A_k^{(t)}) \to Y^{-1}(L_k) \quad \mathbb{P}_\gamma \text{ a.e.}$$

(i.e. in the limit allmost everybody will be estimated to be in its correct class).

Proof : Since $\mathbb{P}_\gamma[A\times L] = \int_L \mathbb{P}^\lambda[A]\, d\gamma(\lambda)$ for all $A \in F_t$ and all $L \in L$ we also have $\mathbb{P}_\gamma[A\times L] = \int_L d\gamma(\lambda) \int_A d\mathbb{P}^\circ h_t(\eta,\lambda)$

We now claim that $\mathbb{P}_\gamma$ a.e.

$$\mathbb{E}\,[1_L \circ Y \mid F_t] = \frac{\int_L h_t(\eta,\lambda)\,\gamma(d\lambda)}{\int_\Lambda h_t(\eta,\lambda)\,\gamma(d\lambda)}$$

Indeed

$$\int_{A\times\Lambda} \frac{\int_L h_t(\eta,\lambda)\,\gamma(d\lambda)}{\int_\Lambda h_t(\eta,\lambda)\,\gamma(d\lambda)}\, d\mathbb{P}_\gamma(\eta,\mu)$$

$$= \int_\Lambda d\gamma(\mu) \int_A d\mathbb{P}^\circ\; h_t(\eta,\mu) \frac{\int_L h_t(\eta,\lambda)\,\gamma(d\lambda)}{\int_\Lambda h_t(\eta,\lambda)\,\gamma(d\lambda)}$$

$$= \int_A d\mathbb{P}^\circ \int_L h_t(\eta,\lambda)\,\gamma(d\lambda) = \int_{A\times L} h_t(\eta,\lambda)\, d\mathbb{P}^\circ\, d\gamma(\lambda)$$

$$= \mathbb{P}_\gamma\,[A\times L] = \int_A 1_L\circ Y\; d\mathbb{P}_\gamma.$$

From the theory of martingales and since DxL_k is measurable for $\tilde{F}_\infty$ we obtain for $k \leq K$

$$\frac{\int_{L_k} h_t(\eta,\lambda)\, d\gamma(\lambda)}{\int_{\Lambda} h_t(\eta,\lambda)\, d\gamma(\lambda)} \to 1_{L_k} \circ Y \quad \mathbb{P}_\gamma \quad \text{a.e.}$$

For allmost all $(\eta,\lambda') \in DxL_k$ we then can conclude that

$$\frac{\int_{L} h_t(\eta,\lambda)\, d\gamma(\lambda)}{\int_{\Lambda} h_t(\eta,\lambda)\, d\gamma(\lambda)} \to \begin{matrix} 1 \text{ if } s=k \\ 0 \text{ if } s\neq k \end{matrix}$$

Hence for t large enough and allmost all $w \in DxL_k$ we obtain that $w \in A_k^{(t)} x \Lambda$
Λ

Since A_k^t form a partition we obtain that

$$1_{A_k^t x\Lambda} \to 1_{DxL_k} \qquad \text{a.e.}$$

Remark 1 : Assumption A1) is reasonable if the measures would not be equivalent, we could make a distinction on the risk parameters using the appropriate null sets. Also the distributions $\mathbb{P}^\lambda$ represent the same kind of risks, they only differ in "size"

Remark 2 : Assumption A2) is critical since we need that $L_1 \ldots L_k$ should be in $\tilde{F}_\infty$. It also has a strong consequence.

<u>Proposition</u> : There is a set $\Lambda_o \subset \Lambda$, $\gamma(L_o) = 1$ such that for $\lambda,\mu \in \Lambda_o$ we have $\mathbb{P}^{\mu}$ and $\mathbb{P}^{\lambda}$ are disjoint.

<u>Proof</u> : The mapping $Y: \Omega \to E$ is measurable for $\tilde{F}_{\infty}$. Since $(E,\mathcal{E})$ is countably generated we can consider E as a subset of $[0,1]$ (Try the mapping $E \to [0,1]$ defined as $x \to \sum_k 2^{-k} 1_{E_k}(x)$. Where $E_k, k \geq 1$ generates $\mathcal{E}$)

From standard results on measurability it follows that there is a mapping

$\alpha : D \to [0,1]$ such that α is F_{∞} measurable and

$Y = \alpha_o Z.$ $\mathbb{P}_{\gamma}$ a.e.

Let $\Omega_o \subset \Omega$ be such that $\mathbb{P}_{\gamma}(\Omega_o) = 1$ and $Y = \alpha_o Z$ on Ω_o

Since $\mathbb{P}_{\gamma}(\Omega_o) = 1 = \int_{\Lambda} d\gamma \int 1_{\Omega_o}(\eta,\lambda)\, d\mathbb{P}^{\lambda}$ we find that

$\exists\, \Lambda_o \subset \Lambda, \gamma(\Lambda_o) = 1$ such that of $\lambda \in \Lambda_o : \mathbb{P}^{\lambda}[\eta | (\eta,\lambda) \in \Omega_o] = 1.$

Fix now $\lambda, \mu \in \Lambda_o$ and let

$A = \{\eta \,|\, (\eta,\lambda) \in \Omega_o\}$ $\qquad$ $B = \{\eta \,|\, (\eta,\mu) \in \Omega_o\}$

We claim that $A \cap B = \emptyset$ concluding the proof of the proposition. To see this observe that for $w \in A$ we have $Y(w) = \lambda$ whereas $Y(w) = \mu$ for $w \in B$ and hence $\eta \in A$ implies $\alpha(\eta) = \Lambda$ whereas $\alpha(\eta) = \mu$ for $\eta \in B$.

Remark 3 : Of course we can minimize other expressions then (★). One may think of calculating certain losses due to wrong decisions. This gives expressions of the form

$$\psi = \sum_{k \neq j} 1_{A_k} \, 1_{L_j} \, f_j(w) \text{ when } f_j \text{ is measurable for } F_t.$$

If we want to minimize $\mathbb{E}_\gamma[\psi]$ we obtain

$$\mathbb{E}_\gamma[\psi] = \sum_{k \neq j} \mathbb{E}_\gamma[1_{A_k} \, 1_{L_j} \, f_j]$$

$$= \sum_{k=1}^{K} \sum_{j \neq k} \int_{A_k} d\mathbb{P}^\circ f_j(\eta) \int_{L_j} h_t(\eta, \lambda) \, d\gamma(\lambda)$$

$$= \sum_{k=1}^{K} \sum_{j \neq k} \int_{A_k} d\mathbb{P}^\circ f_j(\eta) \, p_j^{(t)}(\eta)$$

$$= \sum_{k=1}^{K} \int_{A_k} q_k(\eta) d\mathbb{P}^\circ \quad \text{with } q_k(\eta) = \sum_{j \neq k} f_j(\eta) \, p_j^{(t)}(\eta)$$

The optimal choice is

$$A_k^{(t)} = \{\eta \mid q_k(\eta) < q_s(\eta) \quad \text{for } s < k \; ; \; q_k(\eta) \leq q_s(\eta)$$

$$\text{for } s > k \}$$

Remark 4 : The approach given here is in essential the one introduced by Neyman Pearson in the theory of testing hypotheses.

Remark 5 : If we have a weighted Markov process as in the previous paper then we obtain that $p_j^{(t)}(\eta)$ will be measurable for Z_t, hence the decision will only depend on the observation at time t. (This is precisely the meaning of a sufficient statistic)

Remark 6 : A classical example in collective risk theory is the automobile insurance. The introduction of the Bonus Malus system in Belgium took place around 1970. The system contains 18 classes but to make it Markov we need the consideration of 30 classes. A computer simulation give as a result that after an infinite time the separation of these 30 classes is very poor. Low risks with bad luck overlap with high risks having good luck, Moreover most of the drivers tend to be in the lower classes (a fact already observed).
A computer simulation using the same parameters but taking as class number the number of accidents caused during the last 5 years gave better results (the optimal choice according to paragraph 3).

WEIGHTED MARKOV PROCESSES WITH AN APPLICATION TO RISK THEORY

F. Delbaen
Vrije Universiteit Brussel

J. Haezendonck
Universitaire Instelling
Antwerpen

Summary. We give necessary and sufficient conditions for a weighted Markov process to be Markov.

Keywords. risk theory, weighted distributions, Markov processes

F. Delbaen, Vrije Universiteit Brussel, Departement Wiskunde, Pleinlaan 2 - F7, B-1050 Brussel. Belgium

F. de Vylder et al. (eds.), Premium Calculation in Insurance, 121–131.

§ 1. Notations and basic assumptions

In risk theory one often deals with weighted processes (see [1]). More precisely in collective risk theory one gives a family of processes X_t^λ, indexed by a parameter λ. The parameter λ itself is distributed according to a "structure distribution" γ. To describe a collective risk one first chooses λ according to the distribution γ and then one considers the process X_t^λ (see [6] and [2] for an application to a model of automobile insurance).
In many cases one can assume that the individual processes are Markov. We will give necessary and sufficient conditions under which the weighted process is still Markovian. The result is related to a result of Salminen [7], however our approach is more general. Before giving mathematical details we first need some notation.

For all definitions and notions about Markov processes we refer to Dynkin [4].

The measurable space $(E,\mathcal{E})$ will be the space of states of the different processes. The set E and the σ-algebra $\mathcal{E}$ will be fixed in the rest of the paper. To avoid "technical" difficulties we assume that $\mathcal{E}$ is countably generated i.e. $\mathcal{E}$ is generated by a sequence of subsets of E. Of course for practical applications this is no restriction at all.

The set D will denote the set of all realisations or trajectories (f.i. if $E = \mathbb{R}$ then D will be the set of all functions from the positive real numbers $\mathbb{R}_+$ into E, which are right continuous and have left limits for all $t > 0$). The mappings $X_t : D \to E$ defined for $t \geq 0$ will describe the evolution in time of the stochastic processes. By F_t we mean the σ-algebra generated by all X_s with $s \leq t$ i.e. $F_t = \sigma(X_s, s \leq t)$. With F_∞ we mean $\sigma(X_t, t \geq 0)$.

The measurable space $(\Lambda, \mathcal{L})$ is the space of parameters used to describe the weighted process. For each $\lambda \in \Lambda$ there is given a probability measure $\mathbb{P}^\lambda$ on F_∞. The family $\mathbb{P}^\lambda$ satisfies the regularity condition : $\forall A \in F_\infty$ the mapping $\lambda \mapsto \mathbb{P}^\lambda(A)$ is measurable for the σ-algebra $\mathcal{L}$. By a structure distribution we simply mean a probability measure defined on $\mathcal{L}$.

On $\Omega = D \times \Lambda$ we put the σ-algebras $G_t = F_t \otimes \mathcal{L}$. Given γ on $(\Lambda, \mathcal{L})$ we denote by $\mathbb{P}_\gamma$ the unique probability on G_∞ defined by the rule

$$\forall A \in F_\infty \; ; \; \forall L \in \mathcal{L} \; : \; \mathbb{P}_\gamma(A \times L) = \int_L \mathbb{P}^\lambda(A) \; \gamma(d\lambda).$$

The assumptions on $\mathbb{P}^\lambda$ guarantee the existence of $\mathbb{P}_\gamma$ (see [5]).

On Ω there are two "natural" projections $Z : \Omega \to D$ and $Y : \Omega \to \Lambda$. The process Z_t is defined on Ω as $Z_t = X_t \circ Z$. The σ-algebras generated by Z_t are denoted by $\tilde{F}_t$, i.e.

$$\tilde{F}_t = Z^{-1}(F_t) = \sigma(Z_s \; ; \; s \leq t) = F_t \otimes \{\phi, \Lambda\}.$$

Since different probability distributions are involved we also make the assumption that, unless stated otherwise, all expectations, conditional expectations ..., are performed on Ω with the probability $\mathbb{P}_\gamma$.

§ 2. The main result

Theorem : Let γ be a probability on $(\Lambda,\mathcal{L})$. Suppose that for γ allmost all λ the process $(D,(F_t)_{t\geq 0},\mathbb{P}^\lambda,(X_t)_{t\geq 0} : D \to E)$ is Markov. If for all $t \geq 0$ and all $a : \Omega \to \mathbb{R}$, $\widetilde{F}_t$ measurable and bounded :

$$\mathbb{E}[a|Z_t,Y] = \mathbb{E}[a|Z_t] \quad \mathbb{P}_\gamma \text{ a.e.} \qquad (\star)$$

then the weighted process $(\Omega,(\widetilde{F}_t)_{t\geq 0},\mathbb{P}_\gamma,(Z_t)_{t\geq 0}:\Omega\to E)$ is Markov.

Conversely if the weighted process is Markov, if for γ-allmost all λ the individual process is Markov and if moreover for all $t \geq 0$ the σ-algebra $\sigma(Y)$ is (up to sets of $\mathbb{P}_\gamma$-measure zero) included in $\sigma(Z_s, s \geq t)$, then condition $(\star)$ holds.

Remark : Condition $(\star)$ has an immediate interpretation : it says that conditionally to the present situation Z_t, the past does not depend on the parameter λ. The condition $(\star)$ is related to the condition of sufficiency in statistics. The extra condition in the converse part of the theorem says that by observing the process "at infinity" one can discover the value λ. For practical applications it is clear that if a parameter λ is introduced it should be possible to discover the value by observing the process. The above condition is a little bit more restrictive : one can discover λ by observing the process starting at any time t.

Lemma : Let $d : D \to \mathbb{R}$ be bounded and F_∞ measurable. Suppose $t \geq 0$ and let $f : E \times \Lambda \to \mathbb{R}$ be the (bounded) function such that $\mathbb{E}[d \circ Z | Z_t, Y] = f(Z_t, Y)$. The function f then satisfies for γ-allmost all λ :

$$f(X_t, \lambda) = \mathbb{E}_\lambda[d | X_t] \quad \mathbb{P}^\lambda \text{ a.e.}$$

($\mathbb{E}_\lambda$ denotes the expectation with respect to $\mathbb{P}^\lambda$).

<u>Proof of the lemma</u> : First of all remark that the existence of f is an easy exercise on measurability (see [5]). Since $(E, \mathcal{E})$ is countably generated there is a countable set $(h_n)_{n \geq 1}$ of bounded functions $h_n : E \to \mathbb{R}$ such that for all $\lambda \in \Lambda$:

$$h_n \circ X_t \text{ is dense in } L^2(D, \sigma(X_t), \mathbb{P}^\lambda).$$

The conditional expectation of d is uniquely determined by the following equation

$$\int_D |d - \mathbb{E}_\lambda[d | X_t]|^2 \, d\mathbb{P}^\lambda =$$

$$\min_{h \in L^2(D, \sigma(X_t), \mathbb{P}^\lambda)} \int_D |d - h|^2 \, d\mathbb{P}^\lambda = m(\lambda).$$

By the density of the sequence $h_n \circ X_t$ we can replace this relation by

$$\int_D |d-\mathbb{E}_\lambda[d|X_t]|^2 \, d\mathbb{P}^\lambda = \inf_{n\geq 1} \int_D |d-h_n(X_t)|^2 \, d\mathbb{P}^\lambda.$$

For each n let L_n be the measurable set defined as

$$L_n = \{\lambda \mid \int_D |d-f(X_t,\lambda)|^2 \, d\mathbb{P}^\lambda > \int_D |d-h_n(X_t)|^2 \, d\mathbb{P}^\lambda\}$$

Suppose that for some n we have $\gamma(L_n) > 0$. Then we put

$$k(e.\lambda) = f(e,\lambda) \quad \text{for } \lambda \notin L_n$$
$$= h_n(e) \quad \text{for } \lambda \in L_n$$

We obtain :

$$\int_\Omega |doZ - k(Z_t.Y)|^2 \, d\mathbb{P}_\gamma =$$

$$\int_{L_n} \gamma(d\lambda) \int_D |d-h_n(X_t)|^2 \, d\mathbb{P}^\lambda + \int_{L_n^c} \gamma(d\lambda) \int_D |d-f(X_t.\lambda)|^2 \, d\mathbb{P}^\lambda$$

$$< \int_{L_n} \gamma(d\lambda) \int_D |d-f(X_t.\lambda)|^2 \, d\mathbb{P}^\lambda + \int_{L_n^c} \gamma(d\lambda) \int_D |d-f(X_t.\lambda)|^2 \, d\mathbb{P}^\lambda$$

$$< \int_\Omega |doZ - f(Z_t,Y)|^2 \, d\mathbb{P}_\gamma$$

which is a contradiction to $f(X_t,Y) = \mathbb{E}[doZ|Z_t,Y]$ a.e.

It follows that for all n : $\gamma(L_n) = 0$ and hence for γ-allmost all λ :

$$\int_D |d - f(X_t,\lambda)|^2 \, d\mathbb{P}^\lambda = m(\lambda)$$

i.e. $f(X_t,\lambda) = \mathbb{E}_\lambda[d|X_t]$ $\mathbb{P}^\lambda$ a.e.

Proof of the theorem : Let g be bounded and $\mathcal{E}$ measurable $g : E \to \mathbb{R}$. Let a be an arbitrary bounded function $a : \Omega \to \mathbb{R}$ measurable for $\widetilde{F}_t$. By the definition of $\widetilde{F}_t$ there is a F_t measurable function $b : D \to \mathbb{R}$ such that $a = b \circ Z$. For $s > t$ we obtain

$\mathbb{E}[\mathbb{E}[g(Z_s)|Z_t].a]$

$= \mathbb{E}[g(Z_s)\ \mathbb{E}[a|Z_t]]$

$= \mathbb{E}[g(Z_s)\ \mathbb{E}[a|Z_t,Y]]$ by condition (★)

$= \mathbb{E}[a\ \mathbb{E}[g(Z_s)|Z_t,Y]]$

$= \mathbb{E}[b\circ Z.f(Z_t,Y)]$ for some $f : E \times \Lambda \to \mathbb{R}$

$= \int \gamma(d\lambda) \int d\mathbb{P}^\lambda\ b\ .\ \mathbb{E}_\lambda[g(X_s)|X_t]$ by the lemma

$= \int \gamma(d\lambda) \int d\mathbb{P}^\lambda\ b\ .\ \mathbb{E}_\lambda[g(X_s)|F_t]$ since the individuals are Markov

$= \int \gamma(d\lambda) \int d\mathbb{P}^\lambda\ b\ .\ g(X_s)$

$= \int b \circ Z.g \circ Z_s\ d\mathbb{P}_\gamma$

$= \mathbb{E}[a\ g(Z_s)]$

$= \mathbb{E}[a\ \mathbb{E}\ g(Z_s)|\widetilde{F}_t]]$

Since the equations hold for all a we have for s > t

$$\mathbb{E}[g(Z_s)|Z_t] = \mathbb{E}[g(Z_s)|\widetilde{F}_t] \quad \mathbb{P}_\gamma \text{ a.e.}$$

which establishes the Markov property.

Proof of the converse :

Let $f : \Omega \to \mathbb{R}$ be bounded and $\widetilde{F}_t$ measurable and let $g : \Omega \to \mathbb{R}$ be bounded and measurable for $\sigma(Z_s, s \geq t)$. By the extra assumption we have (up to sets of measure zero)

$$\sigma(Z_s, s \geq t) = \sigma(Z_s, s \geq t ; Y)$$

hence $L^\infty(\Omega, \sigma(Z_s, s \geq t), \mathbb{P}_\gamma) = L^\infty(\Omega, \sigma(Z_s, s \geq t, Y), \mathbb{P}_\gamma)$.

An easy application of the Markov property gives us that

$$\mathbb{E}[f.g|Z_t] = \mathbb{E}[f|Z_t].\mathbb{E}[g|Z_t]$$

The same property for the individual processes together with the lemma gives :

$$\mathbb{E}[f.g|Z_t, Y] = \mathbb{E}[f|Z_t, Y].\mathbb{E}[g|Z_t, Y] \quad \mathbb{P}_\gamma \text{ a.e.}$$

Combining the equalities yields :

$$\begin{aligned}\mathbb{E}[f.g|Z_t] &= \mathbb{E}[\mathbb{E}[f.g|Z_t\ Y]|Z_t] \\ &= \mathbb{E}[\mathbb{E}[f|Z_t, Y].\mathbb{E}[g|Z_t\ Y]|Z_t] \\ &= \mathbb{E}[f|Z_t].\mathbb{E}[g|Z_t].\end{aligned}$$

Hence

$$\mathbb{E}[\mathbb{E}[f|Z_t].g] = \mathbb{E}[\mathbb{E}[f|Z_t\ Y].g]$$

or $$\mathbb{E}[g(\mathbb{E}[f|Z_t]-\mathbb{E}[f|Z_t,Y])] = 0$$

for all $g \in L^{\infty}(\sigma(Z_s,\ s\geq t)) = L^{\infty}(\sigma(Z_s,\ s\geq t\ ;\ Y))$.

This certainly implies $\mathbb{E}[f|Z_t] = \mathbb{E}[f|Z_t,Y]$ $\mathbb{P}_{\gamma}$ a.e.

§ 3. Examples

(1) Suppose that D is the set of all increasing right continuous step functions from $\mathbb{R}_+$ into $\mathbb{R}_+$ i.e. D is the set of functions α satisfying

1) $\alpha(0) = 0$
2) $\forall t \geq 0 \ \ \exists \varepsilon > 0$ such that $\alpha(t) = \alpha(s)$ for all $t \leq s \leq t+\varepsilon$
3) for $s < t$: $\alpha(s) \leq \alpha(t)$.

The mappings X_t are nothing else but point evaluations $X_t(\alpha) = \alpha(t)$. The set $(\Lambda,\mathcal{L})$ is the set of positive real numbers with the Borel σ-algebra. For each $\lambda > 0$ let $\mathbb{P}^{\lambda}$ be the probability on D such that X_t is a homogeneous Poisson process with parameter $\lambda > 0$, i.e. X_t has independent increments and

$\mathbb{P}[X_t=n] = e^{-\lambda t}\frac{(\lambda t)^n}{n!}$. An easy calculation shows that condition (*) is fulfilled for all probability measures on $(\Lambda,\mathcal{L})$.

We also remark that for all γ one has $\lambda = \lim_{t\to\infty}\frac{X_t}{t}$ $\mathbb{P}_{\gamma}$ a.e. and hence the extra condition is also fulfilled. It turns out that for all γ the weighted Poisson process is a Markov process.

(2) Same notation as above but this time $\Lambda = \{1,2\}$ and $\mathbb{P}^1$ and $\mathbb{P}^2$ are defined so that

(i) $\mathbb{P}^1$ gives a homogeneous Poisson process with parameter $\lambda > 0$.

(i') $\mathbb{P}^2$ gives up to $t_0 > 0$ a homogeneous Poisson process with parameter $\mu \neq \lambda$.
For $t > t_0$ the process $X_t - X_{t_0}$ is independent from F_{t_0} and is an homogeneous Poisson process with parameter λ.

Condition (★) and the extra condition are not fulfilled but nevertheless one can easily show that for $\mathbb{P} = \frac{1}{2}\mathbb{P}^1 + \frac{1}{2}\mathbb{P}^2$ we obtain a Markov process.

(3) Same situation as in (1), this time we denote by N_t the homogeneous Poisson process (for the measure $\mathbb{P}^\lambda$). Let $(U,\mathcal{U},\mathbb{Q})$ be an auxiliary probability space and let X_t be defined as

$$X_t = \sum_{k=1}^{N_t} S_k$$

Here S_k is a sequence of independent identically distributed strictly positive variables defined on $(U,\mathcal{U},\mathbb{Q})$. On D x U we put the family of distributions $\mathbb{P}'^\lambda = \mathbb{P}^\lambda \otimes \mathbb{Q}$. If $\mathbb{E}[S_k] < \infty$ then one easily sees that $\lim_{t\to\infty} \frac{X_t}{t} = \lambda.\mathbb{E}[S_k]$ and hence the extra condition is fulfilled. It follows that condition (★) is necessary and sufficient to conclude that the so-obtained weighted compound Poisson process is Markov. Unfortunately even for very reasonable S_k

the condition (*) is not fulfilled.
Since the weighted process is also a risk process in the sense of [3] one can apply the results obtained there. It turns out that weighted compound processes are (except in very trivial cases) never Markov.

References

[1] Buhlmann, H. (1970), Mathematical Methods in Risk Theory, Springer Verlag, Berlin

[2] Delbaen, F., Depauw, J., De Rouck, M., Haezendonck, J., Van Coethem, P. (1980), Etude Mathématique de l'assurance automobile belge. Journal de l'ARAB, 21-68

[3] Delbaen, F., Haezendonck, J. (1983), On Risk Processes with the Markov Property and with Independent Increments. To appear in : Insurance Mathematics and Economics

[4] Dynkin, E.B. (1960), Foundations of the theory of Markov processes, Pergamon Press, Oxford

[5] Neven, J. (1964), Bases Mathématiques du Calcul des probabilités, Masson, Paris

[6] Norberg, R. (1976), A credibility theory for Automobile Bonus Systems, Scand. Actuarial J., 92-107

[7] Salminen, P. (1983), Mixing Markovian Laws ; with an application to Path Decompositions, Stochastics, Vol. 9, 223-231

PRACTICAL MODELS IN CREDIBILITY THEORY, INCLUDING PARAMETER ESTIMATION (*)

F. de Vylder

Catholic University of Louvain,
Louvain-la-Neuve, Belgium

ABSTRACT

In credibility theory, an unobservable random vector Y is approximated by a random vector $\hat{Y}$ in a pre-assigned set $\mathcal{A}$ of admitted estimators for Y. The credibility approximation $\hat{Y}$ for Y is best in the sense that it is the vector $V \in \mathcal{A}$ minimizing the distance d(V,Y) between V and Y.

Several models in credibility theory (depending on $\mathcal{A}$, d and the set of all involved random variables) have been developed in recent years. The starting pointof modern credibility theory is the model by Bühlmann (1967). This model has been generalized in several directions. We only retain the models having already proved their practical usefulness and, among the newer models, those that we believe to be most promising for future applications. These are :

- The time-homogeneous model with unweighted observations (Bühlmann, 1967).
- The time-homogeneous model with weighted observations (Bühlmann and Straub, 1970).
- The regression model (Hachemeister, 1975)
- The regression model with scalar credibility weights (De Vylder, to be published)
- A multiplicative model (De Vylder, submitted for publication)

(*) Invited lecture at the 14th European Meeting of Statisticians (Wroclaw, 1981).

F. de Vylder et al. (eds.), Premium Calculation in Insurance, 133–150.

Our approach has a geometrical character. The vector Y shall be a point in a Hilber space $\mathcal{H}$ and $\mathcal{A}$ shall be a closed convex set in $\mathcal{H}$. Then $\hat{Y}$ shall be the orthogonal projection of Y on $\mathcal{A}$.

The credibility estimator $\hat{Y}$ depends on observable random variables but also on unknown structure parameters. In the practical models, the latter can be estimated from realizations of the observable random variables.

1. HILBERT SPACE OF RANDOM VECTORS

1.1. Abstract Hilbert space

Our development of credibility theory is based on the following rather elementary parts of Hilbert space theory.

1.1.1. Definition

A <u>real Hilbert space</u> $\mathcal{H}$ is a vector space over R equipped with a real scalar product, i.e. a real function $\langle X,Y\rangle$ defined for $X,Y\in\mathcal{H}$ and having the following properties.
$\langle X,X\rangle = o \Rightarrow X=o$, $\langle X,X\rangle \geqslant o$, $\langle X+Y,Z\rangle = \langle X,Z\rangle + \langle Y,Z\rangle$, $\langle aX,Y\rangle = a\langle X,Y\rangle$ $(a\in R)$.

This scalar product defines a norm by the relation $\|X\| = \langle X,X\rangle^{1/2}$ $(X\in\mathcal{H})$. By definition of Hilbert space, $\mathcal{H}$ is also supposed to be complete in the topology induced by this norm.

1.1.2. Subspaces and translated subspaces

A <u>subspace</u> $\mathcal{L}$ of the real Hilbert space $\mathcal{H}$ is a subset of $\mathcal{H}$ containing all (finite) linear combinations of its elements. A <u>translated subspace</u> $\mathcal{T}$ of $\mathcal{H}$ is a subset of $\mathcal{H}$ that can be displayed as $\mathcal{T} = X+\mathcal{L} = \{X+V/V\in\mathcal{L}\}$, where $X\in\mathcal{H}$ and where $\mathcal{L}$ is a subspace of $\mathcal{H}$. The subset $\mathcal{T}$ of $\mathcal{H}$ is a translated subspace iff it contains all (finite) normed linear combinations of its elements.

Finite-dimensional translated subspaces of $\mathcal{H}$ (in particular, finite-dimensional subspaces) are closed subsets of $\mathcal{H}$.

Translated subspaces (in particular, subspaces) are convex sets.

1.1.3. Orthogonality and projection

Let $X\in\mathcal{H}$ and let $\mathcal{A}$ be a closed convex set in the real Hilber space $\mathcal{H}$. Then there is a unique point $P\in\mathcal{A}$ minimizing the distance $d(X,P) = \|X-P\|$. The point P is called the (orthogonal) <u>projection of</u> X <u>on</u> $\mathcal{A}$ and is denoted by $Pro(X/\mathcal{A})$.

$X,Y \in \mathcal{H}$ are said to be orthogonal (notation $X \perp Y$) if $<X,Y> = o$.

If $\mathcal{A}$ is a supspace of $\mathcal{H}$, then the point $P \in \mathcal{A}$ is the projection of X on $\mathcal{A}$ iff $X-P \perp V$ for all $V \in \mathcal{A}$. More generally, if $\mathcal{A}$ is a translated subspace of $\mathcal{H}$, then $P \in \mathcal{A}$ is the projection of X on $\mathcal{A}$ iff $X-P \perp U-V$ for all $U,V \in \mathcal{A}$.

If $\mathcal{A}_1, \mathcal{A}_2$ are translated closed subspaces of $\mathcal{H}$ with $\mathcal{A}_1 \subset \mathcal{A}_2$, then $\mathrm{Pro}(X/\mathcal{A}_1) = \mathrm{Pro}(\mathrm{Pro}(X/\mathcal{A}_2)/\mathcal{A}_1)$ (theorem on iterated projections).

1.2. Hilbert space of random vectors

1.2.1. General notations

Random quantities are represented by capital letters or (in case of estimators) by small lettersbearing one or two circumflex accents. Non-random quantities are represented by small letters (without circumflex accents).

Usual (no boldface) letters are used for vectors and matrices. The indication $\binom{n}{m}$ accompanying a matrix means that that matrix has m rows and n columns. For instance $X \binom{1}{t}$ shall be a column with t random components. The setof $\binom{n}{m}$ matrices with real (non-random) elements is denoted by R^n_m.

As a general rule we assume that all displayed quantities (expectations, covariances, inverses of matrices, products of matrices, ...) exist and are finite. However, in most cases this existence shall result from precisely stated assumptions.

If V is a random matrix, we denote by $V° = V-EV$ the corresponding centered random matrix. For $X\binom{1}{m}$, $Y\binom{1}{n}$, the covariance matrix $\mathrm{Cov}(X,Y')$ is defined to bethe matrix $E(X°Y°') = E(XY') -EX.EY' \binom{n}{m}$. The covariance matrix $\mathrm{Cov}(X,X')$ of X with itself is more simply written as Cov X.

The unit matrix is denoted by 1, its dimensions always being clear from the context.

1.2.2. Definition of the space $\mathcal{H}_t(p)$

All random variables considered inthe sequel are supposed to be defined on a fixed probability space. Let t be a positive integer and let $\mathcal{H}_t$ be the set of random vectors $X \binom{1}{t}$ with square-integrable components, two such vectors being considered as identical if they are equal a.s. Let $p \binom{t}{t}$ be a positive definite matrix, called weighting matrix. The relation $<X,Y> = E\ X'pY$ defines a scalar product on $\mathcal{H}_t$. We denote by $\mathcal{H}_t(p)$

the space $\mathcal{H}_t$ equipped with the just defined scalar product. It is easily verified that $\mathcal{H}_t(p)$ is a Hilbert space (In order to prove completeness, diagonalize p and use the well-known theorem stating that $\mathcal{H}_1$, with p irrelevant in that case, is complete).

2. THE REGRESSION MODEL IN CREDIBILITY THEORY (Hachemeister, 1975)

2.1. Case of one observable vector

2.1.1. Definition of the model

In this subsection 2.1 we consider a fixed vector $X \in \mathcal{H}_t$, called the <u>observable vector</u> and a fixed random variable Θ, called the <u>structure variable</u> (in fact all following arguments are valid if Θ is any family of random variables). We make the following assumptions :

(i) Regression assumption : $E(X/\Theta) = y\beta(\Theta)$ where $y \in R_t^g$ is a given <u>design matrix</u> of full rang $g < t$ and where $\beta(\Theta) \in \mathcal{H}_g$.

(ii) Covariance assumption : $Cov(X/\Theta) = \sigma^2(\Theta).v$ where $v \in R_t^t$ is a given positive definite matrix and where $E\sigma^2(\Theta) < \infty$.

The interpretation of Θ is the following. In fact we initially assume that the distribution of the observable vector X depends on an unknown parameter Θ. This parameter is considered as being the unknown realization of some random variable Θ.

Very roughly stated, the problem in this regression model is to find approximations for $\beta(\Theta)$.

2.1.2. Initial space and contracted space

The space $\mathcal{H}_t$ containing the observable vector X shall be called the <u>initial space</u> and the space $\mathcal{H}_g$ containing the vector $\beta(\Theta)$, the <u>contracted space</u>. The vectors in $\mathcal{H}_g$ are shorter than those in $\mathcal{H}_t$.

Corresponding Hilbert space $\mathcal{H}_t(p)$ and $\mathcal{H}_g(q)$ shall also be considered. Here $p \in R_t^t$ and $q \in R_g^g$ are arbitrary weighting matrices. In some cases it may be interesting to assume that they are related by $q=y'py$. Indeed, to each vector $U \in \mathcal{H}_g(q)$ corresponds a vector $yU \in \mathcal{H}_t(p)$. If (U,V) is a couple of vectors in $\mathcal{H}_g(q)$ and if (yU,yV) is the corresponding couple in $\mathcal{H}_t(p)$, then the scalar product $\langle U,V \rangle$ in $\mathcal{H}_g(q)$ equals the scalar product $\langle yU,yV \rangle$ in $\mathcal{H}_t(p)$ if $q=y'py$.

To each closed convex set $\mathcal{A}$ in $\mathcal{H}_g(q)$ corresponds a closed convex set $y\mathcal{A} = \{yU/U \in \mathcal{A}\}$ in $\mathcal{H}_t(p)$. If $q=y'py$ and $V = Pro(U/\mathcal{A})$, then $yV = Prof(yU/y\mathcal{A})$.

2.1.3. Notations

1. Non-random quantities

$b = E\beta(\Theta)\ \binom{1}{g}$, $a = \mathrm{Cov}\ \beta(\Theta)\ \binom{g}{g}$, $s^2 = E\ \sigma^2(\Theta)$ (scalar),

$c = \mathrm{Cov}\ X\ \binom{t}{t}$, $u = (y'v^{-1}y)^{-1}\ \binom{g}{g}$, $z = a(a+s^2u)^{-1}\ \binom{g}{g}$.

a,b,s^2 are the <u>structure parameters</u>. z is the <u>credibility matrix</u>.

2. Vectors in the contracted space

$$\hat{\mathfrak{b}} = (y'py)^{-1}y'pX\ \binom{1}{g},$$

$$\mathfrak{b} = (y'v^{-1}y)^{-1}y'v^{-1}X = (y'c^{-1}y)^{-1}y'c^{-1}X\ \binom{1}{g},$$

$$B = z\mathfrak{b} + (1-z)b\ \binom{1}{g}.$$

We notice that $\hat{\mathfrak{b}}$ depends on the matrix p defining the scalar product in $\mathcal{H}_t(p)$. The vectors $\mathfrak{b}$,B do not depend on the weighting matrices p,q.

The equality in the definition of $\mathfrak{b}$ results from the two following lemma's. If $p = v^{-1}$ or $p = c^{-1}$, then $\hat{\mathfrak{b}} = \mathfrak{b}$.

3. Subsets of the contracted space

$\mathcal{L}_{con.} = R^1_g$ (subspace of constant vectors),

$\mathcal{L}_{hom.} = \{\zeta X / \zeta \varepsilon R^t_g\ ,\ \zeta y = 1\}$ (translated subspace of homogeneous unbiased combinations of the observable variables),

$\mathcal{L}_{gen.} = \{\eta + \zeta X / \eta \varepsilon R^1_g,\ \zeta \varepsilon R^t_g\}$ (subspace of general - unbiased or not, homogeneous or not - linear combinations of the observable random variables),

$\mathcal{L}(\Theta) = \{f(\Theta)\ /\ f$ such that $f(\Theta) \varepsilon \mathcal{H}_g\}$ (subspace of functions of Θ).

of course, the observable random variables are the components of the observable vector X.

The unbiasedness condition $\zeta y=1$ in the definition of $\mathcal{L}_{hom.}$ implies that $E\zeta X = b = E\beta(\Theta)$ for all $\zeta X \varepsilon \mathcal{L}_{hom.}$.

4. Subsets of the initial space

$\mathcal{L}^*_{con.} = R^1_t$ (subspace of constant vectors),

$\mathcal{L}^*_{gen.} = \{\eta + \zeta X\ /\ \eta \varepsilon R^1_t, \zeta \eta \varepsilon R^t_t\}$ (subspace of general linear combinations of the observable random variables),

$\mathcal{L}^\star(\Theta) = \{f(\Theta)\ /\ f \text{ such that } f(\Theta)\varepsilon\mathcal{H}_t\}$ (subspace of functions of Θ).

In the initial space we shall also consider the sets $y\mathcal{L}_{con.}$ (subspace), $y\mathcal{L}_{hom.}$ (translated subspace) and $y\mathcal{L}(\Theta)$ (subspace).

2.1.4. Lemma

$$E\ X = yb,\ E\tilde{b} = E\hat{b} = EB = b,$$

$$c = s^2v+yay',\ Cov(\beta(\Theta),X') = ay',\ Cov(\hat{b},\beta'(\Theta)) = a$$

$$Cov(\hat{b},X') = a+s^2u)y',\ Cov\ \hat{b} = a+s^2u$$

Demonstration. Immediate from general relations such as $Cov(rX,(sY)') = r\ Cov(X,Y')\ s'$, $EX = EE(X/\Theta)$, $Cov(X,Y') = Cov(E(X/\Theta),E(Y/\Theta)) + E\ Cov(X,Y'/\Theta)$.

2.1.5. Lemma

If c,v,a,y are any matrices with dimensions, positiveness and rank properties of the matrices defined before and if these matrices are related by c=v+yay', then

$$(y'c^{-1}y)^{-1}y'c^{-1} = (y'v^{-1}y)^{-1}y'v^{-1}\ ,\ (y'c^{-1}y)^{-1} = (y'v^{-1}y)^{-1} + a.$$

Demonstration. The verification of these relation, starting with the last one, is rather immediate.

2.1.6. Theorem

In the initial space $\mathcal{H}_t(p)$, the following projections hold.

$$Pro(X/\mathcal{L}^\star(\Theta)) = y\beta(\Theta), \quad (1)$$
$$Pro(X/y\mathcal{L}(\Theta)) = y\beta(\Theta), \quad (2)$$
$$Pro(X/\mathcal{L}^\star_{con.}) = yb, \quad (3)$$
$$Pro(X/y\mathcal{L}_{con.}) = y\tilde{b}, \quad (4)$$
$$Pro(X/y\mathcal{L}_{hom.}) = y\hat{b}, \quad (5)$$
$$Pro(y\beta(\Theta)/\mathcal{L}^\star_{gen.}) = yB. \quad (6)$$

In the contracted spece $\mathcal{H}_g(q)$, the following projections hold.

$$Pro(\beta(\Theta)/\mathcal{L}_{con.}) = b, \quad (7)$$
$$Pro(\beta(\Theta)/\mathcal{L}_{hom.}) = \hat{b}, \quad (8)$$
$$Pro(b/\mathcal{L}_{hom.}) = \hat{b}, \quad (9)$$
$$Pro(\beta(\Theta)/\mathcal{L}_{gen.}) = B. \quad (10)$$

Demonstration. In order to prove these relations it is sufficient to verify the corresponding orthogonality relations (see 1.1.3). As an illustration, we consider (10), the verification of the other relations being similar (and much simpler in most cases).

We first notice that indeed $B\varepsilon\mathcal{L}_{gen.}$. Ten it is sufficient to verify that

$$B-\beta(\Theta) \perp \eta + \xi X \quad (\eta\varepsilon R_g^1,\ \xi\varepsilon R_g^t).$$

But

$$B-\beta(\Theta) = z\hat{b}+(1-z)b-\beta(\Theta) = z(\hat{b}-b)-(\beta(\Theta)-b) = z\hat{b}^\circ-\beta^\circ(\Theta)$$

and so it is successively sufficient to verify that

$$z\hat{b}^\circ-\beta^\circ(\Theta) \perp \eta + \xi X\ ,\ z\hat{b}^\circ-\beta^\circ(\Theta) \perp \xi X,$$

$$z\hat{b}^\circ-\beta^\circ(\Theta) \perp \xi X^\circ\ ,\ E\ X^{\circ\prime}\xi' q(z\hat{b}^\circ-\beta^\circ(\Theta)) = o.$$

In order to transform the latter equality we use the cyclical property of the trace of a matrix, the fact that a scalar equals its trace, the fact that the trace and expectation operators commute, the covariance relations in 2.1.4. Then we have to verify that

$$\mathrm{tr}\ E\ q(z\hat{b}^\circ-\beta^\circ(\Theta))X^{\circ\prime}\xi' = o$$

or

$$\mathrm{tr}\ q(z(a+s^2u)y'-ay')\xi' = o$$

or

$$\mathrm{tr}\ q(z(a+s^2u)-a)y'\xi' = o.$$

The last relation results from the definition of z.

2.1.7. Comments and definitions

It is worth while to notice that none of the projections in the preceding theorem, except $y\hat{b}$ (5), depends on the weighting matrices p,q.

Each projection is the solution of a minimization problem. Most authors in credibility theory work (implicitly) in the initial space and obtain a relation equivalent to (6).

The following definitions are based on the projections in the contracted space.

According to (7) we call b the best constant estimator for $\beta(\Theta)$. According to (8) we call $\hat{b}$ the best homogeneous estimator for $\beta(\Theta)$. The vector B (10) is the credibility estimator for $\beta(\Theta)$. From the inclusions $\mathcal{L}_{con.}$ $\mathcal{L}_{gen.}$, $\mathcal{L}_{hom.}$ $\mathcal{L}_{gen.}$ results that B is closer to $\beta(\Theta)$ than b and $\hat{b}$.

The expression $B=z\hat{b}+(1-z)b$ shows that the credibility estimator can be considered as being a weighted average of the homogeneous estimator $\hat{b}$ and the constant estimator b (with matricial coefficients z and 1-z).

Looking back at the definitions of $\hat{b}$ and z (2.1.3), we see that B is expressed in terms of the known matrices v,y, the observable vector X and the structure parameters a,b,s^2. In practical work the latter are replaced by estimates (obtained in next section).

2.2. Case of several observable vectors

2.2.1. Definition of the model

We now consider $k > 1$ observable vectors $X_j \varepsilon \mathcal{H}_{t_j}$ and corresponding structure variables Θ_j $(j=1,2,\dots,k)$.
Notice that the length t_j of X_j may depend on j.

For each couple (X_j,Θ_j) we adopt a regression assumption and a covariance assumption (see 2.1.1). For all j, the results in 2.1 are valid. All notations introduced there must now be completed by a lower subscript j.

We make the following supplementary assumptions.

(j) Independence assumption: the couples $(X_1,\Theta_1),(X_2,\Theta_2),\dots,(X_k,\Theta_k)$ are independent

(jj) Invariability of structural parameters assumption : the structure parameters b_j (and then also the length g_j of that vector, a_j, s_j^2 do not depend on j.

These structure parameters arethen denoted by $b\binom{1}{g}$, $a\binom{g}{g}$ s^2 (scalar).

The credibility estimator for $\beta_j(\Theta_j)$ is B_j. We shall see how the involved structure parameters a, b, s^2 can be estimated from the observable vectors X_j.

Remark. The following assumptions (k), (kk) imply (jj).

(k) $\Theta_1,\Theta_2,\dots,\Theta_k$ are identically distributed.

(kk) The functions $\beta_j(.) = \beta(.)$, $\sigma_j^2(.) = \sigma^2(.)$ do not depend on j.

Usually one adopts (j), (k), (kk) but only (j) and (jj) are used in the sequel.

2.2.2. Notation

If w_j (j=1,2,...,k) is a finite sequence of matrices (in particular vectors or scalars) of same dimensions, we define w_Σ by

$$w_\Sigma = \sum_{j=1}^{k} w_j$$

If moreover w_j is square, the pre-normed sequence $\bar{w}_j$ (j=1,2,...,k) is defined by

$$\bar{w}_j = (w_\Sigma)^{-1} w_j$$

where it is understood then that w_Σ is invertible.

2.2.3. Estimation of b.

From the relation $E\hat{b}_j = b$ results that $\hat{b}_j$ is an unbiased estimator for b. Then, for all w_j ($\overset{g}{g}$) (j=1,2,...,k),

$$\hat{b}_w = \sum_{j=1}^{k} \bar{w}_j b_j$$

is an unbiased estimator for b (w occurring in the first member if of course a fixed subscript related to the w_j). What are the best w_j ? From the following theorem results that

$$\hat{b}_z = \sum_{j=1}^{k} \bar{z}_j \hat{b}_j ,$$

where $z_j = a(a+s^2u_j)^{-1}$ is the jth credibility matrix, is optimal in some sense.

THEOREM. In $_g(q)$ let us consider the set

$$\mathcal{L}_{hom} = \{ \sum_{j=1}^{k} \xi_j X_j \ / \ \xi_j \varepsilon R_g^{t_j} (j=1,2,\dots,k) \ , \ \sum_{j=1}^{k} \xi_j y_j = 1 \}$$

(translated subspace of linear homogeneous unbiased cominations of the observable random variables).

Then

$$\text{Pro}(b/\mathcal{L}_{hom.}) = \hat{b}_z \text{ (independent of q).}$$

Demonstration. The vector $\hat{b}_z$ equals

$$\hat{b}_z = \sum_{j=1}^{k} \bar{z}_j (y_j' v_j^{-1} y_j)^{-1} y_j' v_j^{-1} X_j .$$

The last member is 1 ($\overset{g}{g}$) if X_j is replaced by y_j. This proves that $\hat{b}_z \varepsilon \mathcal{L}_{hom.}$. Then it remains to prove the orthogonality

$$\hat{b}_z - b \perp \sum_j (\xi_j - \eta_j) X_j \quad (\xi_j, \eta_j \varepsilon R_g^{t_j}, \ \sum_j \xi_j y_j = \sum_j \eta_j y_j = 1),$$

or

$$\sum_i \bar{z}_i \hat{b}_i^\circ \perp \sum_j (\xi_j - \eta_j) X_j^\circ$$

or

$$\sum_{ij} E\, X_j^{\circ\prime} (\xi_j' - \eta_j') q \bar{z}_i \hat{b}_i^\circ = o$$

or (independence assumption, properties of the trace)

$$\sum_j \operatorname{tr} E\, q z_j \hat{b}_j^\circ X_j^{\circ\prime} (\xi_j' - \eta_j') = o$$

or (2.1.4, definition of z_j)

$$\sum_j \operatorname{tr} qa(a+s^2 u_j)^{-1} (a+s^2 u_j) y_j' (\xi_j' - \eta_j') = o$$

or

$$\operatorname{tr} qa(\sum_j \xi_j y_j - \sum_j \eta_j y_j)' = o.$$

Remark. Taylor (1977) (and before, Bühlmann and Straub, 1970, in their particular model) obtained the estimator $\hat{b}_z$ for b in an implicit way. Next theorem is a variant of the result by Taylor. In that theorem $\hat{\mathcal{L}}_{hom.}$ is the translated subspace defined in the preceding theorem and $\hat{B}_j = z_j \hat{b}_j + (1-z_j)\hat{b}_z$ is the vector $B_j = z_j \hat{b}_j + (1-z_j) b$ where b has been replaced by its estimator $\hat{b}_z$.

THEOREM. In $\mathcal{H}_g(q)$,

$$\text{Pro}(\beta_j(\Theta_j)/\mathcal{L}_{hom.}) = \hat{B}_j \text{ (independent of q).}$$

Demonstration. Left to the reader (use the orthogonality obtained in the preceding theorem).

2.2.4. Estimation of s^2

$$\hat{s}_j^2 = \frac{1}{t_j - g} (X_j y_j \hat{b}_j)' v_j^{-1} (X_j - y_j \hat{b}_j)$$

is an unbiased (classical) estimator for s^2 based only on the observable vector X_j.

Then, for all scalars w_j $(j=1,2,\ldots,k)$,

$$\hat{s}_w^2 = \sum_{j=1}^k \bar{w}_j \hat{s}_j^2,$$

is an unbiased estimator for s^2. In particular

$$\hat{s}^2 = \frac{1}{k} \sum_{j=1}^k \hat{s}_j^2$$

is an unbiased estimator for s^2. It has an optimal character under normal assumptions. See DeVylder (to be published).

Remark. In some situations it may be interesting to drop the assumption that s_j^2 does not depend on j. Then $\hat{s}_j^2$ is an unbiased estimator for s_j^2.

2.2.5. Estimation of a

The relations

$$\mathrm{Cov}(\mathfrak{b}_z,\mathfrak{b}_j') = \mathrm{Cov}(\mathfrak{b}_j,\mathfrak{b}_z') = \mathrm{Cov}(\mathfrak{b}_z,\mathfrak{b}_z') = z_\Sigma^{-1} a = a\, z_\Sigma'^{-1} ,$$

where $\mathfrak{b}_z$ is the vector defined in 2.2.3, are easy consequences of the independence assumption and the relation $\mathrm{Cov}\,\mathfrak{b}_j = a+s^2u_j$. Then, centering the vectors $\mathfrak{b}_j, \mathfrak{b}_z$ it is verified that

$$\hat{a} = \frac{1}{k-1} \sum_{j=1}^{k} z_j(\mathfrak{b}_j-\mathfrak{b}_z)(\mathfrak{b}_j-\mathfrak{b}_z)'$$

is an unbiased estimator for a.

An unbiased estimator of another type is

$$\hat{a}_w = \frac{1}{w_\Sigma w_\Sigma - \sum_{j=1}^{k} w_j w_j} \left(\frac{1}{2} \sum_{ij=1}^{k} w_i w_j (\mathfrak{b}_i-\mathfrak{b}_j)(\mathfrak{b}_i-\mathfrak{b}_j)' - \hat{s}^2 \sum_{j=1}^{k} w_j(w_\Sigma - w_j)u_j\right),$$

where $w_j (j=1,2,\ldots,k)$ is a sequence of scalars.

2.2.6. The model in practice

If an estimator $\hat{a}_w$ corresponding to known weights w_j (say $w_j = \mathrm{tr}\, y'v_j^{-1}y$) is adopted, then

$$\hat{s}^2, \mathfrak{b}_j (j=1,2,\ldots,k), \hat{a}_w, z_j (j=1,2,\ldots,k), \mathfrak{b}_z, B_j (j=1,2,\ldots,k)$$

are calculated successively. At each step, the involved structure parameters are replaced by their already obtained estimates.

The situation is different if the estimator $\hat{a}$, depending on the credibility matrices z_j, is adopted. Indeed, then the estimator $\hat{a}$ depends on the parameter a to be estimated. In that case an initial value $a_{(o)}$ is chosen and successive improvements $a_{(1)}$, $a_{(2)}, \ldots$ are calculated from the formulae. In most cases it is observed numerically that the sequence of improvements tend to a limit independent from the initial $a_{(o)}$. After some corrections (symmetrization; the estimate must be semidefinite positive) the limit is adopted as estimate for a.

General families of estimators for the structure parameters have been defined by Norberg (1980) who, moreover, proved asymptotical properties.

3. TWO TIME-HOMOGENEOUS MODELS

3.1. Definition of the models

In this section, we develop the time-homogeneous model with weighted observations due to Bühlmann and Straub (1970). It is the regression model with several observable vectors X_j $(j=1,2,\ldots,k)$ particularized as follows (see 2.2) :

$t_j = t$ independent of j, $g = 1$, $y' = (1,1,\ldots,1)$,

$v_j = \operatorname{diag}(w_{j1}^{-1}, w_{j2}^{-1}, \ldots, w_{jt}^{-1})$,

where w_{js} $(j=1,2,\ldots,k\ ;\ s=1,2,\ldots,t)$ is the natural weight (given) of the observation X_{js}, the s-th component of the vector X_j.

For $w_{js} = 1$, we obtain the time-homogeneous model with unweighted observations by Bühlmann (1967).

The time-homogeneous models are particularly useful in insurance. The model by Bühlmann (1967) opened a new era in credibility theory. It contains, in the simplest form, the basic ideas of the more sophisticated models.

3.2. Notations

All quantities displayed further in this section 3 are scalars. The structure parameters now are

$$b = E\beta_j(\Theta_j) = EX_{js}\ ,\ s^2 = E\sigma_j^2(\Theta_j)\ ,\ a = \operatorname{Var}\beta_j(\Theta_j).$$

We denote $(j=1,2,\ldots,k\ ;\ s=1,2,\ldots,t\ ;\ w,z,\Sigma$ are fixed subscripts)

$$w_{j\Sigma} = \sum_{s=1}^{t} w_{js},\ w_{\Sigma\Sigma} = \sum_{j=1}^{k} w_{j\Sigma},\ z_j = \frac{aw_{j\Sigma}}{s^2+aw_{j\Sigma}},\ z_\Sigma = \sum_{j=1}^{k} z_j,$$

$$X_{jw} = \sum_{s=1}^{t} \frac{w_{js}}{w_{j\Sigma}} X_{js},\ X_{zw} = \sum_{j=1}^{k} \frac{z_j}{z_\Sigma} X_{jw},\ X_{ww} = \sum_{j=1}^{k} \frac{w_{j\Sigma}}{w_{\Sigma\Sigma}} X_{jw}.$$

z_j is the j-th credibility weight.

3.3. Credibility estimator

The credibility estimator for $\beta_j(\Theta_j) = E(X_{js}/\Theta_j)$ now equals

$$B_j = z_j X_{jw} + (1-z_j)b, \quad (j=1,2,\dots,k).$$

3.4. Parameter estimation

The estimator $\hat{b}_z$ for b here equals $\hat{b}_z = X_{zw}$ (see 2.2.3).

The estimator $\hat{s}^2$ (2.2.4.) equals

$$\hat{s}^2 = \frac{1}{k(t-1)} \sum_{j=1}^{k} \sum_{s=1}^{t} w_{js}(X_{js}-X_{jw})^2.$$

The estimator $\hat{a}$ (2.2.5) here becomes

$$\hat{a} = \frac{1}{k-1} \sum_{j=1}^{k} z_j(X_{jw}-X_{zw})^2.$$

The Bühlmann-Straub unbiased estimator for a is equivalent to the following

$$\hat{\hat{a}} = \frac{w_{\Sigma\Sigma}}{w_{\Sigma\Sigma}^2 - \sum_j w_{j\Sigma}^2} \left(\sum_{j=1}^{k} w_{j\Sigma}(X_{jw}-X_{ww})^2 - \hat{s}^2(k-1)\right).$$

4. THE REGRESSION MODEL WITH SCALAR CREDIBILITY WEIGHTS

(De Vylder, to be published)

4.1. Motivation and definition

Here the assumptions and notations are the same as in the regression model (see 2.1, 2.2). The parameter estimation problem is the same as in that regression model and so we can restrict ourselves to one observable vector X.

The credibility approximation $yB = yz\hat{b}+y(1-z)b$ for $y\beta(\Theta)$ (see (6)) is not necessarily componentwise between the homogeneous approximation $y\hat{b}$ and the constant approximation yb. This condition, sometimes required in practical situations, is satisfied if z is a scalar between o and 1. The optimal such scalar z results from the following theorem.

4.2. Theorem

Let $\mathcal{L}$ be the following translated subspace of $\mathcal{H}_g(q)$.

$$\mathcal{L} = \{\zeta\hat{b} + (1-\zeta)b \,/\, \zeta\in R\} .$$

Then

$$\text{Pro}(\beta(\Theta) \,/\, \mathcal{L}) = z\hat{b} + (1-z)b$$

where

$$z = \text{tr}(qa)/\text{tr}(q(a+s^2u)).$$

This z is between 0 and 1

Demonstration. The orthogonality relation is easily verified from the covariance relations in 2.1.4. The last statement of the theorem results from the fact that the trace of a product of two semidefinite positive matrices is nonnegative.

4.3. Choice of the weighting matrix q.

Notice that the scalar z depends on q. A rather natural choice for p in the definition of $\mathcal{H}_t(p)$ is $p = c^{-1} = \text{Cov}^{-1}X$. The corresponding q (see 2.1.2 and 2.1.5) is

$$q = y'c^{-1}y = (s^2(y'v^{-1}y)^{-1} + a)^{-1} = (a+s^2u)^{-1}.$$

Then the scalar z of the theorem in 4.2 equals

$$z = \frac{1}{g}\,\text{tr}(a(a+s^2u)^{-1}).$$

For a numerical illustration with such scalar credibility weights, see De Vylder (to be published).

More constrained credibility matrices have been considered by De Vylder and Sundt (submitted for publication).

5. A MULTIPLICATIVE MODEL (De Vylder, submitted for publication)

5.1. Problem

In this model, also useful in insurance (in particular for the estimation of reserves for IBNR claims; IBNR = incurred but not reported), we consider a square array of random variables X_{js} ($j=1,2,\ldots, k$; $s=1,2,\ldots,t$).

The variables X_{js} are partitioned in two subsets : the <u>observable variables</u> and the <u>variables to be predicted</u>.

X_{11}	X_{21}	X_{31}	X_{41}	X_{51}
X_{12}	X_{22}	X_{32}	X_{42}	X_{52}
X_{13}	X_{23}	X_{33}	X_{43}	X_{53}
X_{14}	X_{24}	X_{34}	X_{44}	X_{54}

Table 1 : Observable random variables (above staircase line) and variables to be predicted (under that line).

For instance, see table 1.

For all $s=1,2, \ldots,t$, we denote by K_s the set of subscripts j such that x_{js} is observable. For all $j=1,2, \ldots, k$, we denote by T_j the set of subscripts s such that X_{js} is observable. The number of elements in K_s, T_j is k_s, t_j respectively. It is assumed that no K_s, T_j is empty and that at least one T_j contains more than one element.

We denote by X_j $\binom{1}{t}$ the column

$$X_j = (X_{j1}, X_{j2}, \ldots, X_{jt})'.$$

Based on realizations of the observable variables, we shall have to make forecasts concerning the variables to be predicted.

5.2. Definition of the model

We consider that the distribution of X_j $(j=1,2, \ldots, k)$ depends on a (multidimensional) parameter θ_j that we interpret as an unknown realization of a (multidimensional) structure variable Θ_j. We adopt the following assumptions.

(I) Independence assumption : the couples $(\Theta_1,X_1),(\Theta_2,X_2),\ldots(\Theta_k,X_k)$ are independent.

(II) Multiplicative assumption $(j=1,2,\ldots,k)$: $X_j = Y_j.\beta(\Theta_j)$, where $Y_j = Y_{j1},Y_{j2},\ldots,Y_{jt})'$ is a $\binom{1}{t}$ vector independent of Θ_j and $\beta(.)$ a scalar function.

(III) Distribution assumptions on Y_j $(j=1,2,\ldots,k)$:

$$E\, Y_j = y = (y_1,y_2,\ldots,y_t)', \text{ independent of } j.$$

$$\text{Cov } Y_j = \frac{r^2}{w_j}.1 \binom{t}{t}, \text{ where}$$

r^2 is a scalar independent of j,
w_j is the natural weight of X_j,
1 is the $\binom{1}{1}$ unit matrix.

(iv) Equidistribution assumption : the structure variables $\Theta_1, \Theta_2, \ldots, \Theta_k$ are identically distributed.

5.3. Structure parameters

The natural weights w_j are supposed to be given, but we shall have to estimate the following structure parameters :

$$y,\ b = E\beta(\Theta),\ s^2 = r^2\ E\beta^2(\Theta_j)\ ,\ a = \mathrm{Var}\beta(\Theta_j),$$

all scalars, except y $\binom{1}{t}$.

5.4. Corresponding regression model in credibility theory

We shall approximate the variance X_{js} to be predicted by

$$E(X_{js}/\Theta_j) = y_s\ \beta(\Theta_j),$$

where $\beta(\Theta_j)$ shall be approximated by a credibility estimator using only the observable variables.

We shall use the following notation. If V_j (depending on j or not) is a $\binom{1}{t}$ vector (random or not), we denote by V_j^c (c is a fixed superscript) the vector V_j restricted to its components V_{js} with subscript s in T_j. For instance, in case of table 1 :

$$X_4^c = (X_{41}, X_{42})'\ ,\ Y_3^c = (Y_{31}, Y_{32}, Y_{33})'\ ,\ Y_4^c = (y_1, y_2)'.$$

Then it results from the assumptions that :

(m) $(\Theta_1, X_1^c), (\Theta_2, X_2^c), \ldots, (\Theta_k, X_k^c)$ are independent.

(mm) $E(X_j^c/\Theta_j) = y_j^c\ \beta(\Theta_j)\ ,\ (j=1,2,\ \ldots,k)$.

(mmm) $\mathrm{Cov}(X_j^c/\Theta_j) = \sigma^2(\Theta_j)\ v_j^c\ ,\ (j=1,2,\ \ldots,k)$,

where $\sigma^2(.) = r^2\ \beta^2(.)\ ,\ v_j^c = \frac{1}{p_j}\ 1_j^c$

and 1_j^c is the unit matrix with t_j columns.

(mv) $\Theta_1, \Theta_2, \ldots, \Theta_k$ are identically distributed.

This means that the assumptions of the regression model are satisfied for the couples (Θ_j, X_j^c), $(j=1,2,\ \ldots,\ k)$. (See 2.1.1 and 2.2.1). Of course, here g=1.

The credibility estimator for $\beta(\Theta_j)$ equals $B_j = z_j \hat{b}_j + (1-z_j)b$

where

$$\hat{b}_j = (y_j^{c\prime} y_j^c)^{-1} y_j^{c\prime} X_j^c = (\sum_{s\in T_j} y_s X_{js}) / (\sum_{s\in T_j} y_s^2) ,$$

$$z_j = a/(a+s^2 u_j) , \quad u_j = \frac{1}{w_j}(y_j^{c\prime} y_j^c)^{-1} = 1/(w_j \sum_{s\in T_j} y_s^2) .$$

5.5. Parameter estimation

From the multiplicative assumption results that we may divide Y_j by any number and then multiply $\beta(\Theta_j)$ by the same number. That means that we can take b=1 (assuming of course b≠0). Then $y_s = E\,Y_{js} = E\,X_{js}$ and each X_{js} is an unbiased estimator for y_s. Then any weighted average of such unbiased estimators is again an unbiased estimator for y_s. We suggest to estimate y_s by

$$\hat{y}_s = (\sum_{j\in K_s} w_j X_{js}) / (\sum_{j\in K_s} w_j) , \quad (s=1,2,\ldots,t).$$

The suggested estimator for s^2 is (see 2.2.4)

$$\hat{s}^2 = \frac{1}{k_o} \sum_j \frac{w_j}{t_j-1} (X_j^c - y_j^c \hat{b}_j)'(X_j^c - y_j^c \hat{b}_j)$$

$$= \frac{1}{k_o} \sum_j \frac{w_j}{t_j-1} \sum_{s\in T_j} (X_{js} - y_s \hat{b}_j)^2 ,$$

where the summation in j is over the subscripts with $t_j > 1$ and where k_o is the number of such subscripts.

From the relation $E(\hat{b}_j - b)^2 = a+s^2 u_j$ (see 2.1.4), the expression for z_j and the relation b=1, it is clear that

$$\hat{a} = \frac{1}{k} \sum_{j=1}^{k} z_j (\hat{b}_j - 1)^2$$

is an unbiased estimator for a (to be used iteratively, i.e. by a method of successive approximations; see 2.2.6).

REFERENCES

Bühlmann H. (1967), Experience rating and credibility. ASTIN Bulletin 4, p. 199-207.
Bühlmann H. & Straub E. (1970). Glaubwürdigkeit für Schadensätze. Mitt. der Ver. Schweiz. Vers. Math., p. 11-133.
De Vylder F. (1976). Geometrical credibility. Scandinavian Actuarial Journal, p. 121-149.
De Vylder F. (to be published). Practical credibility theory with emphasis on optimal parameter estimation. ASTIN Bulletin.

De Vylder F. (to be published). Regression model with scalar credibility weights. Mitt. Ver. Schweiz. vers. Math.

De Vylder F. (submitted for publication). Estimation of IBNR claims by credibility theory. Insurance.

De Vylder F. & Sundt B. (submitted for publication). Constrained credibility estimators in the regression model. Scandinavian Actuarial Journal.

Hachemeister C.A. (1975). Credibility for regression models with application to trend. In Kahn, p. 129-163.

Kahn P.M. (Editor), (1975). Credibility : Theory and Applications. Academic Press.

Norberg R. (1979). The credibility approach to experience rating. Scandinavian Actuarial Journal. p. 181-221.

Norberg R. (1980). Empirical Bayes credibility. Scandinavian Actuarial Journal, p. 177-194.

RATE MAKING AND SOCIETY'S SENSE OF FAIRNESS

G.W. de Wit and J. van Eeghen

Nationale-Nederlanden N.V. - Research Department
P.O. Box 796, 3000 AT Rotterdam, NL

Premium differentiation on the basis of observable risk factors has always been one of the main activities of the rate making actuary. Conflicts may arise between the differentiation proposed by the actuary and society's general sense of fairness. This may be a reason for a company to deliberately ignore in its tariff part of the risk information available. The consequences of doing so are examined and it is shown that increased solvency requirements will in the end prove to be ineffective. The only solution is a voluntary transfer of premium between companies. The situation is illustrated by an example of health insurance in the Netherlands, where proposals to arrive at such transfers are presently being discussed.

Actuaries have always been in search of ways to determine premiums which match the risks insured as closely as possible. They do this by differentiating between them on the basis of observable risk factors. In practice, many examples of such risk factors are being used: age and sex for life insurance; location, type of building etc. for fire insurance. Motor insurance is perhaps the most characteristic branch with respect to this phenomenon: in tariffs we find factors like weight, price or cylinder capacity of the car, age of the driver, area of residence, past claims experience (bonus/malus), annual mileage etc.

Outsiders may not always be very positive about such a refined premium differentiation. The basis of insurance, they say, should be solidarity among insureds; premium differentiation is basically opposed to this.

F. de Vylder et al. (eds.), Premium Calculation in Insurance, 151–169.

Another statement heard in the field is: "Premium differentiation ultimately results in letting every individual pay his own claims, it is the end of insurance".
Much confusion arises during discussions about this subject, especially between actuaries and non-actuaries. It will therefore be useful to further define what we are talking about.

FORMS OF SOLIDARITY

If no insurance is purchased, the situation can be briefly summarized as follows:

	Carried by insured	Carried by insurer
risk	X	0
expected risk	E(X)	0
variance of risk	Var(X)	0

By the insurance transaction, the risk is transferred from the insured to the insurer, the result of which is [1])

	Carried by insured	Carried by insurer
risk	E(X)	X-E(X)
expected risk	E(X)	0
variance of risk	0	Var(X)

It has been recognized however that risks are like leaves in a tree: similar, but never identical. We therefore say that the risk of an individual is characterized by a distribution P_θ, where θ differs from one insured to another. θ is unobservable and is in turn looked upon as a realization of a random variable Θ, whose distribution is characteristic for the market. Thus the risk process is divided into two parts: first the "θ-lottery" [2]) (realization of Θ), then the "claim lottery" ruled by the probability law P_θ .
Suppose for a moment that θ is observable and that each insured is charged a premium equal to the expected value (given θ) which corresponds to his own risk $E(X|\Theta)$. After the insurance transaction the total division of risk is as follows:

	Carried by insured	Carried by insurer
risk	$E(X\|\Theta)$	$X-E(X\|\Theta)$
expected risk	$E(X)$	0
variance of risk	$Var\{E(X\|\Theta)\}$	$E\{Var(X\|\Theta)\}$

While in the first example, the insured transferred his full risk to the insurer (X is replaced by E(X)), he now keeps part of the risk for himself, for his premium $E(X|\Theta)$ is a random variable.

We may now define actual solidarity as the variance of the risk transferred to the insurer (i.e. shared among insureds). Full solidarity is achieved in the first example:

$$S= Var(X)$$

In the second example the actual solidarity remains restricted to the "purely probabilistic" part of it:

$$S_p= E\{Var(X|\Theta)\}$$

The part of the variance which is caused by Θ, the "risk solidarity", remains with the insured

$$S_r= Var\{E(X|\Theta)\}$$

The subdivision is complete now, for it can easily be checked that

$$S= S_p + S_r$$

S_p is the minimum and S is the maximum amount of actual solidarity that can be achieved.

As we have said before, θ is unobservable; the probability distribution of an individual risk is never known. However, we do have some information on the differences in distribution of the risks in our portfolio by means of observable risk factors. These risk factors can be viewed as a (possibly vector-valued) random variable F. Mathematically, F satisfies the following:

$$\text{for all sets } A: \Pr\{X \in A \mid \Theta, F\} = \Pr\{X \in A \mid \Theta\} \qquad (1)$$

i.e. the conditional distribution of the risk given Θ does not depend on F.

If each insured is charged a premium $E(X|F)$ 3) (i.e. information on risk factors is taken into account in pricing), the result of the insurance transaction is as follows:

	Carried by insured	Carried by insurer
risk	$E(X\|F)$	$X-E(X\|F)$
expected risk	$E(X)$	0
variance of risk	$\mathrm{Var}\{E(X\|F)\}$	$E\{\mathrm{Var}(X\|F)\}$

Now we can write:

$$\mathrm{Var}(X|F) = E(X^2|F) - E^2(X|F)$$

(because of (1))

$$= E\{E(X^2|\Theta)|F\} - E^2\{E(X|\Theta)|F\}$$

$$= E\{\mathrm{Var}(X|\Theta)|F\} + \mathrm{Var}\{E(X|\Theta)|F\}$$

Hence

$$E\{\mathrm{Var}(X|F)\} = E\{\mathrm{Var}(X|\Theta)\} + E\{\mathrm{Var}[E(X|\Theta)|F]\}$$

$$= \qquad S_p \qquad + \qquad S_{ur}$$

S_{ur} can be interpreted as the part of risk solidarity S_r that remains unknown after the information contained in F has been taken into account. It may therefore be called the "unknown-risk solidarity".

Similarly, Var{E(X|F)} can be viewed as the part of S_r that becomes known through F. It is therefore called the "known-risk solidarity", S_{kr} .
Evidently we have:

$$S = S_p + S_r = S_p + S_{ur} + S_{kr} \quad 4)$$

The result of the insurance transaction (with premiums equal to E(X|F)) can therefore be rewritten as:

	Carried by insured	Carried by insurer
risk	E(X|F)	X-E(X|F)
expected risk	E(X)	0
variance of risk	S_{kr}	S_p+S_{ur}

The endeavours of the rate making actuary can now be represented as follows:

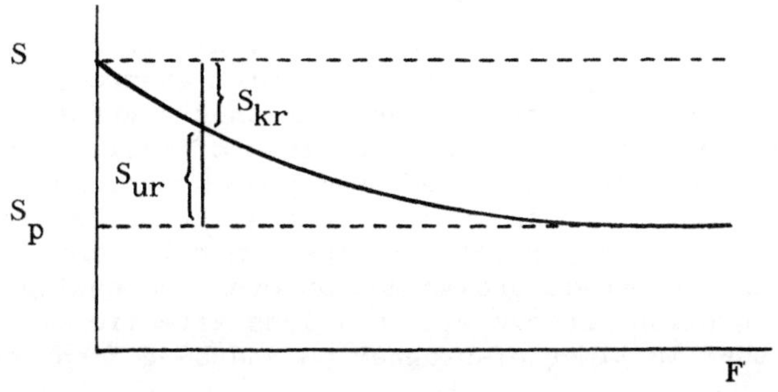

Figure 1. Solidarity shared among insureds

As F moves "towards full information about Θ" (i.e. towards the right in the above graph) - which is what the actuary tries to achieve - the unknown risk solidarity S_{ur} will tend towards 0, and the solidarity shared among insureds will remain restricted to S_p. If no F is used for rate making purposes, i.e. F is at the origin of the graph, then there will be full solidarity.

SOCIETY'S SENSE OF FAIRNESS

In the past, it used to be very difficult to discover risk factors both in a qualitative and in a quantitative sense. Solidarity was therefore - unavoidably - considerable.
But recent developments have changed this situation:

- with the help of computers it has become possible to make thorough risk analyses, and consequently to arrive at further premium differentiation,
- the consumer's attitude towards tariffs became more critical. He requested more information and if he was a good risk objected to pay the same premium as the bad ones.

Both developments have their own special character.
The first one shows that in recent years the actuary has been successive in his travel to the right of the F-axis of figure 1 [5].
With the help of large data files and the possibility to analyse such data in detail, he is on his way to reduce actual solidarity to purely probabilistic solidarity S_p .
This S_p is the smallest possible value of the actual solidarity shared among insureds. It can be considered as a limit-situation in which F contains all information about Θ. Computers have made us some closer to this limit value, so one could say that in insurance computers reduce the solidarity among insureds.
Perhaps a parallel could be drawn. There are other applications of electronics which reduce the contact between people.

The consumer's attitude is of a completely different nature. In the past, he would simply accept the premium charged, but today things are different. Premiums have increased a great deal, coverages have been extended and the risks of society have grown. Because of the relative height of premiums, the consumer has become more sensitive to price differences. Price sensitivity is probably also closely related to the general economic situation. In days of rapid economic growth and an ever increasing level of personal consumption, people will pay less attention to premium differences than in times of stagnation and budget squeezes.

Besides these financial criteria, the general attitude people have towards one another is important too. At all levels of society we can observe alienation between people by both moral and economic causes. People feel less related to one another. Personal assistance is gradually taken over by professional organizations. It has turned our society into an expensive and impersonal one. Natural solidarity is declining. These developments have also influenced insurance. Many people would like to see their ideals ("I pay for myself only") realized by a tariff consisting of pure premiums: $E(X|\Theta)$.

Nevertheless, a changing attitude is starting to become apparent, caused by a critical review of today's society. It is this change that gives rise to the type of statements mentioned in the introduction. People justly start realizing that a certain restauration of human contacts might be desirable. For insurance this would mean the wish to increase solidarity. First of all this seems to apply to branches of insurance which are in the closest relation to people themselves. Therefore: healthy persons will be paying for the less healthy ones. And: should someone who works under circumstances which endanger his physical condition have to pay a higher premium in spite of the fact his work is of vital importance to the economy? The answer to such questions is often determined by the degree of influence a person has on his own risk. Should solidarity be extended to cover those people who harm their own health by their voluntarily chosen way of life? The answer to this question would generally be affirmative, but a non-smoker's discount, for instance, denies this form of solidarity. One might object against this form of solidarity, because it reduces one's own responsibility and has an anti-prevention character. It therefore seems justified to restrict solidarity to factors for which one is not personally responsible. Alternatively, the community can impose solidarity by safety rules (helmets for motorcyclists, safety belts in cars etc.).
But solidarity is not merely related to "personal" branches of insurance, but applies to more material fields also. Should someone who, for economic reasons, lives in a certain area pay a higher motor insurance premium, because of the higher traffic risks? The higher rent he has to pay in such an area may even be compensated by special subsidies. This brings us to a totally different aspect of solidarity. Should premiums be such that everyone can afford insurance? In the past, this question used to be relevant for social insurance only. The reasoning of private insurers was: if you cannot pay for insurance, don't buy it. But times have changed. Many types of insurance have become such common commodities, that they are being considered as basic needs and must therefore be affordable by everyone. If private insurers do not want to see their tasks taken over by social insurance, they should keep this aspect in mind.

One final remark. Where the foregoing considerations have a mainly social character, legal aspects may (or will) also be important, in the form of restrictions which preclude insurance companies from using certain risk factors, even if these factors can be proved to be statistically significant.
We are thinking of:[6)]

- emancipation. It will no longer be allowed to distinguish between men and women for rate making purposes. For the European Community this rule will be laid down in a forthcoming directive;

- discrimination. Tariffs are not allowed to differentiate between racial groups.

What we are saying is that there may be reasons (mostly related to society's sense of fairness) for us not to use certain parts of the risk information available in our tariffs. Although we might be able to push premium differentiation to point A in figure 2, we prefer to stick with situation B. In practice, the difference between A and B can be quite significant.

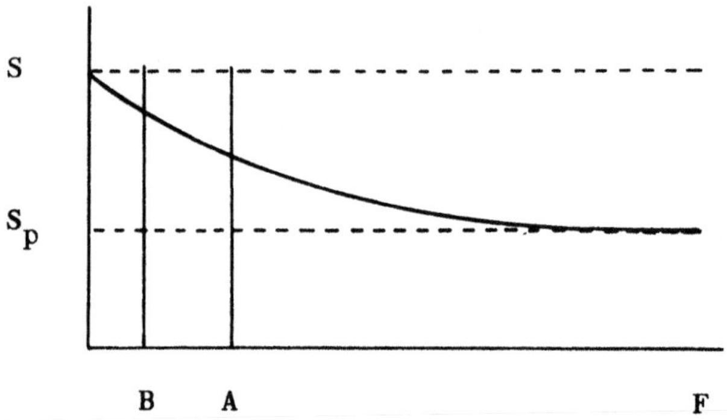

Figure 2.

CONSEQUENCES OF DELIBERATELY IGNORING RISK INFORMATION

If it is decided that not all information available will be used for the rating structure, this policy should be adopted by all companies operating on the same market. It will be clear that, if with one company two different risks have to pay the same premium, while with another one the good risks pay a low premium and the bad risks a higher one, the former will attract relatively more bad risks and thus will incur a loss [7]. This implies that in a market where every company is free to fix its own rating structure (like the Dutch or the British ones) it will be less easy to respond to social pressure for "fairness" than in markets where rating structures are imposed by the authorities (like in Germany or Belgium).

But even in a regulated market there may be problems. If certain aspects of risk information are ignored in the rating structure, they may be used for underwriting purposes. Applications by unwanted risk groups may be refused or "forgotten". By doing so, insurance companies can increase their profitability, while other companies will see their profits reduced.

This may also occur with "special character" companies: in almost every country there are companies (often mutuals) which sell insurance to agriculturers or to civil servants only or which operate in one specific area. These groups may turn out to consist of non-average risks.

To illustrate the effects of such a situation, let us assume that we are dealing with a two-company market. A risk factor F has been discovered but it is not used in the tariffs [8]. So both companies charge a level premium E(X). Suppose that, due to the market mechanism described above, the insureds of company I are characterized by $F \in \mathcal{F}$ and those of company II by $F \notin \mathcal{F}$, where $\mathcal{F}$ is a subset of the possible outcomes of F. The expected risk carried by the companies now becomes:

$$E(X|F \in \mathcal{F}) - E(X) \qquad \text{for company I}$$

and

$$E(X|F \notin \mathcal{F}) - E(X) \qquad \text{for company II}$$

one of which, say for company I, may be positive.

This situation clearly leads to modifications in the profit and loss accounts of companies I and II. In a free market (but still assuming that F remains removed from the tariffs), company I might consider adjusting its overall premium level. But this would mean that F is effectively used as a rating factor, not on a company level but on the market level. Risks characterized by $F \in \mathcal{F}$ would be charged a higher premium (by company I) than risks characterized by $F \notin \mathcal{F}$ (insured by company II). Since this is what we were trying to avoid on grounds of social fairness, the situation is not very satisfactory and seems to call for another solution, especially if the premium differences are very large.

Another possibility is that the companies do not adjust their premiums to reflect the special character of the risks of their portfolio. The resulting positive value of the expected risk for company I is a risk theoretical impossibility: the insurer will soon be ruined. In practice however, premiums contain loadings for security, expenses, profits etc.

This loading will now turn out to be lower than expected, because of the special risk selection represented by $F \varepsilon \mathcal{F}$. It may very well be possible that the company can still live and survive with this smaller loading. But its existence will have become subject to more risk and a larger safety buffer may therefore be required [9].

Let us return to figure 2. It illustrates that by deliberately ignoring some risk information, we find ourselves in situation B instead of A. Solidarity between insureds $S_p + S_{ur}$, which was defined as the variance carried by the insurer, was thus increased. This increase of the variance is a second indication that the solvency margin of an insurer is to be increased when not all possible risk information is used in determining premiums. Such an increase would be based on risk variance grounds and therefore its nature is different from the one which reflects premium inadequacy due to risk selection ($F \varepsilon \mathcal{F}$).

The necessary provisions in situations A and B can thus be written as:

$$R_A = RV_A + RS_A$$

$$R_B = RV_B + RS_B$$

where RV is the variance part of the provision and RS is the risk selection part of the provision.
We have

$$RV_A < RV_B \quad ;$$

$$RS_A < RS_B$$

and hence

$$R_A < R_B$$

It should further be noted that RV expressed as a percentage of premium income, tends to zero when the size of the portfolio increases. This is not true for RS [10]. RS will be zero when all risk information is reflected in premium differentiation.
While much literature exist on the determination of the level of the "variance part" of the reserve, RV, it is hard to say anything general about the level of RS. We will simply mention two of the factors that can influence it:

- the explanatory power of the deliberately omitted risk factor F. The more $E(X|F=f)$ varies with f, the greater will be the premium inadequacy resulting from adverse risk selection. Schematically, the risk pattern in figure 3a is more dangerous than the one in figure 3b.

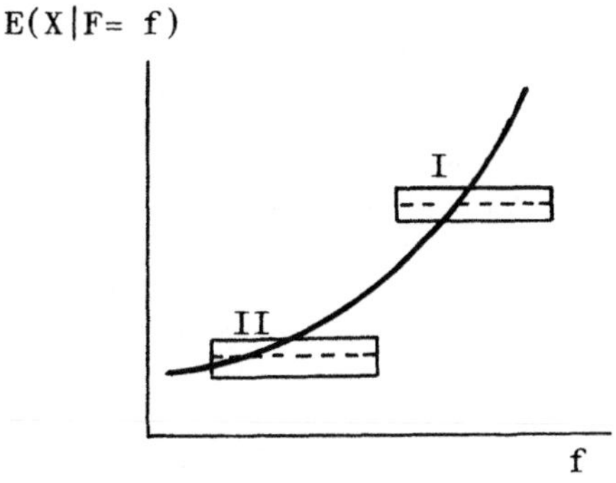

Figure 3a

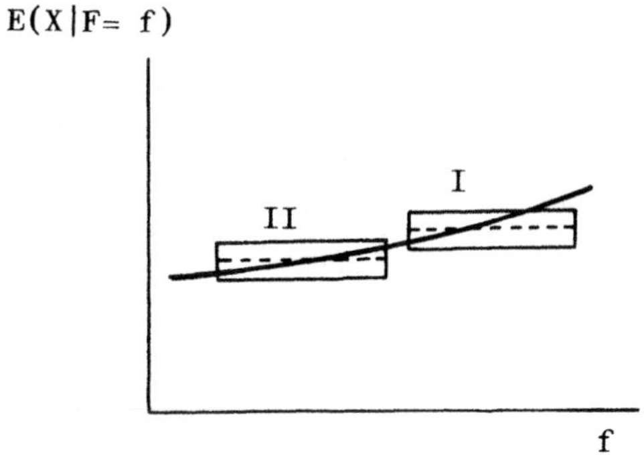

Figure 3b

- the possibility of the market to arrive at an effective risk selection. The most dangerous risk selection would result from a choice of $\mathcal{F}$ such that $E(X|F\in\mathcal{F}) - E(X|F\notin\mathcal{F})$ is maximized. In practice however, not all choices of $\mathcal{F}$ are possible. Social tolerance can be important in this respect. A health insurance company for instance, cannot openly say that it accepts insureds under age 30 only, without being highly controversial. So full risk selection (through age) would be impossible, but some degree of risk selection may be possible by means of carefully planned marketing campaigns.

As we have seen, an extra security buffer may be necessary for protection against risk selection effects in a market where part of the risk information is not reflected in tariffs. This may however not be sufficient. The possibilities of fighting premium inadequacy by setting up extra provisions are limited. The difference between the net premium charged ($E(X)$) and the necessary net premium ($E(X|F\in\mathcal{F})$) may be too large to be financed from the premium loading. In such cases another solution is necessary. Since considerations of social fairness have lead to the decision of non-differentiation of premiums with respect to F, this same sense of social fairness suggests a transfer of premium income from the companies characterized by $F\notin\mathcal{F}$ to those portfolios for which $F\in\mathcal{F}$ holds. Since F is an observable risk factor, with known effect on the expected losses, the level for such a "fair transfer" can be computed.

To realize such transfers in practice, a consensus between companies is necessary. This consensus can only be reached when there is full agreement on what is "socially fair". Public opinion and political pressure can prove to be major factors in the process of reaching a final agreement.
These principles have been applied in the Dutch health insurance market. "Age of the insured" is a highly significant risk factor but it is not used in tariffs. Deliberate and undeliberate risk selection by some companies have had a great impact on the profit and loss accounts of the Dutch companies. As a consequence of the free market mechanism premium levels have been adjusted, leading to highly differing premium levels from company to company. Thus age has factually returned as a rating factor. At present, proposals are being discussed to arrange transfers of premium income between companies, to cure the situation. The theoretically necessary safety provisions could then be reduced to a realistic level. Details can be found in the Appendix.

CONCLUSION

In this contribution we have tried to show that, depending on social circumstances, practical tariffs should not always reflect all the risk information available.
If the reduction of the relationship on a micro-level between risks insured and premiums charged is pushed to an extreme, the nature of insurance industry will change profoundly, the end being a full socialization of insurance, with for instance income-related premiums. The different stages of such a process can be summarized as follows:

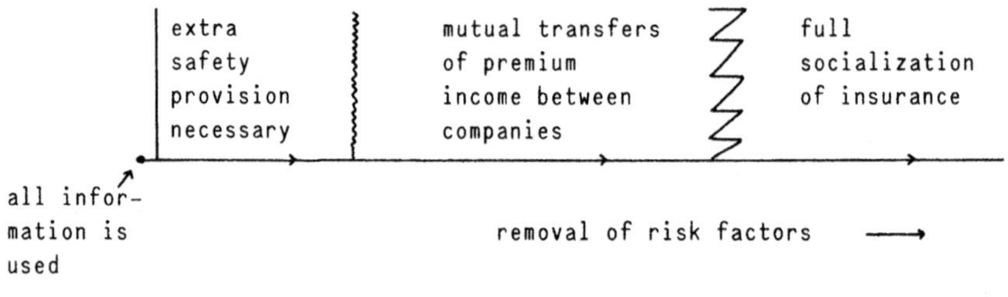

It is not our intention to plead for tariffs from which as many risk factors as possible have been removed, for this would imply a full socialization of insurance. But in modern society, we should be aware of the fact that some aspects of the free market mechanism need to be adjusted. This also applies to insurance, and insurance companies need not feel threatened by these developments.

The intention of this paper is therefore to take better notice - especially in a quantitative sense - of the consequences of solidarity transfers. A better knowledge of these transfers may help the insurance industry to react adequately upon general social developments. If such a reaction should result in a reduced premium differentiation, the insurance industry will have to reach a consensus on whether to increase solvency requirements or to neutralize the effects of risk selection by mutual transfers of premium income.

APPENDIX

HEALTH INSURANCE IN THE NETHERLANDS

In the Netherlands, 30% of the population obtains full health coverage from private insurance companies.
The most important risk factor (F) for this type of risk is age of the insured. The expected loss for an individual of age x is:

$$E(X|F=x) = S(x)$$

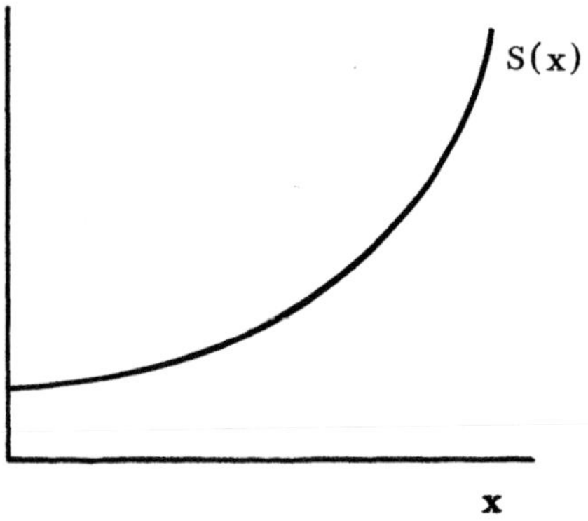

Clearly, age is a very significant risk factor. Premiums however, originally do not depend on age, but only on type of coverage and level of deductible. So if N(x) is the number of insureds of age x in the portfolio, the net premium applicable to all insureds is:

$$P = \left(\sum_x N(x)\, S(x)\right) / \sum_x N(x)$$

No mathematical reserve forming takes place in the process; the financing system used is pay-as-you-go.

This situation is a typical example of a rating structure where an important risk factor has been deliberately ignored, and the features described in the previous paragraph are observable in the market.

Premiums differ widely from company to company, as a function of the age composition of the portfolio. Companies with a "young" portfolio have low premiums and therefore attract the largest number of new (mainly young) insureds. Companies with an "old" portfolio have high premiums, they will therefore not be able to attract enough new insureds, the portfolio therefore grows older and as a result they have to increase their premiums, possibly to an unacceptably high level.
This feature is reinforced by the steepness of the S(x)-curve: S(85) is about 7 times as high as S(20). The situation in the market can therefore well be represented by figure 3a of the main text (for two companies). Due to the free market mechanism, age is de facto used as a rating factor.

An extra provision [11] necessary to protect companies from possible premium inadequacies can be formed as follows.
Suppose there is a portfolio with a given age composition which attracts no more new insureds. The age composition in subsequent years is therefore fully determined by aging and mortality.

If p_x is the one-year survival probability of an x-year old insured, the expected number of insureds of age x+t after t years will be:

$$N_t(x+t) = p_x \cdot p_{x+1} \cdots p_{x+t-1}\, N(x) \;.$$

Suppose the premiums are to be kept constant at a level P, the discounted (interest i= 4%) premium income for the company will be:

$$DP = P \sum_t (1+i)^{-t} \sum_x N_t(x+t)$$

while the discounted yearly claims total is:

$$DC = \sum_t (1+i)^{-t} \sum_x N_t(x+t)\; S(x+t)$$

The difference DC - DP could be considered as a theoretically necessary extra provision [12].
For the average Dutch portfolio this reserve would be 440% of net premium income, an amount which is simply not available [13].

To cope with this problem and with the problem of widely differing premium levels in the market (part of which is due to differing age compositions), the Dutch health insurance companies are in the process of deciding to share the costs of older insureds. Basically the proposal is as follows:

- for each insured of age $x > 55$, the insurer receives $r[S(x)-S(55)]$ where $0 \leqslant r \leqslant 1$;
- for each insured of age $x < 55$, the insurer pays $\lambda[S(55)-S(x)]$,

where λ is determined in such a way that for the market as a whole and for fixed r, the balance of income and expenditure is zero.

The result of these transfers is a new (less steep) curve of expected losses:

$$
\begin{aligned}
S^*(x) &= (1-\lambda)S(x) + \lambda S(55) && x < 55 \\
&= S(55) && x = 55 \\
&= (1-r)S(x) + rS(55) && x > 55
\end{aligned}
$$

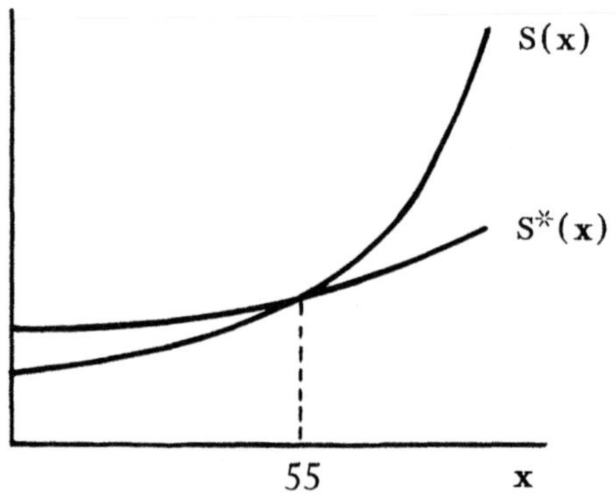

Due to the reduction of the slope of the S-curve, premium differences in the market will decrease, so that the resulting situation might be represented by figure 3b of the main text.
If we compute the level of the necessary safety provision in a similar way as above, but with S(x) replaced by S*(x), we find, for the values of r and λ suggested by the Dutch insurers, a provision of 20% of premium income, which is more in line with the financial position of the Dutch health insurance industry than the previous 440%.

This is a mixed solution. The S^*-curve is not completely horizontal. The factor age continues to be of importance, also because the premium transfers relate only to a specific part of the total health insurance coverage. An extra provision (20%) thus remains necessary. Premium differences due to differences in age composition are strongly reduced. Solidarity between younger and older insureds is thus secured through an agreement between insurance companies which does not interfere with normal, healthy composition.

FOOTNOTES

1) For simplicity's sake we use the expected value premium principle only, and forget about the loading.

2) One might think of the θ-lottery as an underwriting experiment. Each time a risk is accepted a θ is drawn at random from the population Θ .

3) We implicitely assume that a good estimate of $E(X|F)$ is available. For simplicity, we assume that $E(X|F)$ (like $E(X)$) is known.

4) These relations hold only in the cases that premiums are based on the expected value of X (conditionally or not to Θ or F). In social insurance however, premiums may not be related to the risk insured at all. Think e.g. of income related premiums or premiums which have to be paid in spite of the certainty one will never receive a benefit (women having to pay for a widow's pension under the Dutch social insurance). In such cases actual solidarity can be defined as

$$E[E\{(X-P(F))^2 |F\}]$$

(where $P(F)$ is the premium payable)

which may well exceed $S= Var(X)$.

5) See,for instance, G.W. de Wit: "Premium calculation in theory and practice", in General Insurance Bulletin (Australia), nr 34, June 1982; also published in: Bulletin de l'Association Royale des Actuaires Belges 1982, nr 76.

6) See also G.W. de Wit: "Note on premium differentiation", presented to the General Insurance Seminar, Stratford-upon-Avon, October 1982.

7) In reality, the difference has to be substantial before the effect becomes noticeable. Moreover, we simplify by considering the risk process only and by ignoring expenses, marketing and client service aspects.

8) Omitting a risk factor can be seen as a special case of not expressing its full effects on the risk in the premium charged. This may happen in practice when for a certain group of large surcharge is (risk theoretically) necessary but a much smaller one appears in the tariff (for reasons of "social fairness"). The reasoning in the main text applies to such situations as well.

9) Attention will be paid to this subject in a forthcoming paper "Management oriented approach to solvency" to be presented at the 22nd International Congress of Actuaries in Sydney, October 1984 by a group of Dutch actuaries.

10) This shows that the "law of large numbers" is not the whole story of insurance, as is popularly believed.

11) We restrict ourselves to this extra provision only (RS).

12) This comes down to the premium reserve for a fully capitalized health insurance system, without inflation.

13) In these calculations, we take only one type of coverage into account. Moreover, we allow for a small premium adjustment only.

THE IMPACT OF REINSURANCE ON THE INSURER'S RISK

Hans U. Gerber

Ecole des H. E. C., Université de Lausanne

A desirable reinsurance arrangement reduces the profitability but increases the security. The former is measured by the expected profit, the latter by the adjustment coefficient (which is closely connected with the probability of ruin.) For different types of reinsurance coverages these two parameters are calculated. In the special case where the claims are normally distributed, the numerical values of expected profit and adjustment coefficient that result from a stop-loss contract can be obtained from a table.

BUYING REINSURANCE: A COMPROMISE

"Should we buy reinsurance, and if so, what kind and how much?". The insurer who asks this question is confronted with a dilemma: On the one hand, the purchase of reinsurance will reduce the profitability; on the other hand, a meaningful reinsurance arrangement will increase the security (or stability) in some sense.

To quantify this dilemma, we need measures for the profitability and the security. For the "profitability" we use expected profit as a measure; for "security" the measure is probability of ruin. Thus any possible reinsurance arrangement should be judged by two numbers, the resulting expected profit and the resulting probability of ruin.

F. de Vylder et al. (eds.), Premium Calculation in Insurance, 171–181.

THE PROBABILITY OF RUIN IN THE ABSENCE OF REINSURANCE

Let u denote the initial surplus of the insurance company, U_n the surplus at the end of year n, and G_n the companies gain in year n $(n = 1,2, \ldots)$. Then

$$U_n = u + G_1 + G_2 + \ldots + G_n . \tag{1}$$

It is assumed that $G_1, G_2, \ldots$ are independent and identically distributed random variables. "Ruin" is the event that U_n will be negative sometime in the future. The probability of ruin depends on the initial surplus u on the one hand, and on the common distribution of the G_n's on the other hand. We shall denote this probability by $\Psi(u)$. Furthermore, let N denote the time of ruin, with the convention that $N = \infty$, if ruin does not occur.

For assessing the probability of ruin the so-called <u>adjustment coefficient</u> R plays a central role. It is defined as the positive solution of the equation

$$E\,[e^{-rG}] = 1 \tag{2}$$

or,

$$\log E\,[e^{-rG}] = 0 . \tag{3}$$

A classical result is that

$$\Psi(u) = \frac{e^{-Ru}}{E\,[e^{-RU_N} \mid N < \infty]} . \tag{4}$$

Since U_N is by definition negative, it follows that

$$\Psi(u) < e^{-Ru} . \tag{5}$$

For computational simplicity, we shall use the right hand side of this inequality as a measure for the security. Thus it is enough to calculate the adjustment coefficient R in a given situation.

In the absence of reinsurance,

$$G_n = P_n - S_n , \tag{6}$$

where P_n is the premium received, and S_n the claims paid (in year n .)

In general equation (2) has to be solved numerically for R . A notable exception is the case where the common distribution of the G_n's is normal, say with mean $\mu > 0$ and standard deviation σ . Then (2) has an explicit solution:

$$R = \frac{2\mu}{\sigma^2} . \tag{7}$$

Thus in this case the adjustment coefficient is twice the expectation of the annual gain divided by its variance. In the general case this expression can be used as a first approximation. This can be seen from the Taylor series expansion of the left hand side of (3) around $r = 0$.

AN ELEMENTARY PROOF OF FORMULA (4)

While an elegant proof of formula (4) can be obtained from the observation that the sequence $\exp(- RU_n)$ is a martingale, we shall here look at a more elementary proof.

From the definition of R it follows that for any n

$$e^{-Ru} = E\,[e^{-RU_n}] = E\,[e^{-RU_N} I_{(N \leqq n)}] + E\,[e^{-RU_n} I_{(N > n)}] . \tag{8}$$

For $n \to \infty$ the first term on the right hand side converges to

$$E\,[e^{-RU_N} I_{(N < \infty)}] = E\,[e^{-RU_N} \mid N < \infty] \cdot \Psi(u) . \tag{9}$$

Thus formula (4) follows, if we can show that the last term on the right hand side of (8) vanishes for $n \to \infty$.

Let $a = E\,[G_n]$, $b^2 = \mathrm{Var}\,[G_n]$. The last term on the right hand side of (8) can be estimated from above as follows:

$$E\,[e^{-RU_n} I_{(N > n,\ U_n \leqq u + an - b\,n^{2/3})}]$$

$$+ E\,[e^{-RU_n} I_{(N > n,\ U_n > u + an - b\,n^{2/3})}]$$

$$\leqq \Pr(U_n \leqq u + an - b\,n^{2/3}) + e^{-R(u + an - b\,n^{2/3})}]$$

$$\leqq n^{-1/3} + \exp(-R(u + an - b\,n^{2/3}) , \tag{10}$$

where we have used Chebyshev's inequality in the last step. Since the upper bound in (10) vanishes for $n \to \infty$, the last term on the right hand side of (8) has this property, which completes the proof of formula (4).

THE PROBABILITY OF RUIN IN THE PRESENCE OF REINSURANCE

With an appropriate reinterpretation of G_n the previous results can be used. Now

$$G_n = P_n - P_n^{re} - S_n^{ret} , \tag{11}$$

where P_n^{re} is the reinsurance premium paid, and S_n^{ret} the retained claims, i.e., $S_n^{ret} = S_n$ minus (payment of the reinsurer).

Thus, to evaluate a given reinsurance arrangement, one calculates the expected profit,

$$E\,[G_n] = P_n - P_n^{re} - E\,[S_n^{ret}] , \tag{12}$$

and the resulting adjustment coefficient. To obtain the latter, we substitute (11) in (2) or (3).

The methodology will now be illustrated by some examples.

CHOOSING A STOP-LOSS CONTRACT

We assume that S_n has a compound Poisson distribution such that the expected number of claims per year is 1.5 and that an individual claim amount is 1 or 2, with probabilities 2/3 and 1/3, respectively. Thus the mean claim size is 4/3, and $E\,[S_n] = 2$. We suppose that the premium received is $P_n = 2.5$, i.e., that it contains a loading of 25%. Stop-loss coverage with a deductible of $d = 3$, 4, or 5 can be purchased, for a reinsurance premium that contains a loading of 100%.

In the case of no reinsurance $(d = \infty)$ the adjustment coefficient is best obtained from (3). This is the condition that

$$1/3\, e^{2r} + 2/3\, e^{r} - (2.5)r - 1 = 0 . \tag{13}$$

For a stop-loss contract with deductible d we first calculate the reinsurance premium:

$$P^{re} = 2 \sum_{x=d+1}^{\infty} (x-d)\,Pr(S=x)$$

$$= 4 - 2d + 2 \sum_{x=0}^{d} (d-x)\,Pr(S=x)\ . \qquad (14)$$

Then the adjustment coefficient is obtained from the equation

$$\sum_{x=0}^{d-1} e^{r(2.5-P^{re}-x)}\,Pr(S=x) + e^{r(2.5-P^{re}-d)}\,Pr(S\geq d) = 1\ . \quad (15)$$

The results are displayed in Table 1.

Table 1

Assessing a Stop-loss Arrangement

(1) Deductible d	(2) Stop-loss premium, P^{re}	(3) Adjustment coefficient, R	(4) Expected gain per year
3	.676	.25	.162
4	.312	.35	.344
5	.134	.34	.433
∞	0	.28	.500

With respect to security (as measured by the adjustment coefficient) a deductible of 4 is better than one of 5, which in turn is better than no reinsurance. With respect to expected gain this order is reversed. Furthermore, selecting a deductible of 3 would be an irrational decision: It would be worse than no reinsurance both with respect to security and expected gain.

CHOOSING PROPORTIONAL REINSURANCE

We assume that S_n has a compound Poisson distribution such that the expected number of claims per year is 1 and that the individual claim amounts are uniformly distributed between 0 and 1. Thus $E[S_n] = 1/2$. The premium received is supposed to be $P_n = 1$ (which amounts to a loading of 100%). Proportional reinsurance can be purchased, whereby the reinsurer pays aS_n $(0 \leq a \leq 1)$. We consider two cases: one where the reinsurance premium contains a loading of 100% (identical to the original loading), and the other where the loading is 140%. The resulting adjustment coefficients are displayed in Table 2.

Table 2

Adjustment Coefficient for Proportional Reinsurance

	Reinsurance Loading	
a	100%	140%
0	1.793	1.793
.1	1.993	1.936
.2	2.242	2.095
.3	2.562	2.268
.4	2.989	2.436
.5	3.587	2.538
.6	4.483	2.335
.7	5.978	.635
.8	8.966	*
.9	17.933	*
1.0	∞	*

CHOOSING EXCESS OF LOSS REINSURANCE

Assume that S_n has the same compound Poisson distribution as in the preceding example, and that excess of loss coverage can be purchased, whereby the reinsurer pays the excess of any claim over the deductible b . Again the adjustment coefficient has been calculated for two cases, for a reinsurance loading of 100% and for one of 140%. The results are displayed in Table 3.

Table 3

Assessing an Excess of Loss Arrangement

Deductible	Reinsurance Loading	
b	100%	140%
1.0	1.793	1.793
.9	1.833	1.828
.8	1.940	1.920
.7	2.116	2.062
.6	2.378	2.259
.5	2.768	2.518
.4	3.373	2.840
.3	4.400	3.138
.2	6.478	2.525
.1	12.746	*

We observe that if the reinsurance loading is 100% (identical to the loading contained in P_n), the adjustment coefficient is the higher, the higher the degree of reinsurance. If, on the other hand, the reinsurance loading is higher (140% in the example), the adjustment coefficient increases first, but then decreases rapidly to zero. After a certain point the expected annual gain becomes negative; these cases, in which ruin is certain, are marked by asterisk.

COMPARISON OF THE EFFECTS OF PROPORTIONAL AND EXCESS OF LOSS COVERAGE

To compare the results of the last two examples, we look at pairs of a and b, for which the expected payment of the reinsurer is the same. This is the condition that $a = (1-b)^2$. The resulting adjustment coefficients are compared in Table 4.

Table 4

Comparison of Adjustment Coefficients for Proportional and Excess of Loss Reinsurances with Equal Payments Expected form Reinsurer

		Reinsurance Loading			
		100%		140%	
a	b	Proportional	Excess of Loss	Proportional	Excess of Loss
0	1.0	1.793	1.793	1.793	1.793
.01	.9	1.811	1.833	1.807	1.828
.04	.8	1.868	1.940	1.848	1.920
.09	.7	1.971	2.116	1.921	2.062
.16	.6	2.135	2.378	2.030	2.259
.25	.5	2.391	2.768	2.181	2.518
.36	.4	2.802	3.373	2.372	2.840
.49	.3	3.516	4.400	2.535	3.138
.64	.2	4.981	6.478	1.992	2.525
.81	.1	9.438	12.746	*	*
1.00	0	∞	∞	*	*

We observe that if proportional and excess of loss reinsurance can be bought for the same loading, the excess of loss arrangement leads to a higher adjustment coefficient (and therefore to a higher degree of security) than the comparable proportional arrangement. This is no coincidence: A general result says that the excess of loss coverage is optimal among all reinsurance arrangements where the payments are made on a claim by claim basis, and where the reinsurance premium contains the same loading.

THE NORMAL APPROXIMATION AND STOP-LOSS COVERAGE

We assume that the distribution of S_n is normal with $Var[S_n] = 1$; the latter can always be obtained by an appropriate choice of the monetary units. Under a stop-loss arrangement the annual gain G_n has a truncated normal distribution. Thus G_n assumes the value m with probability $\Phi(m-\mu)$ and has a continous distribution for $x > m$ with density $\phi(x-\mu)$.

Tables 5 and 6 show the resulting values of R and $E[G_n]$.

Table 5

Normally distributed claims modified by a stop-loss contract: the adjustment coefficient

m \ μ	- .2	0	.2	.4	.6	.8	1.0
-3	*	.0005	.401	.802	1.20	1.61	2.01
-2.5	*	.004	.407	.810	1.22	1.62	2.04
-2	*	.018	.426	.839	1.25	1.68	2.10
-1.5	*	.066	.492	.925	1.37	1.82	2.28
-1	*	.231	.701	1.19	1.69	2.22	2.76
- .5	.312	.897	1.52	2.18	2.90	3.66	4.48

Table 6

Normally distributed claims modified by a stop-loss contract: the expected gain

m \ μ	- .2	0	.2	.4	.6	.8	1.0
-3	neg	.0004	.200	.400	.600	.800	1.000
-2.5	neg	.002	.201	.401	.600	.800	1.000
-2	neg	.008	.205	.403	.601	.801	1.000
-1.5	neg	.029	.218	.411	.606	.804	1.002
-1	neg	.083	.256	.437	.623	.814	1.008
- .5	.067	.198	.343	.500	.669	.846	1.029

Example: The distribution of S_n is normal with mean .7 and variance 1. To pay for these claims, the first insurer receives $P_n = 1$. A stop-loss coverage with a deductible $d = 1.9$ is available for a premium of $P^{re} = .1$. Find the expected annual gain and the adjustment coefficient a) without reinsurance, and b) with reinsurance.

Solution: a) Without reinsurance the distribution of G_n is normal with mean 1 - .7 = .3 and variance 1. Hence its expectation is .3 , and the resulting adjustment coefficient is $R = .6$, see formula (7).

b) With the proposed stop-loss coverage the annual gain is

$$G_n = \begin{cases} P_n - P^{re} - S_n & \text{if } S_n \leq d \\ P_n - P^{re} - d & \text{if } S_n > d \end{cases},$$

or,

$$G_n = \begin{cases} .9 - S_n & \text{if } S_n \leq 1.9 \\ -1 & \text{if } S_n > 1.9 \end{cases}.$$

Thus G_n has a truncated normal distribution with $\mu = .9 - .7 = .2$ and $m = -1$. From Table 6 we see that the expected annual gain is now reduced to .256 , and from Table 5 we see that the adjustment coefficient has been increased to .701 .

Reference: Bowers, N.L., Gerber, H.U., Hickman, J.C., Jones, D.A. and Nesbitt, C.J.: Risk Theory Study Note, Society of Actuaries, Chicago (1982).

CHAINS OF REINSURANCE

Hans U. Gerber

Ecole des H. E. C., Université de Lausanne

A one-dimensional, hierarchical system of reinsurance is considered. A member of the chain is in direct contact with only two other members: the one from which coverage is bought and the one to which coverage is sold. Exceptions are the first link (which does not sell any reinsurance coverage) and the last link (which does not buy any). The problem is to find the values of the quotas and the loadings that are optimal in some sense. Assuming exponential utility functions and a normal distribution for the claims, an explicit solution is found for two versions of the model.

1. INTRODUCTION

There are different systems of reinsurance. One possibility is to organize a pool, in which the partners play a more or less symmetric role. The notion of Pareto-optimal risk exchange (a popular topic in the last twenty years) falls into this context. Here, we shall consider another extreme system: a chain of reinsurance. This is a one-dimensional, hierarchical system, since it is assumed that each participating company is in direct contact only with two other companies: the one to which coverage is sold, and the one from which coverage is bought.

We consider $n+1$ companies. Company 0 is the first insurer, companies $1, \ldots, n-1$ are the intermediate reinsurers, and company n is the terminal reinsurer. The original claims payable by company 0 are denoted by X. We use the notations

F. de Vylder et al. (eds.), Premium Calculation in Insurance, 183–193.

$\mu = E[X]$, $\sigma^2 = Var[X]$, and $C(t) = \log E[e^{tX}]$ for the cumulant generating function of X . Company i sells coverage to company $i-1$ (for a premium of $(\mu+\lambda_i) f_i$ it reimburses company $i-1$ for a fraction f_i of X) and buys coverage from company $i+1$ (for a premium of $(\mu+\lambda_{i+1}) f_{i+1}$ it will obtain $f_{i+1} X$ from company $i+1$) . The first insurer does not sell any reinsurance coverage, and the terminal reinsurer does not buy any.

The problem is to find the values of f_1 , λ_1, ..., f_n, λ_n that are optimal in some sense. It is assumed that company i will base its decisions on a utility function u_i . Thus company 0 will seek to maximize

$$E[u_0(-(1-f_1)X - (\mu+\lambda_1)f_1)] , \tag{1}$$

company i ($i = 1, ..., n-1$) will seek to maximize

$$E[u_i(-(f_i-f_{i+1})X + (\mu+\lambda_i)f_i - (\mu+\lambda_{i+1})f_{i+1})] , \tag{2}$$

and company n will seek to maximize

$$E[u_n(-f_nX + (\mu+\lambda_n)f_n)] . \tag{3}$$

We shall assume exponential utility functions,

$$u_i(x) = (1 - e^{-a_ix}) / a_i , \tag{4}$$

where a_i is the constant risk aversion of company i , and we shall assume throughout most of this note that X is normally distributed, i. e.

$$C(t) = \mu t + \frac{1}{2} \sigma^2 t^2 . \tag{5}$$

With these assumptions it will be possible to solve the problem explicitly.

To complete the model, we have to specify the <u>bargaining rules</u>. We assume that the buyer of reinsurance will let the seller know how much he will buy as a function of the loading, and that the seller (knowing this function) will determine the loading. Thus, the first insurer chooses the function $f_1 = f_1(\lambda_1)$ in order to maximize (1) and communicates this function to company 1 . Then company 1 chooses λ_1 and f_2 , both functions of λ_2 , in order to maximize its expected utility, and communicates the function $f_2 = f_2(\lambda_2)$ to company 2 , etc. Finally, company $n-1$ communicates to company n the function $f_n(\lambda_n)$, i.e. the amount that it will buy as a function of the loading λ_n . With this information company n chooses the numerical value of λ_n in order to maximize (3). Then the resulting values of f_n, λ_{n-1}, f_{n-1}, ..., λ_1, f_1 can be calculated by successive substitutions.

In the following three sections we shall study the three types of subproblems. Afterwards, the results can be combined to obtain the solution of the original problem for any n .

In section 9 we shall discuss an alternative set of bargaining rules.

2. THE FIRST INSURER'S DEMAND FOR REINSURANCE

Because of (4), maximizing (1) is the same as minimizing

$$C(\ a_0(1-f_1)\ ) + a_0(\mu+\lambda_1)f_1\ . \tag{6}$$

Hence, $f_1 = f_1(\lambda_1)$ is obtained from the condition that

$$-\ C'(\ a_0(1-f_1)\ ) + \mu + \lambda_1 = 0\ . \tag{7}$$

Assuming (5) we obtain an explicit solution for f_1 :

$$f_1 = 1 - \lambda_1\ /\ (\sigma^2\ a_0)\ . \tag{8}$$

Evidently, this solution is hypothetical unless $\lambda_1 \leq \sigma^2\ a_0$.

3. THE SITUATION OF AN INTERMEDIATE REINSURER

We consider company i $(i = 1, \ldots, n-1)$. Knowing the function $f_i(\lambda_i)$, it chooses λ_i and f_{i+1} in order to maximize (2) for any given value of λ_{i+1} . Then the resulting function $f_{i+1}(\lambda_{i+1})$ is communicated to company $i+1$.

Because of (4), maximizing (2) is the same as minimizing

$$C(\ a_i(f_i-f_{i+1})\ ) - a_i(\mu+\lambda_i)f_i + a_i(\mu+\lambda_{i+1})f_{i+1}\ . \tag{9}$$

Setting the partial derivatives with respect to f_{i+1} and λ_i equal to zero we obtain the equations

$$-\ C'(\ a_i(f_i-f_{i+1})\ ) + \mu + \lambda_{i+1} = 0 \tag{10}$$

and

$$f_i'(\lambda_i)\ C'(\ a_i(f_i - f_{i+1})\ ) - f_i - (\mu+\lambda_i)\ f_i'(\lambda_i) = 0\ . \tag{11}$$

Combining the two equations, we see that the optimal value of λ_i is obtained from the condition that

$$f_i'(\lambda_i)\ (\lambda_{i+1} - \lambda_i) = f_i(\lambda_i)\ . \tag{11}$$

It is remarkable that the risk aversion a_i of company i does not enter in this formula.

We shall now assume that f_i is a linear function,

$$f_i = A_i - B_i\ \lambda_i\ . \tag{12}$$

A justification of this assumption will be given later. Then, (11) can be solved explicitly, and we find that

$$\lambda_i = \frac{1}{2}\ \lambda_{i+1} + \frac{1}{2}\ A_i/B_i\ . \tag{13}$$

Finally, we assume (5). From (10) we obtain

$$f_i - f_{i+1} = \lambda_{i+1}\ /\ (\sigma^2\ a_i)\ , \tag{14}$$

which is an appealing expression for the excess of insurance sold over insurance purchased. Substituting (12) and (13), we find that

$$f_{i+1} = A_{i+1} - B_{i+1}\ \lambda_{i+1}\ , \text{ where}$$
$$A_{i+1} = \frac{1}{2}\ A_i\ ,\ B_{i+1} = \frac{1}{2}\ B_i + (\sigma^2\ a_i)^{-1}\ . \tag{15}$$

Thus linearity of $f_i(\lambda_i)$ implies linearity of $f_{i+1}(\lambda_{i+1})$. Since f_1 is linear, see (8), we conclude that all f_i's are linear functions, and that the parameters can be calculated recursively from (15), starting with $A_1 = 1$, $B_1 = (\sigma^2\ a_0)^{-1}$.

4. THE SITUATION OF THE TERMINAL REINSURER

Because of (4), maximizing (3) is the same as minimizing

$$C(\ a_n f_n\ ) - a_n(\mu+\lambda_n)f_n\ . \tag{16}$$

Thus, the optimal value of λ_n is obtained from the condition that

$$f_n'(\lambda_n)\ C'(\ a_n f_n\ ) - f_n - (\mu+\lambda_n)\ f_n'(\lambda_n) = 0\ . \tag{17}$$

We assume (5). It follows that $f_n = A_n - B_n\ \lambda_n$, and from (17) we see that

$$\lambda_n = \frac{A_n}{B_n}\,\frac{1+\sigma^2\ a_n\ B_n}{2+\sigma^2\ a_n\ B_n} \tag{18}$$

Then the resulting portion of reinsurance sold is

$$f_n = A_n\ /\ (\ 2+\sigma^2\ a_n\ B_n\ )\ . \tag{19}$$

Formula (18) can be written as

$$\lambda_n = \frac{1}{2}\,\frac{A_n}{B_n} + \frac{1}{2}\,\frac{A_n}{B_n}\,\frac{\sigma^2\ a_n\ B_n}{2+\sigma^2\ a_n\ B_n}\ , \tag{20}$$

and, because of (15), as

$$\lambda_n = \frac{A_{n-1}}{B_{n-1}}\,\frac{\sigma^2\ a_{n-1}\ B_{n-1}}{2+\sigma^2\ a_{n-1}\ B_{n-1}}\,\frac{1+\sigma^2\ a_n\ B_n}{2+\sigma^2\ a_n\ B_n}\ . \tag{21}$$

We shall use the last two formulas in section 6.

5. EXAMPLES

If $n = 1$, there is only one transaction, and we write simply λ instead of λ_1 and f instead of f_1 . Since $A_1 = 1$, $B_1 = (\sigma^2\ a_0)^{-1}$, we obtain from (18) and (19) immediately the desired solution:

$$\lambda = \sigma^2\ a_0\,\frac{a_0 + a_1}{2a_0 + a_1} \tag{22}$$

$$f = \frac{a_0}{2a_0 + a_1}\ . \tag{23}$$

While it is not surprising that f is increasing in a_0 and decreasing in a_1 , it is noteworthy that f is always less than one half.

If $n = 2$ (one intermediate reinsurer), we substitute $A_2 = 1/2$, $B_2 = (2\sigma^2 a_0)^{-1} + (\sigma^2 a_1)^{-1}$ in (18) and (19) to obtain

$$\lambda_2 = \frac{\sigma^2}{a_0^{-1} + 2a_1^{-1}} \frac{a_0^{-1} + 2a_1^{-1} + 2a_2^{-1}}{a_0^{-1} + 2a_1^{-1} + 4a_2^{-1}} \tag{24}$$

$$f_2 = \frac{a_2^{-1}}{a_0^{-1} + 2a_1^{-1} + 4a_2^{-1}} . \tag{25}$$

From (13) we obtain $\lambda_1 = \lambda_2 / 2 + \sigma^2 a_0 / 2$, and finally, $f_1 = 1 - \lambda_1 / (\sigma^2 a_0)$!

If we compare the results for $n = 1$ and $n = 2$, the question arises if $\lambda_1 \leq \lambda$, i. e. if the addition of the last company (with risk aversion a_2) is to the first insurer's advantage. In the following section we shall see that the answer to this and a more general question is yes.

6. THE ADDITION OF A LAST LINK

We consider first the situation of n companies with risk aversion a_0 , a_1 , ... , a_{n-1} and denote the resulting loadings by $\lambda_i^{(n-1)}$, $i = 0, 1, \ldots, n-1$. Now we add company n (with risk aversion a_n) at the end of the chain, and we denote the resulting loadings by $\lambda_i^{(n)}$, $i = 0, 1, \ldots, n$. Is this to the advantage of the original chain, in the sense that

$$\lambda_i^{(n)} \leq \lambda_i^{(n-1)} \quad \text{for} \quad i = 0, 1, \ldots, n-1 \, ? \tag{26}$$

We shall show that the answer is yes. In view of the recursive formula (13) it is enough to derive the inequality for $i = n-1$.

In the formula $\lambda_{n-1}^{(n)} = \lambda_n^{(n)} / 2 + A_{n-1} / (2 B_{n-1})$ we replace $\lambda_n^{(n)}$ by the right hand side of (21), and we substitute for $\lambda_{n-1}^{(n-1)}$ using (20), with n replaced by $n-1$, to see that

$$\lambda_{n-1}^{(n-1)} - \lambda_{n-1}^{(n)} = \frac{1}{2} \frac{A_{n-1}}{B_{n-1}} \frac{\sigma^2 a_{n-1} B_{n-1}}{2 + \sigma^2 a_{n-1} B_{n-1}} \frac{1}{2 + \sigma^2 a_n B_n} \tag{27}$$

Thus, the difference is positive, and the proof is completed. The expression for the difference shows also that the difference is a decreasing function of the risk aversion of the company that has been added.

Inequality (26) shows that $\lambda_i^{(n)}$ has a limit for $n \to \infty$. From (13) we gather that this limit is

$$\frac{1}{2} A_i/B_i + \frac{1}{4} A_{i+1}/B_{i+1} + \frac{1}{8} A_{i+2}/B_{i+2} + \dots \quad (28)$$

under fairly general conditions.

7. LIMITING CASES

So far, the a_i's were assumed to be positive numbers. The limits $a_i = 0$ and $a_i = \infty$ have an interpretation and are of some interest, too.

If $a_i = \infty$ for some i ($i = 1, \dots, n-1$), the risk aversion of company i is extreme. It follows from (14) that $f_i = f_{i+1}$, i. e., that the coverage bought is identical to the coverage sold (but note the difference of the premiums!). Thus company i plays the role of a broker.

If $a_i = 0$, company i has a linear utility function and does not buy any reinsurance. Thus the chain does not go beyond such a company.

As an illustration, let us look at the case where $a_0 = a > 0$, $a_1 = \dots = a_{n-1} = \infty$, $a_n = 0$. Thus the first insurer and the reinsurer (in the proper sense) are separated by a series of $n-1$ brokers. In this case the solution has a simple form. First,

$$A_i = 2^{-i+1}, \quad B_i = (\sigma^2 a)^{-1} 2^{1-i} \quad \text{for } i = 1, \dots, n. \quad (29)$$

From (19) and (14) we see that

$$f_i = 2^{-n} \quad \text{for } i = 1, \dots, n. \quad (30)$$

Finally, we substitute these expressions in (12) and solve for λ_i to obtain

$$\lambda_i = \sigma^2 a \left(1 - 2^{-n-1+i} \right) \quad \text{for all } i. \quad (31)$$

It is interesting that the amount of reinsurance traded does not depend on a, see (30). If n is large, the situation is neither realistic nor desirable.

8. DEPARTURE FROM NORMALITY

While the results were derived under the assumption that X is normally distributed, they can be used in the general case as approximations, provided that $a_0, a_1, \ldots, a_n$ are sufficiently close to zero. This follows from the observation that

$$C'(at) = \mu + \sigma^2 \ at + \text{terms with } a^2 \tag{32}$$

and a glance at formulas such as (7), (10), (11), and (17).

In the general case, it seems to be difficult to find explicit results. We outline the starting point if $n = 1$, where we set (as before) $\lambda_1 = \lambda$ and $f_1 = f$. Thus (7) tells us that

$$\mu + \lambda = C'(\, a_0(1-f)\,) , \tag{33}$$

which defines the relation between λ and f. In the following, we prefer to consider λ as a function of f, symbolically, $\lambda = \lambda(f)$. If we differentiate (33), we obtain

$$\lambda'(f) = -\, a_0 \, C''(\, a_0(1-f)\,) . \tag{34}$$

If we differentiate (16), we get

$$C'(\, a_1 f\,) - \lambda'(f)\, f - (\, \mu + \lambda\,) = 0 . \tag{35}$$

We use (33) and (34) to eliminate λ' and λ to obtain

$$C'(\, a_1 f\,) + a_0 \, f\, C''(\, a_0(1-f)\,) - C'(\, a_0(1-f)\,) = 0 . \tag{36}$$

This is the equation for the desired value of f.

Finally, we would like to point out an interpretation of formulas (7) and (10) which holds in the general case. For example, (10) can be written as

$$C'(\, a_i(f_i - f_{i+1})\,) = \mu + \lambda_{i+1} . \tag{37}$$

This means that company i buys that portion of coverage, such that the Esscher premium (parameter a_i) for the claims in the retention equals the quoted reinsurance premium for full coverage.

9. ALTERNATIVE BARGAINING RULES

One can show that under our previous assumptions

$$f_1 \leqq \frac{1}{2} \text{ , and } \lambda_1 \geqq \frac{1}{2} \sigma^2 a_0 \text{ ,} \tag{38}$$

with equality holding only if $a_1 = 0$. (See formulas (22) and (23) for an illustration). Thus in our model the first insurer would retain at least 50 percent of the original claims. Since this does not always correspond to the reality, we will now discuss a modified model which explains higher degrees of reinsurance.

The new assumption is that the seller of reinsurance will let the buyer know how much he is willing to sell (as a function of the loading), and that the buyer will determine the loading on the basis of this information. Thus, the terminal reinsurer chooses the function $f_n = f_n(\lambda_n)$, which is communicated to company n-1 . This company chooses λ_n and f_{n-1} , both functions of λ_{n-1} , and communicates the function $f_{n-1}(\lambda_{n-1})$ to company n-2 , etc. . Finally, company 1 communicates the function $f_1(\lambda_1)$ to the first insurer, who then selects the value of λ_1 . Now the resulting values of $f_1, \lambda_2, f_2, \ldots, \lambda_n, f_n$ can be calculated in this order.

The calculations are analoguous and somewhat simpler than in the original model. Therefore, we will limit ourselves to a short derivation of the results.

The terminal reinsurer chooses f_n in order to minimize (16). Thus, $f_n = f_n(\lambda_n)$ is obtained from the condition that

$$C'(a_n f_n) - (\mu + \lambda_n) = 0 \text{ .} \tag{39}$$

Assuming (5), we obtain

$$f_n = \lambda_n / (\sigma^2 a_n) \text{ ,} \tag{40}$$

which is the counterpart of (8).

Consider now company i (i = 1, ..., n-1) . Knowing the function $f_{i+1}(\lambda_{i+1})$, it chooses f_i and λ_{i+1} to minimize (9) for given λ_i . Setting the partial derivatives equal to zero, we obtain the equations

$$C'(a_i(f_i - f_{i+1})) - (\mu + \lambda_i) = 0 \tag{41}$$

and

$$- f'_{i+1} \; C'(\; a_i(f_i - f_{i+1}) \;) + f_{i+1} + (\mu + \lambda_{i+1}) \; f'_{i+1} = 0. \quad (42)$$

It follows that

$$f'_{i+1}(\lambda_{i+1}) \; (\lambda_i - \lambda_{i+1}) = f_{i+1}(\lambda_{i+1}) \; , \quad (43)$$

from where λ_{i+1} is obtained. Assuming that f_{i+1} is proportional to λ_{i+1}, i. e. $f_{i+1} = B_{i+1} \; \lambda_{i+1}$, we conclude from (43) that

$$\lambda_{i+1} = \frac{1}{2} \lambda_i \; . \quad (44)$$

Using all this and (5) in (42) we see that

$$f_i = B_i \; \lambda_i \; , \text{ with } \; B_i = \frac{1}{2} B_{i+1} + (\sigma^2 \; a_i)^{-1} \; . \quad (45)$$

Thus, the f's are indeed proportional to the corresponding λ's , and the B_i's can be calculated recursively, starting with $B_n = (\sigma^2 \; a_n)^{-1}$, see (40). One finds that

$$\sigma^2 \; B_i = a_i^{-1} + \frac{1}{2} a_{i+1}^{-1} + \ldots + 2^{i-n} \; a_n^{-1} \; , \quad (46)$$

for $i = 1, \ldots, n$.

Finally, the first insurer chooses λ_1 to minimize (6). Thus the condition is that

$$- f'_1 \; C'(\; a_0(1-f_1) \;) + f_1 + (\mu + \lambda_1) \; f'_1 = 0 \; , \quad (46)$$

which is the counterpart of (17). Using (5) and $f_1 = B_1 \; \lambda_1$ we obtain

$$\lambda_1 = \frac{\sigma^2 \; a_0}{2 + B_1 \; \sigma^2 \; a_0}$$

$$= \sigma^2 \; (\; 2 \; a_0^{-1} + a_1^{-1} + \frac{1}{2} a_2^{-1} + \ldots + 2^{1-n} \; a_n^{-1} \;)^{-1} \quad (47)$$

Then $f_1 = B_1 \; \lambda_1$, or,

$$f_1 = \frac{a_1^{-1} + \frac{1}{2} a_2^{-1} + \ldots + 2^{1-n} \; a_n^{-1}}{2 \; a_0^{-1} + a_1^{-1} + \frac{1}{2} a_2^{-1} + \ldots + 2^{1-n} \; a_n^{-1}} \; . \quad (48)$$

For $i = 2, \ldots, n$ we use (44) to see that

$$\lambda_i = 2^{1-i} \lambda_1 . \tag{49}$$

Since $f_i = B_i \lambda_i$, it follows that

$$f_i = \frac{2^{1-i} a_i^{-1} + 2^{-i} a_{i+1}^{-1} + \ldots + 2^{1-n} a_n^{-1}}{2 a_0^{-1} + a_1^{-1} + \frac{1}{2} a_2^{-1} + \ldots + 2^{1-n} a_n^{-1}} . \tag{50}$$

We see that our modified assumptions have led to a solution that is substantially different. For f_1 all values between 0 and 1 are possible, and for given a_0 the loading λ_1 can be arbitrarily close to 0 . This is in contrast to the original model, see (38). For example, if $a_0, a_1, \ldots, a_{n-1}$ are given positive numbers, $a_n \to 0$ implies that $f_i \to 1$ and $\lambda_i \to 0$ for all i .

Observations similar to those of section 8 can be made. For example, in the special case $n = 1$ the desired value of $f_1 = f$ is obtained from the condition that

$$- C'(a_0(1-f)) + C'(a_1 f) + a_1 f C''(a_1 f) = 0 , \tag{51}$$

which should be compared with condition (36) of the original model.

10. CONCLUSION

While chains of reinsurance are frequently encountered in practice (and definitely more so than in the literature), there remains the question how to model them mathematically. In this note, two models have been submitted to the discussion. Of course, the final answer will have to come from the practitioners, if and when they decide to use a model ...

NET STOP-LOSS ORDERING AND RELATED ORDERINGS

M.J. Goovaerts
K.U. Leuven
Dekenstraat 2
B 3000 Leuven

ABSTRACT

Some invariance properties of net-stop-loss ordering are given. In particular net-stop-loss ordering is examined for addition and mixing of risks as well as for compounding risks. It is shown how ordering of claim size and of claim frequencies are related to ordering of total claim sizes. Also some generalizations of stop-loss ordering, namely stop-loss dominance and the related economical ordering, called stochastic dominance are considered.

1. INTRODUCTION

In ref. [5] W.S. Jewell gave some general characteristics of the basic risk models in actuarial applications. Traditionally a distinction has been made between two classes of models : those used in life insurance companies and those which arise in non-life applications. As W.S. Jewell explained in [5] the basic mathematical models common to all branches of insurance have three key elements :

(1) One or more random variables which characterize the major dimensions of the risk, such as duration, size and number.
(2) A set of well-defined states of nature, separated by observable transition events or epochs, together with a deterministic law of motion between the states.
(3) An economic function, associated with the underlying random variables and for the states and transition events, which may also be deterministic of random, but is most often linked to uncontrollable economic externalities, such as

F. de Vylder et al. (eds.), Premium Calculation in Insurance, 195–234.

market growth, inflation, currency risk etc., but also to economic performance under the control of the company, such as profit margin, portfolio performance etc..

1.1. Life contingencies

The basic random variable appearing in life contingencies is T, the remaining lifetime from moment of underwriting until death. For practical premium calculation in life insurances one considers the risk T/H, the remaining lifetime given the information H, usually the individual's age, sex, health etc.. The probability distribution of T/H is usually constructed by means of an empirically observed mortality table. The economic functions of interest are the associatated present values. If the force of interest is δ per year the present value of the lump sum is $A(t) = \exp(-\delta t)$ and the present value of the continuous annuity is $a(t) = \frac{1}{\delta}(1-e^{-\delta t})$. The link to the underlying risk process is through the durations of these forms. Let the information H be the age of the insured head (x), then the present value of the monetary unit paid at the death of (x) is the variable $A(T_x)$ and the present value of the monetary unit per year payable until the death of (x) is the variable $a(T_x) = \frac{1}{\delta}(1-e^{-\delta T_x})$

A premium principle is a rule that assigns a premium to the economic functions (risks) $A(T_x)$, $a(T_x)$. In practice the fair premiums which are of interest to the life actuary are nothing more than the mean values of these random variables

$$\bar{A}_x = E(A(T_x)) \tag{1.1}$$

$$\bar{a}_x = E(a(T_x)) \tag{1.2}$$

Multiple contingencies can also be attacked from the point of view of random variables. Consider the reversionary continuous annuity to (x) after (y). Suppose the joint density of (T_x,T_y) the remaining lifetimes is known. Let $T_{xy} = \min(T_x,T_y)$ then the new economic function for this multiple contingency is given by

$$a(T_{xy},T_x) = a(T_x) - a(T_{xy}) \tag{1.3}$$

From this relation it follows that (1.3) is thrue for random variables, and for the mean as is well known. From (1.3) it follows that the hypothesis of independence between the two lifes is not necessary provided that the distribution of T_{xy} is calculated properly.

1.2. The aggregate claims model

The basic model in casualty insurance is the accumulated claim or aggragate claim model in which basically two random variables are recognized, namely the claim number or claim intensity N and the claimsize of each claim $\{X_1, X_2, \ldots\}$. The economic function of interest is the total claim size which is a random sum of random variables.

$$S = X_1 + X_2 + \ldots + X_N \tag{1.4}$$

Usually two assumptions are made in this model, namely :

(1) Given N = n, the random variables $\{X_1, X_2, \ldots, X_n\}$ are mutually independent and identically distributed.
(2) Their common distribution does not depend upon n.

The important practical question that arises then reads : what theoretical premium should we charge for this economic function.

Remark
The word theoretical is added to premium because premium calculation principles provides us with measures of risk, rather than with commercial premiums (which include some commercial loadings). The only kind of loading taken into account in applying a premium calculation principle is a safety loading.

If F is a distribution on $\mathbb{R}^+$ and α locally bounded on $\mathbb{R}^+$, the convolution product $\alpha * F$ is the function on $\mathbb{R}^+$ defined by

$$(\alpha * F)(x) = \int_0^x (x - y)\, dF(y) \tag{1.5}$$

If $\beta = F - G$ is the difference of two distributions on $\mathbb{R}^+$ then $\alpha * \beta$ is defined by

$$\alpha * \beta = \alpha * F - \alpha * G \tag{1.6}$$

It is readely seen that this definition does not depend on the particular representation $\beta = F - G$ of β. If F and G are distributions on $\mathbb{R}^+$, then both $F * G$ and $G * F$ are defined and are equal. It follows that $\alpha * \beta = \beta * \alpha$ if α and β are linear combinations of distributions. If F, G and H are distributions on $\mathbb{R}^+$, then

$$(F * G) \quad H = F \quad (G * H) \tag{1.7}$$

and this associativity is trivially extended to the case of linear combinations of distributions. If α is a linear combination of distributions (in particular a distribution) then $\alpha^{*n} (n = 0,1,2, \ldots)$ is defined on $\mathbb{R}^+$ by

$$\alpha^{0} = 1 \quad \alpha^{*(n+1)} = \alpha^{*n} * \alpha \ (n = 0,1,2,\ldots) \tag{1.8}$$

Notice that $\alpha^{*0} = H(x)$ where H now denotes the Heaviside function.

The distribution function F_S in (1.4) is connected with the distribution function F_X and with theclaim intensity distribution $\{p_0, p_1, \ldots\}$ with $p_n = \text{prob}(N = n)$ as follows :

$$F_S(x) = \sum_{n=0}^{\infty} p_n \, F_X^{*n}(x) \tag{1.9}$$

A distribution of the kind (1.9) is called a compound distribution.

In order to explicate some certain non-obvious variations in observed outcomes from a portfolio, or collective of risks mixtures of distributions are considered. If F_Y is a distribution function with support N_Y and $F_X^{(y)}$, $y \in N_Y$ is a family of distribution functions such that $F_X^{(y)}$ is Borel-measurable as a function of y for all x, then the function F_X defined by

$$F_X = \int_{-\infty}^{+\infty} F_X^{(y)} \, dF_Y(y) \tag{1.10}$$

is again a distribution function, called a mixture of distributions. Of course $F_X^{(y)}$ can be interpreted as the distribution of the conditional random variable X/Y with

$$F_{X/Y}(y,x) = F_X^{(y)}(x) \tag{1.11}$$

Hence :

$$F_X(x) = \int F_{X/Y}(\ ,x) \, dF_Y(y) \tag{1.12}$$

Of course in case densities exist one has :

$$f_X(x) = \int f_{X/Y}(y,x) f_Y(y) \tag{1.13}$$

In fact

$$F_{XY}(x,y) = \int_0^y F_{X/Y}(y,x) \, dF_Y(Y)$$

denotes the joint distribution of the two random variables (X,Y)

$$F_{XY}(x,y) = \text{prob}(X \leqslant x, Y \leqslant y) \tag{1.14}$$

If h(x,y) is a function of two variables (x,y) then Z = h(X,Y) is a random variable with expectation

$$E(Z) = \int h(x,y) dF_{XY}(x,y) \tag{1.15}$$

Of course we have the following results for expectations of dependent variables (X,Y).

$$E(Y) = E(E(Y/X)) \tag{1.16}$$

$$= \int_{\mathbb{R}_+} dF_X(x) \int_{\mathbb{R}} y d_y F_{Y/X}(x,y) \tag{1.17}$$

and of course :

$$Var(Y) = E(Var(Y/X)) + Var(E(Y/X)) \tag{1.18}$$

Example

Let

$$F_{S/Y}(y,x) = \sum_{n=0}^{\infty} e^{-y} \frac{y^n}{n!} F_X^{*n}(x) \tag{1.19}$$

and

$$prob(Y \leqslant y) = F_Y(y)$$

then :

$$F_S(x) = \sum_{n=0}^{\infty} p_n F_X^{*n}(x) \tag{1.20}$$

with :

$$p_n = \int_0^{\infty} e^{-y} \frac{y^n}{n!} dF_Y(y) \tag{1.21}$$

The distribution function (1.19) still depends on the so-called risk parameter, while the distribution function (1.20) describes a collectivety of risks. The expression (1.19) is called a compound Poisson distribution, while (1.20) is referred to as a weighted compound Poisson distribution.

From (1.11) it follows that

$$M_X(t) = E_Y(M_{X/Y}(t)) \tag{1.22}$$

in case convergent expressions are obtained in both members. Let us examine next some properties of compound distributions. Let

$$P(u) = \sum_{i=0}^{\infty} p_i u^i \tag{1.23}$$

then, one easily verifies :

$$M_S(t) = P(M_X(t)) \tag{1.24}$$

Of course :

$$E(S) = E(N).E(X) \tag{1.25}$$

$$\mathrm{Var}(S) = \sigma^2(S) = \mathrm{Var}(N)E(X)^2 + E(N).\mathrm{Var}(X) \tag{1.26}$$

which are obtained from (1.24) by taking the derivative with respect to t twice.

Example
Let

$$p(N = n) = e^{-\lambda}\,\frac{\lambda^n}{n!},$$

then

$$P(u) = e^{-\lambda+\lambda u} \tag{1.27}$$

such that

$$M_S(t) = e^{-\lambda+\lambda M_X(t)} \tag{1.28}$$

as well as :

$$E(S) = \lambda E(X) \tag{1.29}$$

$$\mathrm{Var}(S) = \lambda E(X^2) \tag{1.30}$$

2. NET-STOP LOSS ORDERING FOR CLAIM DISTRIBUTIONS

Definiton 1

Let G,H be any distributions on the real line. Then we say that G H if

i) $$\int_0^\infty x\,dG(x) < \infty \tag{2.1}$$

ii) $$\int_t^\infty (x-t)\,dG(x) \leqslant \int_t^\infty (x-t)\,dH(x) \quad -\infty < t < +\infty \tag{2.2}$$

Remarks

1. The reason for calling this ordering a net-stop-loss ordering is due of course to the condition (2.2). This condition simply means that for any retention limit t the net-stop-loss premium for a risk whose c.d.f. is G is not higher than the one of a risk whose c.d.f. is H.
2. The condition (2.2) can be transformed by means of one partial integration into the following equivalent condition.

$$\int_t^\infty (1-G(x))\,dx \leqslant \int_t^\infty (1-H(x))\,dx \qquad -\infty < t < +\infty$$

3. In what follows we will restrict ourselves to distributions of bounded random variables, $X \in D\,[o,a\,]$. However most of the results still hold in case of distributions on $\mathbb{R}^+$ and on $\mathbb{R}$.

Theorem 1

If $G \prec H$, then

$$\int_0^{\infty} v(x)dG(x) \leq \int_0^{+\infty} v(x)dH(x) \tag{2.4}$$

$\forall v$ which have a continuous $v''(x) > 0$, $\forall x \in \mathbb{R}^+$, $v'(o) \geq 0$.

Proof

We have :

$$v(x) = \int_0^{\infty} (x-t)_+ v''(t)dt + xv'(o) + v(o) \tag{2.5}$$

consequently :

$$\int_0^{\infty} v(x) \; [dG(x)-dH(x)] =$$

$$\int_0^{\infty} v''(t)dt \int_0^{\infty} (x-t)_+ [dG(x)-dH(x)] + v'(o) \int_0^{\infty} x [dG(x)-dH(x)]$$

Because of (2.2) we have

$$\int_0^{\infty} v(x) [dG(x)-dH(x)] \leq 0$$

Remark :

If in addition $\int_0^{\infty} xdG(x) = \int_0^{\infty} xdH(x)$ the condition $v'(o)$ can be dropped.

Theorem 2

Let $\varphi(\nu)$ with $\nu = 0, \ldots, n$ denotea finite sequence such that

$$\Delta^2\varphi(\nu) \geq 0 \qquad \nu = 0,1,2, \ldots, n-2$$

then

$$\sum_{\nu=0}^{n} \varphi(\nu)a_\nu \geq 0 \tag{2.6}$$

if and only if :

$$\sum_{\nu=0}^{n} \varphi(\nu-k)_+ a_\nu \geq 0 \qquad k = 0,1, \ldots, n-1 \tag{2.7}$$

and

$$\begin{cases} \sum_{\nu=0}^{n} a_\nu = 0 \\ \sum_{\nu=0}^{n} \nu a_\nu = 0 \end{cases} \tag{2.8}$$

Proof

The conditions (2.7) and (2.8) are certainly necessary, because

$$\varphi(\nu) = (\nu-x)_+ \text{ and } \varphi(\nu) = \pm 1, \pm\nu$$

are such that $\Delta^2\varphi(\nu) \geqslant 0$.
Let us next prove the sufficiency.
We consider a function $\psi(t)$ such that

$$\psi(t) = \sum_{j=1}^{n-1} \alpha_j (t-j)_+ + \beta_0 + \beta_1 t \tag{2.9}$$

and we determine the coefficients in order that

$$\psi(\nu) = \varphi(\nu) \qquad \nu = 0, \ldots, n$$

We find : $\varphi(0) = \beta_0$
$\varphi(1) = \beta_0 + \beta_1$
$\varphi(2) = \alpha_1 + \beta_0 + 2\beta_1$

Consequently $\beta_0 = \varphi(0)$
$\beta_1 = \Delta\varphi(0)$
$\alpha_1 = \Delta^2\varphi(0)$

In general we get

$$\varphi(p+1) = \alpha_p + \sum_{j=0}^{p-1} \alpha_j + \varphi(p) + \beta_1$$

$$\varphi(p+2) = \alpha_{p+1} + \sum_{j=0}^{p} \alpha_j + \varphi(p+1) + \beta_1$$

and hence $\alpha_{p+1} = \Delta^2\varphi(p)$

As a final result for the interpolation formula we get

$$\psi(t) = \varphi(o) + \Delta\varphi(o)\, t + \sum_{j=1}^{n-1} [\Delta^2 \varphi(j-1) \quad (t-j)_+ \tag{2.10}$$

But of course $\sum_{\nu=o}^{n} \varphi(\nu)\, a_\nu = \sum_{=o}^{n} \psi(\nu) a_\nu$, such that finally

$$\sum_{\nu=o}^{n} \varphi(\nu) a_\nu = \sum_{j=1}^{n=1} \Delta^2\varphi(j-1) \sum_{\nu=o}^{n} (\nu-j)_+ \, a_\nu \geqslant 0$$

Remarks
1. This theorem follows as a special case from a result obtained by Karlin and Studden. We like to give this explicit demonstration because of its simplicity.

2. There is no need to restrict ourselves to the case of finite series of the form (2.6). Indeed the theorem still holds in case $n \to \infty$, when convergent series arises.

Theorem 3

Let $\varphi_o(n) = \int_\alpha^\infty (x-\alpha)dF^{*n}(x)$

with $v' > o$, $v'' \geqslant o$, then $\varphi_o(n)$ denotes a convex sequence.

Proof

We thus have to prove that :

$$\varphi_o(n+2) - 2\varphi_o(n+1) + \varphi_o(n) \geqslant 0 \qquad (2.11)$$

which is of course equivalent with

$$\varphi_o(n+1) \leqslant \frac{1}{2}\varphi_o(n) + \frac{1}{2}\varphi_o(n+2)$$

Firstly we limit ourselves to the special case n=0. We then have to show :

$$\int_\alpha^\infty (x-\alpha)dF(x) \leqslant \frac{1}{2}\int_\alpha^\infty (x-\alpha)dF^o(x) + \frac{1}{2}\int_\alpha^\infty (x-\alpha)dF^{*2}(x)$$

This inequality can still be cast into the form :

$$\sum_{i=1}^{2} E((X_i-\alpha)_+) \leqslant 1\ (-\alpha)_+ + E((\sum_{i=1}^{2} X_i-\alpha)_+)$$

where X_1 and X_2 are distributed according to F with F(o) = 0.

Next we prove that the inequality (2.11) implies the following

$$\varphi_0(n+3)-2\varphi_0(n+2) + \varphi_0(n+1) \geqslant 0$$

But of course $_0(n)$ can be cast into the form

$$\varphi_0(n) = \int_\alpha^\infty (1-F^{*n}(x))dx$$

Consequently :

$$\begin{aligned}\varphi_0(n+2) &= \int_\alpha^\infty (1-F^{*(n+2)}(x))dx \\ &= \int_\alpha^\infty (1-F(x))dx + \int_\alpha^\infty (F(x)-F^{*(n+2)}(x))dx \qquad (2.12)\end{aligned}$$

Write $E(\alpha)$ for the first term on the r.h.s., then

$$\begin{aligned}\varphi_0(n+2) &= E(\alpha) + \int_\alpha^\infty \{F(x) - \int_0^x F^{*(n+1)}(x-y)dF(y)\}dx \\ &= E(\alpha) + \int_\alpha^\infty \int_0^x (1-F^{*(n+1)}(x-y))dF(y)dx\end{aligned}$$

Reversal of the order of integration gives :

$$\varphi_0(n+2) = E(\alpha) + \{\int_0^\alpha \int_\alpha^\infty + \int_\alpha^\infty \int_y^\infty \quad (1-F^{*(n+1)}(x-y))dxdF(y) \qquad (2.13)$$

But because (2.11) holds one still has thefollowing inequality

$$\int_\beta^\infty (1-F^{*(n+1)}(x-y))dx \leq \frac{1}{2}\int_\beta^\infty (1-F^{*(n+2)}(x-y)dx + \frac{1}{2}\int_\beta^\infty (1-F^{*n}(x-y))dx \qquad (2.14)$$

Inserting the r.h.s. of (3.14) into the r.h.s. of (2.13) gives :

$$\varphi_0(n+2) \leq E(\alpha) + \frac{1}{2}\{\int_0^\alpha \int^\infty + \int_\alpha^\infty \int_y^\infty\}(1-F^{*(n=2)(x-y))dxdF(y)$$

$$+ \frac{1}{2}\{\int_0^\alpha \int_\alpha^\infty + \int_\alpha^\infty \int_y^\infty\}(1-F^{*n}(x-y))dxdF(y)$$

$$= \frac{1}{2}\varphi_0(n+3) + \frac{1}{2}\varphi_0(n+1)$$

<u>Theorem 4</u>

Let $F_{S_1}(x) = \sum_{n=0}^{\infty} p_n F^{*n}(x)$

$$F_{S_2}(x) = \sum_{n=0}^{\infty} p'_n F^{*n}(x)$$

$$G(x) = \sum_{n=0}^{[x]} p_n, \quad H(x) = \sum_{n=0}^{[x]} p'_n$$

If $G \prec H$, $F_{S_1} \prec F_{S_2}$

<u>Conclusion</u>

When two claim number distributions are ordered according to i) and ii) the corresponding compound distributions are ordered in the same sense.

<u>Theorem 5</u>

Let (G_n), (H_n) be sequences of distributions, and let (p_n) be a discrete probability distribution. If $G_n \prec H_n$ for all n and if $\sum_n p_n \int x dG_n(x) \prec \infty$, then

$$\sum_n p_n G_n \prec \sum_n p_n H_n \qquad (2.15)$$

<u>Proof</u> : trivial

Theorem 6

Let F be a distribution with $\int_0^\infty x dF(x) < \infty$. If G H ten

$$G^*F \prec H^*F \tag{2.16}$$

Proof : Condition i) is certainly satisfied because

$$\int_0^\infty x dG^*F(x) = \int_0^\infty x dG + \int_0^\infty x dF < \infty$$

To prove the validity of ii) we observe that

$$\begin{aligned}\int_t^\infty (1-G^*F(x))dx &= \int_t^\infty \int_{-\infty}^{+\infty} (1-G(x-s))dF(x)dx \\ &= \int_{-\infty}^{+\infty} \int_{t-s}^{\infty} (1-G(y))dydF(s) \\ &\leqslant \int_{-\infty}^{+\infty} \int_{t-s}^{\infty} (1-H(y))dydF(s) \\ &= \int_t^\infty (1-H^*F(x))dx\end{aligned}$$

Consequently, by means of one partial integration

$$\int_t^\infty (x-t)dG^*F(x) \leqslant \int_t^\infty (x-t)dH^*F(x) \qquad \text{q.e.d.}$$

Theorem 7

If $G_i \prec H_i$, $(i = 1,2, \ldots, n)$, then

$$G_1^*G_2^* \ldots ^*G_n \prec H_1^*H_2^* \ldots ^*H_n \tag{2.17}$$

Proof : Repeated application of theorem 6 gives

$$G_1^*G_2^*G_3^* \ldots ^*G_n \prec H_2^*G_2^*G_3^* \ldots ^*G_n$$

$$H_1^*G_2^*G_3^* \ldots ^*G_n \prec H_1^*H_2^*G_3^* \ldots ^*G_n$$

$$H_1^*H_2^*G_3^* \ldots ^*G_n \prec H_1^*H_2^*H_3^* \ldots ^*G_N \qquad \text{etc.}$$

Theorem 8

Let $S_1 = X_1 + X_2 + \ldots + X_{N_1}$

$S_2 = Y_1 + Y_2 + \ldots + Y_{N_2}$

let further

$$F_{N_1} \prec F_{N_2}\ ,\ F_{X_i} \prec F_{Y_i} \qquad \forall i$$

$$F_{S_1} \prec F_{S_2} \tag{2.18}$$

<u>Proof</u>

Because

$$F_{S_1}(x) = \sum_{n=0}^{\infty} p(N_1=n)\, F^{*}_{X_1} F^{*}_{X_2} \cdots {}^{*}F_{X_n}(x)$$

$$F_{S_2}(x) = \sum_{n=0}^{\infty} p(N_2=n)\, F^{*}_{Y_1} F^{*}_{Y_2} \cdots {}^{*}F_{Y_n}(x)$$

we find by theorem 7

$$F_{S_1} \prec F_{S_3}$$

where

$$F_{S_3}(x) = \sum_{n=0}^{\infty} p(N_1=n)\, F^{*}_{Y_1} F^{*}_{Y_2} \cdots {}^{*}F_{Y_n}(x)$$

From theorem 2 and 3 we can conclude that

$$F_{S_3} \prec F_{S_2}$$

Hence

$$F_{S_1} \prec F_{S_2}$$

<u>Remark</u>

If $F_{X_1} \prec F_{X_2}$ one also writes $X_1 \prec X_2$ for simplicity

<u>Applications</u>

1. Let S_1 be a Poisson risk, with parameter λ, claim size distrituition $F_{X_1}(x)$ with mean $\mu = E(x)$ and $F_{X_1}(M) = 1$.

   On the other hand we consider a Poisson risk S_2 with parameter $\Lambda = \lambda \frac{\mu}{M}$ and claim size distribution $F_{X_2}(x) = H(x-M)$
   We will show now

$$S_1 \prec S_2$$

Indeed

$$E(e^{-sS_2}) = e^{-\lambda\frac{\mu}{M} + \lambda\frac{\mu}{M} e^{-sM}}$$
$$= e^{-\lambda + \lambda[1 - \frac{\mu}{M} + \frac{\mu}{M} e^{-sM}]}$$

which means that S_2 can be interpretted as a compound Poisson variable with Poisson parameter λ and claim size distribution given by a Bernouilli variable

$$p(X = 0) = 1 - \frac{\mu}{M}$$
$$p(X = M) = \frac{\mu}{M}$$

On the other hand we have :

$$\int_t^\infty (x-t)dF_{X_1}(x) \leqslant \int_M^M \frac{x-t}{x} .xdF_{X_1}(x)$$
$$\leqslant \int_t^M \max_{x\varepsilon[t,M]} \{ \frac{x-t}{x} \} xdF_{X_1}(x)$$
$$= \frac{M-t}{M} \int_t^M xdF_{X_1}(x) \leqslant \frac{M-t}{M} \int_0^M xdF_{X_1}(x)$$
$$= \frac{M-t}{M} \mu$$

Hence, application of theorem 7, 5 and 4 provides us with the desired results.

2. Let us consider next a Poisson risk S_1 with parameter λ, claim size distribution $F_X(x)$ with mean $E(x) = \mu$, $F_{X_1}(M) = 1$. On the other hand we consider a Poisson risk S_2 with parameter λ, claim size distribution $F_{X_2}(x) = H(x-\mu)$. We will show that

$$S_1 \succ S_2$$

Indeed, in case :

$$t > \mu : \int_t^M (x-t)dF_{X_1}(x) \geqslant 0 = \int_t^M (x-t)dF_{X_2}(x)$$
$$t < \mu : \int_t^M (x-t)dF_{X_1}(x) = \mu - \int_0^t (1-F_{X_1}(x))dx$$
$$= \mu-t + \int_0^t F_X(x)dx$$

Hence $\int_t^M (x-t)dF_{X_1}(x) \geq \mu - t = \int_t^M (x-t)dF_{X_2}(x)$

Consequently, application of theorems 7, 5 and 4 provides us with the desired results.

3. Now we compare the two distributions :

$$F_1 = (1-q)F^{*0} + qF \qquad 0 \leq q \leq 1$$

$$F_2 = \sum_{n=0}^{\infty} p_n F^{*n} \qquad \text{where} \qquad 0 \leq p_n \leq 1$$

$$\sum_{n=0}^{\infty} p_n = 1$$

Suppose further that $F(0) = 0$, and $\int_0^{\infty} x dF(x) < \infty$

If in addition

$$q = \sum_1^{\infty} np_n$$

then :

$$F_1 \prec F_2$$

Of course, because of theorem 8, it is sufficient to show that $G \prec H$ where :

$$G(x) = \begin{cases} 0 & x < 0 \\ 1-q & 0 \leq x < 1 \\ 1 & 1 \leq x \end{cases}$$

$$H(x) = \begin{cases} 0 & x < 0 \\ \sum_{n=0}^{[x]} p_n & x \geq 0 \end{cases}$$

For $t > 1$ we get

$$\int_t^{\infty} (x-t)dG(x) = 0 \leq \sum_{n=[t]+1}^{\infty} (n-t)p_n$$

where $[t]$ denotes the largest integer contained in t.

For $0 < t < 1$ we have to show that :

$$(1-t)q \leq \sum_{n=1}^{\infty} (x-t)p_n$$

or $q \geq 1-p$.

which is certainly satisfied because :

$$p = \sum_{n=1}^{\infty} np_n \geq \sum_{n=1}^{\infty} p_n = 1-p.$$

Theorem 9

Let N_1 and N_2 be two random variables describing the claim numbers of two-compound variables

$S_1 = \sum_{j=1}^{N_1} X_j$, $S_2 = \sum_{j=1}^{N_2} X_j$ then a sufficient condition for

$S_1 \prec S_2$

is

$$\sum_{i=0}^{k} \sum_{j=0}^{i} p(N_1=j) \leq \sum_{i=0}^{k} \sum_{j=0}^{i} p(N_2=j) \qquad \forall k = 0,1,\ \ldots \tag{2.19}$$

in case $E(N_1) = E(N_2)$

Proof

In case $E(N_1) = E(N_2)$ we restrict ourselves to random variables describing the number of claims which have the same expectation. Of course our theorem 8 of this section can be applied in order to give :

$$F_{N_1} \prec F_{N_2} \Longrightarrow F_{S_1} \prec F_{S_2} \tag{2.20}$$

Hence it is sufficient to prove that (2.19) is a sufficient condition for $F_{N_1} \prec F_{N_2}$ to hold. Of course a necessary and sufficient condition is given by

$$\sum_{j=k+1}^{\infty} (j-k)p(N_1=j) \leq \sum_{j=k+1}^{\infty} (j-k)p(N_2=j)$$

or

$$\sum_{j=k+1}^{\infty} (j-k)\ (p(N_1=j) - p(N_2=j)) \leq 0$$

which is still equivalent with

$$\sum_{j=0}^{k} (k-j)(p(N_1=j) - p(N_2=j)) \leq 0$$

or

$$\sum_{j=0}^{k} \sum_{i=j+1}^{k} (p(N_1=j)-p(N_2=j)) \leq 0 \tag{2.21}$$

Hence :

$$\sum_{i=1}^{k} \sum_{j=0}^{i=1} (p(N_1=j)-p(N_2=j) \leq 0 \qquad \text{q.e.d.} \tag{2.22}$$

<u>Application</u>

1. Let $0 \leq p_i \leq 1$, $i = 1, \ldots, n$, $p_i + q_i = 1$.

   Let $T = [p_{ij}]$ be a double stochastic matrix (i.e. $p_{ij} \geq 0$, $\sum_{i=1}^{n} p_{ij} = \sum_{j=1}^{n} p_{ij} = 1$) and let further

$$p_i' = \sum_{j=1}^{n} p_{ij}\, p_j, \quad i = 1, \ldots, n \tag{2.23}$$

   Let b_k be defined by

$$\sum_{0}^{n} b_k x^k = \prod_{1}^{n} (q_i+p_i x) \tag{2.24}$$

   and

$$\sum_{0}^{n} b_k' x^k = \prod_{1}^{n} (q_i'+p_i' x) \tag{2.25}$$

   Of course the product of two double stochastic matrices is again a double stochastic matrix.
   We will show next that an arbitrary double stochastic matrix can be written as the product of a finite number of special matrices of the type

$$T_{k\ell}(t) = \begin{array}{c} \\ \\ k \\ \\ \ell \\ \\ \\ \\ \end{array} \begin{bmatrix} 1 & 0 & & & & & 0 & & 0 \\ 0 & 1 & 1 & & & & & & 0 \\ & & & t & 1 & & 1-t & & \\ & & & & & 1 & & & \\ & & 1-t & & & 1 & t & & \\ & & & & & & & 1 & 0 \\ & & & & & & & & 1 \\ 0 & 0 & & & & & & & \end{bmatrix} \tag{2.26}$$

   and of course, one easily verifies

$$T_{k\ell}(a) . T_{k\ell}(t) = [\delta_{ij}],$$

   in case $t = \frac{a}{2a-1}$ or $a = \frac{t}{2t-1}$

   Consequently we find :

$$p' = \begin{bmatrix} p'_1 \\ p'_n \end{bmatrix} = T \begin{bmatrix} p_1 \\ p_n \end{bmatrix} \quad (= [\,p\,])$$

and of course

$$\begin{bmatrix} p'_1 \\ p'_n \end{bmatrix} = TT_{k\ell}(a) \begin{bmatrix} p''_1 \\ p''_n \end{bmatrix}$$

Let $d([\,p'-p\,])$ denote the discrepancy of $[\,p'\,]$ and $[\,p\,]$, i.e. the number of the differences p'_i-p_i which are not zero. If $d([\,p'-p\,]) = r > 0$, we will show that t can be choosen such that $d([\,p'-p''\,]) \leqslant r-1$. In case $d([\,p'-p\,]) = 0$ one has of course $p'_i = p_i$. Repeated application of multiplication with such special matrices finaly will lead to $T = \prod_{k,\ell} T_{k\ell}\,(a_{k\ell})$.

Without loss of generality we may assume

$$p'_1 \geqslant p'_2 \geqslant \ldots \geqslant p'_n$$

$$p_1 \geqslant p_2 \geqslant \ldots \geqslant p_n$$

Of course we find

$$\sum_{j=1}^{m} p'_j = \sum_{j=1}^{m} \sum_{i=1}^{n} p_{ji}p_i$$

$$= \sum_{i=1}^{n} \sum_{j=1}^{m} p_{ji}p_i$$

But $\sum_{j=1}^{m} p_{ji} = k_i \leqslant 1$ and $\sum_{i=1}^{n} k_i = m$,

hence

$$p'_1+p'_2 + \ldots + p'_m \leqslant k_1p_1 + \ldots + k_{m-1}p_{m-1}+(m-k_1-\ldots-k_{m-1})p_m$$

$$\leqslant (p_1-p_m) + \ldots + (p_{m-1}-p_m) + mp_m$$

$$= p_1 + p_2 + \ldots + p_m$$

This results in :

$$\sum_{j=1}^{m} (p'_j-p_j) \leqslant 0$$

Such that in case $p'_1 = p_1, \ldots, p'_{k-1} = p_{k-1}$, and $p'_k \neq p_k$ we necessarily have $p'_k - p_k < 0$.

Because $\sum_{i=1}^{n} (p_i'-p_i) = 0$ a negative difference

$p_k'-p_k < 0$ necessary implies a positive difference

$p_\ell'-p_\ell > 0$.

We obtain

$$\begin{bmatrix} p_1' \\ \vdots \\ p_k' \\ \vdots \\ p_\ell' \\ \vdots \\ p_n' \end{bmatrix} = TT_k\ (a) \begin{bmatrix} p_1 \\ \vdots \\ tp_k+(1-t)p_\ell \\ \vdots \\ (1-t)p_k+tp_\ell \\ \vdots \\ p_n \end{bmatrix}$$

But also the following set of inequalities can be deduced :

$$p_\ell < p_\ell' < p_k' < p_k$$

Hence we can choose $t = t_0$ such that

$$tp_k + (1-t_0)p_\ell = p_k'$$

or $\quad (1-t_0)p_k + t_0 p_\ell = p_\ell' \quad$ (or eventually both)

Hence by the transformation $T_{k\ell}(t)$ we get a set of numers

$$p_i'' = p_i \qquad i \neq k,\ell$$
$$p_k'' = tp_k + (1-t)p_\ell$$
$$p_\ell'' = (1-t)p_k + t_{p\ell}$$

such that $d([\,p''-p\,]) \leqslant r-1$

Hence a finite number of transformations $T_{k\ell}(t)$ give

$$\begin{bmatrix} p_1' \\ \\ p_n' \end{bmatrix} = T \prod_{k,\ell} T_{k\ell}(a) \begin{bmatrix} p_i' \\ \\ p_n' \end{bmatrix}$$

Hence $T \prod_{k,\ell} T_{k\ell}(a) = [\,\delta_{ij}\,]$,

such that

$$T = \prod_{k,\ell} T_{k,\ell}(t_{k\ell})$$

Let us consider the original problem again. We want to prove that

$$\sum_{j=0}^{k} \sum_{i=0}^{j} (b_i'-b_i) \geq 0$$

and consequently that

$$\sum_{k=0}^{n} \varphi_o(k)(b_k'-b_k) \geq 0$$

for every convex sequence $\varphi_o(k)$

The preceeding results unable us to consider a transformation

$$p_1' = tp_1 + (1-t)p_2$$

$$p_2' = (1-t)p_1 + tp_2 \qquad 0 < t < 1$$

$$p_i' = p_i$$

where t is a parameter, we get

$$b_k'-b_k = (p_1'p_2' - p_1p_2)\ (R_{k-2} - 2R_{k-1} + R_k)$$

with $R_{-1} = R_{-2} = 0$
and

$$\sum_{0}^{\infty} R_k x^k = \prod_{i=3}^{n} (q_i+p_ix)$$

of course $R_{n-1} = R_n = \ldots = 0$. It follows that :

$$\sum_{j=0}^{k} \sum_{i=0}^{j} (b_i'-b_i) = (p_1'p_2' - p_1p_2)R_k$$

which is nonnegative since R_k clearly satisfies $R_k \geq 0$ while :

$$p_1'p_2'-p_1p_2 = t(1-t)(p_1^2+p_2^2 - 2p_1p_2) \geq 0$$

Consequently if :

$$p(N_1 = k) = b_k$$

where b_k is defined by (2.23) and

$$p(N_2 = k) = b_k'$$

where b_k' is defined by (2.24) and where p_i' is defined by (2.22), then :

$$S_1 \prec S_2$$

where $S_1 = \sum_{j=1}^{N_1} X_j$, $S_2 = \sum_{j=1}^{N_2} X_j$.

2. Let $0 \leqslant p_i \leqslant 1$, $i=1, \ldots, n$; $p_i+q_i = 1$ and let b_k be defined by

$$\sum_{k=0}^{n} b_k x^k = \prod_{i=1}^{n} (a_i+p_i x)$$

Let further

$$d_k = \binom{n}{k} p^k (1-p)^{n-k} \text{, i.e.}$$

$$\sum_{k=0}^{n} d_k x^k = (q+px)^n$$

where $p = \frac{1}{n}(p_1 + \ldots + p_n)$ and $p+q = 1$, then, as a particular case of the first application we find that

$$\sum_{k=0}^{n} b_k \varphi(k) \leqslant \sum_{k=0}^{n} d_k \varphi(k)$$

for all convex $\varphi(k)$. Consequently

$$S_1 \prec S_2$$

where $S_1 = X_1 + \ldots + X_{N_1}$

$S_2 = X_1 + \ldots + X_{N_2}$

with $p(N_1=k) = b_k$, and $p(N_2=k) = \binom{n}{k} p^k (1-p)^{n-k}$ is a binomial variable.

3. As a limiting case of the first application of this section we consider the following example. Consider an infinite sequence of independent binomial trials with probabilities of success $p_1, p_2, \ldots$; $0 < p_i < 1$. Suppose further

$$\sum_{i=1}^{\infty} p_i = \lambda < \infty$$

The probability a_k of k successes can be calculated from the generating function

$$\sum_{k=0}^{\infty} a_k x^k = \prod_{i=1}^{\infty} (q_i+p_i x), \qquad q_i+p_i = 1$$

We then have

$$\sum_{k=0}^{\infty} a_k \varphi(k) \leqslant \sum_{k=0}^{\infty} \frac{\lambda^k}{k!} e^{-\lambda} \varphi(k)$$

for every convex φ. Hence

$$S_1 \prec S_2$$

where $p(N_1=k) = a_k$ and $S_1 = \sum_{j=1}^{N_1} X_j$

and S_2 is a compound Poisson process with parameter λ and claim size variable x. Of course

$$\sum_{k=1}^{\infty} p_k = \sum_{k=0}^{\infty} k a_k$$

Hence $E(N_1) = \lambda$.

Until now we considered compound distribution functions of the type

$$F_S(x) = \sum_{n=0}^{\infty} p(N=n)\, F_X^{*n}(x)$$

We now consider as a special cas the weighted compound Poisson distribution

$$F_S(x) = \sum_{n=0}^{\infty} \int_0^{\infty} e^{-w} \frac{w^n}{n!}\, dF_W(w)\, F_X^{*n}(x)$$

We will prove.

Theorem 10

$$W \prec W' \Longrightarrow S \prec S'$$

Proof

Let

$$p_n = \int_0^{\infty} e^{-w} \frac{w^n}{n!} \quad dF_W(w) \qquad (2.26)$$

and

$$p'_n = \int_0^{\infty} e^{-w} \frac{w^n}{n!} \quad dF_W(w) \qquad (2.27)$$

denote the probabilities $p(N=n) = p_n$ and $p(N'=n) = p'_n$. It is sufficient to prove that $N \prec N'$. We therefore consider the difference

$$p_n - p'_n = \int_0^{\infty} e^{-w} \frac{w^n}{n!} \; [\, dF_W(w) - dF'_W(w)\,] \qquad (2.28)$$

By means of two successive partial integrations the r.h.s. of (2.28) can be cast into the form

$$p_n - p'_n = \int_0^{\infty} \left(\frac{d^2}{dw^2}\, e^{-w} \frac{w^n}{n!}\right) \varphi(w)\, dw$$

where :

$$\varphi(w) = \int_{w}^{\infty} (1-F_W(y))dy - \int_{w}^{\infty} (1-F_{W'}(y))dy$$

Because

$$\frac{d^2}{dw^2} e^{-w} \frac{w^n}{n!} = e^{-w} \left[\frac{w^n}{n!} - 2 \frac{w^{n-1}}{(n-1)!} + \frac{w^{n-2}}{(n-2)!} \right]$$

one obtain

$$\sum_{k=0}^{n} \sum_{v=0}^{k} \left(\frac{d^2}{dw^2} e^{-w} \frac{w^v}{v!}\right) = \sum_{k=0}^{n} \left[\frac{w^k}{k!} - \frac{w^{k-1}}{(k-1)!} \right] e^{-w}$$

$$= \frac{w^n}{n!} e^{-w}$$

Consequently

$$\sum_{k=0}^{n} \sum_{v=0}^{k} (p_v - p'_v) = \int_{0}^{\infty} e^{-w} \frac{w^n}{n!} \varphi(w)dw < 0$$

Now the l.h.s. of this last inequality can successively be transformed as follows

$$= \sum_{k=0}^{n} \sum_{v=0}^{k} (p_v - p'_v)$$

$$= \sum_{v=0}^{n} (n-v)\ (p_v - p'_v)$$

$$= \sum_{v=0}^{\infty} (n-v)(p_v - p'_v) + \sum_{v=0}^{n} (n-v)(p_v - p'_v)$$

$$= \sum_{v=n+1}^{\infty} (v-n)(p_v - p'_v)$$

Hence :

$$\sum_{v=n+1}^{\infty} (v-n)p_v < \sum_{v=n+1}^{\infty} (v-n)p'_v$$

3. THE NOTION OF DANGEROUS DISTRIBUTIONS

<u>Definition 1</u>

A distribution H is called more dangerous than a distribution G if

i) the first moment μ_G, μ_H exist and $\mu_H \leqslant \mu_H$

ii) there is a constant β such that

$$\begin{aligned} G(x) &\leqslant H(x) \quad \text{for } x < \beta \\ G(x) &\geqslant H(x) \quad \text{for } x \geqslant \beta \end{aligned} \tag{3.1}$$

Theorem 1

If H is more dangerous than G, then $G \prec H$

Proof

From i) we conclude that μ_G exists. In case $\alpha \geqslant \beta$ we have

$$1-G(x) \leqslant 1-H(x) \qquad \forall x \geqslant \alpha$$

Hence

$$\int_\alpha^\infty (1-G(x)dx \leqslant \int_\alpha^\infty (1-H(x)dx \qquad \forall\alpha \geqslant \beta \tag{3.2}$$

and of course, by means of one partial integration

$$\int_\alpha^\infty (x-\alpha)dG(x) \leqslant \int_\alpha^\infty (x-\alpha)dH(x) \quad \forall\alpha \geqslant \beta$$

In case $\alpha < \beta$ one proceeds as follows :

$$\int_a^\infty (1-G(x))dx - \int_\alpha^\infty (1-H(x))dx = \int_\alpha^\infty (H(x)-G(x))dx$$

$$\leqslant \int_{-\infty}^{+\infty} (H(x)-G(x))dx = \mu_G-\mu_M \leqslant 0 \qquad \text{q.e.d.}$$

and

$$H(x) = \begin{cases} 0 & x < a \\ \dfrac{b-\mu_F}{b-a} & a \leqslant x < b \\ 1 & x \geqslant b \end{cases}$$

Then F is more dangerous than G, and H is more dangerous than F. The proof of these results are trivial.

Theorem 2

Let $\{a_\nu : \nu = 0, \ldots, n\}$ be a set of real numbers, such that

$$\sum_{\nu=0}^n \nu a_\nu \leqslant 0, \quad \sum_{\nu=0}^n a_\nu = 0 \tag{3.3}$$

$$\sum_{\nu=0}^k a_\nu \leqslant 0 \quad k = 0, \ldots, \mathcal{K}$$

$$\sum_{\nu=0}^k a_\nu \geqslant 0 \quad k = \mathcal{K}+1, \ldots, n$$

then

$$\sum_{\nu=0}^n (\nu-r)_+ \, a_j \leqslant 0 \qquad r = 0,1, \ldots, n \tag{3.4}$$

Proof

One has :

$$\sum_{\nu=0}^{n} (\nu-r)_+ a_\nu = \sum_{\nu=r}^{n} (\nu-r)a_\nu$$

We have to distinguish two cases, namely $r \geqslant \mathcal{H} + 1$ and $r \leqslant \mathcal{H}$.

i) $r \geqslant \mathcal{H} + 1$.

In this case we have

$$\sum_{\nu=r}^{n} (\nu-r)a_\nu = \sum_{\nu=r+1}^{n} a_\nu + \sum_{\nu=r+2}^{n} a_\nu + \ldots + \sum_{\nu=n}^{n} a_\nu$$

$$- \sum_{\nu=0}^{r} a_\nu - \sum_{\nu=0}^{r+1} - \ldots - \sum_{\nu=0}^{n} a_\nu \leqslant 0$$

ii) $r \leqslant \mathcal{H}$

In this case one obtains successively :

$$\sum_{\nu=r}^{n} (\nu-r)\, a_\nu = \sum_{\nu=r}^{n} \nu a_\nu - r \sum_{\nu=r}^{n} a_\nu$$

$$= \sum_{\nu=0}^{n} \nu a_\nu + \sum_{\nu=0}^{r-1} (r-\nu)a_\nu$$

$$= \sum_{\nu=0}^{n} \nu a_\nu + \sum_{\nu=0}^{r-1} a_\nu + \sum_{\nu=0}^{r-2} a_\nu + \ldots + \sum_{\nu=0}^{\nu=0} a_\nu \leqslant 0 \qquad \text{q.e.d.}$$

Remark

Also in the present case there is no need to restrict ourselves to finite series. This theorem still holds for $n \to \infty$, when convergent series arises.

Definition 2

A claim frequency distribution H is called more dangerous than a claim frequency distribution G if the expected claim numbers μ_G and μ_H exist and $\mu_G \leqslant \mu_H$ and if there is a constant β such that

$$G(x) \leqslant H(x) \qquad x < \beta$$
$$G(x) \geqslant H(x) \qquad x \geqslant \beta$$

Theorem 3

If a claim frequency distribution H is more dangerous than a claim frequency distribution G then $G \prec H$

Proof : Immediately from the preceeding theorem
In order to be able to generalize the notion of dangerous distributions we consider two lemma's. We suppose that G and H are not identical distribution functions.

Lemma 1

$dG(x) - dH(x)$ exhibits at least 1 sign change

Proof : trivial

Lemma 2

If $\int_a^b x(dG(x)-dH(x)) = 0$ then $dG(x) - dH(x)$ exhibits at least two strict sign changes.

Proof : Suppose $dG(x) - dH(x)$ possesses only 1 sign change, then

$$(dG(x) - dH(x))(dG(y)-dH(y)) \leq 0 \qquad x\varepsilon [a,x_0] \quad y\varepsilon [x_0,b]$$

We then consider

$$w(x) = (1-G(x_0) - (1-H(x_0))(x-x_0)$$

Because G and H can not be the degenerate distribution in the point x_0, we have

$$\int_a^b w(x)(dG(x) - dH(x)) > 0 \qquad (3.5)$$

because $w(x)(dG(x) - dH(x)) \geq 0$

This contradicts the hypothesis.

Lemma 3

If G and H are two distributions on some interval $[a,b]$, if $dG(x)-dH(x)$ possesses exactly one sign change on (a,b) and is nonnegative and nonnull on some interval extending to the endpoint b then

$$\int_a^b \varphi(x)dG(x) \geq \int_a^b \varphi(x)dH(x)$$

for every non decreasing function φ.

Proof : trivial

Lemma 4

If G and H are two distributions on some interval $[a,b]$, if in addition $\mu_G = \mu_H$, and if $dG(x) - dH(x)$ possesses exactly 2 sign changes on (a,b) and is nonnegative and nonnull on some interval to the endpoint b, then

$$\int_a^b \varphi(x)dG(x) \geq \int_a^b \varphi(x)dH(x)$$

for every convex function φ.

Proof

Let J_0, J_1, J_2 be the subdivision of (a,b) such that

$$\begin{aligned} dG(x)-dH(x) &\geq 0 \quad && \forall x \in J_0 \\ dG(x)-dH(x) &\leq 0 && \forall x \in J_1 \\ dG(x)-dH(x) &\geq 0 && \forall x \in J_2 \end{aligned}$$

with strict inequality holding at least for some x-values in each of the three indicated intervals J_0, J_1, J_2. Define t_0, t_1, by $t_i = \sup\{t \mid t \in J_i\}$, $i=0,1$ and let

$$\Theta(t) = \begin{matrix} 1 & 1 & 1 \\ t_0 & t_1 & t \\ \varphi(t_0) & \varphi(t_1) & \varphi(t) \end{matrix}$$

Expanding this determinant we get

$$\Theta(t) = \begin{vmatrix} t_0 & t_1 \\ \varphi(t_0) & \varphi(t_1) \end{vmatrix} - t \begin{vmatrix} 1 & 1 \\ \varphi(t_0) & \varphi(t_1) \end{vmatrix} + \varphi(t) \begin{vmatrix} 1 & 1 \\ t_0 & t_1 \end{vmatrix}$$

such that

$$\int_a^b \Theta(t)(dG(t)-dF(t)) = \begin{vmatrix} 1 & 1 \\ t_0 & t_1 \end{vmatrix} \int_a^b \varphi(t)(dG(t)-dF(t))$$

However because $\varphi(t)$ is convex, we have in (a,b)

$$\Theta(t)(dG(t)-dH(t)) \geq 0,$$

Hence

$$\int_a^b \varphi(t)(dG(t)-dH(t)) \geq 0 \qquad \text{q.e.d.}$$

Theorem 3

Let X, Y be two risks on [a,b], describing the claimsize, possessing densities f_X, f_Y
If

$$f_X(x) \leq f_Y(x) \qquad \forall x \in [a,c]$$
$$f_X(x) \geq f_Y(x) \qquad \forall x \in [c,b]$$

with strict inequality holding at least for some x values in both indicated regions, then

$$Y \prec X$$

Proof : Trivial by application of one of the preceeding lemma's.

Theorem 4

Let X,Y be two risks on [a,b], describing the claimsize possessing densities f_X, f_Y, such that E(X) = E(Y). If

$$f_X(x) \geq f_Y(x) \qquad \forall x \in [a,c]$$
$$f_X(x) \leq f_Y(x) \qquad \forall x \in [c,d]$$
$$f_X(x) \geq f_Y(x) \qquad \forall x \in [d,b]$$

with strict inequality holding at least for some x values in the three indicated regions, then

$$Y \prec X$$

Proof : Trivial by application of one of the preceeding lemma's.

Theorem 5

Let N, N' be two claim frequency variables, such that p(N=k) = p_k,p(N'=k)=p'_k. If

$$p_k \leq p'_k \qquad k = 0,1,2, \ldots, r$$
$$p_k \geq p'_k \qquad k = r+1, \ldots, n$$

with strict inequalities holding atleast once in each of the indicated regions then

$$N' \prec N$$

Proof

Because of a preceeding lemma it is sufficient to prove that

$$\varphi_0(n) = E(\max(0, \sum_{i=1}^{n} X_i-\alpha))$$

is an increasing function of n. This is certainly the case for positive risks X_i, i=1, n q.e.d.

Theorem 6

Let N, N' be two claim frequency variables, such that p(N=k) = p_k, p(N'=k) = p'_k. If

$$\begin{array}{ll} p_k \geqslant p'_k & 0 \leqslant k \leqslant r \\ p_k \leqslant p'_k & r+1 \leqslant k \leqslant s \\ p_k \geqslant p'_k & s+1 \leqslant k \leqslant n \end{array}$$

with strict inequalities holding at least once in each of the indicated regions, then

$$N' \prec N$$

Proof : Because of the convexity of $\varphi_0(n)$ the result follows immediately from a preceeding lemma.

Applications

1. Let G(x) $q_0F^{*0}(x) + q_1F^{*1}(x) + q_2F^{*2}(x)$

$$H(x) = \sum_{n=0}^{\infty} p_nF^{*n}(x)$$

where of course

$$\sum_{n=0}^{\infty} p_n = 1$$

$$q_0+q_1+q_2 = 1$$

$$q_1+2q_2 = \sum_{n=0}^{\infty} np_n$$

Choose q_i such that

$$\begin{array}{l} p_0 > q_0 \\ p_1 < q_1 \\ p_2 > q_2 \end{array}$$

then

$$G \prec H$$

To illustrate this application numerically we consider

$$p_n = e^{-\mu} \frac{\mu^n}{n!} \quad \text{with } \mu = 1.1$$

Consequently we get

$$q_0 + q_1 + q_2 = 1$$
$$q_1 + 2q_2 = \mu$$

Then inequalities then become

$$0.3329 > q_0$$
$$0.3662 < 2-1.1-2q_0$$
$$0.2014 > 0.1+q_0$$

Hence it is sufficient to choose $q_0 < 0.1014$ with

$$q_1 = - +2-2q_0$$
$$q_2 = q_0+\mu-1$$

to obtain a lower bound.

4. DOMINANCE CRITERIA

4.1. Stochastic dominance

Stochastic dominance ranking of risks uses repeated integrals of the distribution function F_X of a risk X.

$$F_X(x) = \int_0^x dF_X(y)$$

$$_1F_X(x) = \int_0^x F_X(y)dy$$

and in general

$$_kF_X(x) = \int_0^x {}_{k-1}F_X(y)dy \tag{4.1}$$

In what follows we consider risks of a bounded range, say $[a,b]$.

Definition 1

The condition for first degree stochastic dominance of X over $Y(Y \overset{1}{<} X)$ is that

$$F_X(x) \leqslant F_Y(x) \qquad \forall x \varepsilon [a,b]$$

The condition for second degree stochastc dominance of X over $Y(Y \overset{2}{<} X)$ is

$$ {}_1F_X(x) \leq {}_1F_Y(x) \qquad \forall x \varepsilon [a,b] $$

The conditions for third degree stochastic dominance of X over Y($Y \overset{3}{<} X$) are :

$$ E(X) \leq E(Y) $$
$$ {}_2F_X(x) \leq {}_2F_Y(x) \qquad \forall x \varepsilon [a,b] $$

The conditions for n-th order stochastic dominance ($Y \overset{n}{<} X$) are:

$$ {}_kF_X(b) \leq {}_kF_Y(b) \qquad \text{for } k = 1,2,\ldots, n-2 $$

and

$$ {}_{n-1}F_X(x) \leq {}_{n-1}F_Y(x) \qquad \forall x \varepsilon [a,b] $$

W.H. Jean shows that n-th order stochastic dominance implies higher order ($> n$) stochastic dominance (see reference [21]). Of course, in case $F_X(0) = 0$, ${}_nF_X(x)$ can be cast into the form :

$$ {}_nF_X(x) = \frac{1}{n!} \int_0^x (x-t)^n \, dF_X(t) $$

Also in ref. [21] it is shown that stochastic dominance implies geometric mean dominance. Indeed

$$ Y \overset{n}{<} X \Rightarrow \exp\{E(\ell nY)\} \leq \exp\{E(\ell nX)\} $$

which follows immediately from the more general result

Theorem 1

$$ Y \overset{n}{<} X \Rightarrow E(v(Y)) \leq E(v(X)) $$

for every function v on [a,b] such that :

$v', v'', \ldots, v^{(n-1)}$ exist and are continuous,
$v^{(n)}$ exists except for at most two points, and is bounded,
$(-1)^{k+1} v^{(k)}(x) \geq 0 \quad \forall k = 1, \ldots, n$ and $\forall x \varepsilon [a,b]$

Proof

We first prove the implication from the right to the left. If $a \leq x_1 < x_2 \leq b$ and if $v(t) = -\frac{1}{n!} [(x_2-t)_+^n - (x_1-t)_+^n]$ then v fulfils the assumptions and :

$$ E(v(X)) = - \int_{x_1}^{x_2} {}_{n-1}F_X(t)dt $$

Now it readily follows that $Y \overset{n}{<} X$.

In order to prove the converse half of the theorem we consider

$$E(v(Y)) = \int_a^b v(y)dF_Y(y)$$

By means of successive partial integrations the following expression is obtained for the l.h.s. :

$$E(v(Y)) = v(b) - v'(b)\,{}_1F_Y(b) + v''(b)\,{}_2F_Y(b) + \dots$$
$$\dots + (-1)^{n-1}v^{(n-1)}(b)\,{}_{n-1}F_Y(b) + (-1)^n \int_a^b v^{(n)}(x)\,{}_{n-1}F_Y(x)dx$$
$$\leqslant v(b) - v'(b)\,{}_1F_X(b) + v''(b)\,{}_2F_X(b) + \dots$$
$$\dots + (n-1)^{n-1}v^{(n-1)}(b)\,{}_{n-1}F_X(b) + (-1)^n \int_a^b v^{(n)}(x)\,{}_{n-1}F_X(x)dx$$

Hence : $E(v(Y)) \leqslant E(v(X))$.

Remark

The special choice $v(x) = \ell nx$ gives as a particular case the result of W.H. Jean (see references).

4.2. Stop-loss dominance

Stop-loss dominance ranking is an extension of stop-loss ordering in analogy to the stochastic dominance discussed in the preceeding section.

In this section we consider without loss of generality bounded risks X with domain [0.b]. We define :

$$\overline{F}_X(x) = 1 - F_X(x)$$
$${}_1\overline{F}_X(x) = \int_x^\infty \overline{F}_X(y)dy$$

and in general

$${}_k\overline{F}_X(x) = \int_x^\infty {}_{k-1}\overline{F}_X(y)dy \tag{4.2}$$

Definition 2

The condition for first degree stop-loss dominance of Y over $X(X \overset{1}{\ll} Y)$ is that :

$$\overline{F}_X(x) \leqslant \overline{F}_Y(x)$$

The condition for second degree stop-loss dominance of Y over $X(X \overset{2}{\ll} Y)$ is that :

$${}_1\overline{F}_X(x) \leqslant {}_1\overline{F}_Y(x)$$

The conditions for third degree stop-loss dominance of Y over $X(X \overset{3}{\ll} Y)$ are :

$$E(X) \leq E(Y)$$

$${}_2\bar{F}_X(x) \leq {}_2\bar{F}_Y(x)$$

The conditions for n-th degree stop-loss dominance of Y over $X(X \overset{n}{\ll} Y)$ are :

$${}_k\bar{F}_X(0) \leq {}_k\bar{F}_Y(0) \qquad \text{for } k = 1,2, \ldots, n-2$$

$${}_{n-1}\bar{F}_X(x) \leq {}_{n-1}\bar{F}_Y(x)$$

Remark 1

First degree stop-loss dominance coïncides with first degree stochastic dominance.

Remark 2.

$${}_1\bar{F}_X(0) \leq {}_1\bar{F}_Y(0) \qquad E(X) \leq E(Y)$$

Remark 3

$${}_k\bar{F}_X(0) = \frac{1}{k!} E(X^k)$$

Theorem 2

The n-th degree stop-loss dominance implies higher degree (> n) stop-loss dominance.

Proof : trivial

Theorem 3

$$X \overset{n}{\ll} Y \qquad E(v(X)) \leq E(v(Y))$$

$$\forall v \text{ with } v'(x) \geq 0,\ v''(x) \geq 0,\ \ldots,\ v^{(n)}(x) \geq 0 \qquad \forall x \in [0,b]$$

Proof

The proof proceeds along the same lines as the one of theorem 1 of this section.

Remark 1

In fact the conditions on the function v can be weakened to

$$v'(0) \geq 0,\ v''(0) \geq 0,\ \ldots,\ v^{(n-1)}(0) \geq 0,\ v^{(n)}(0) \geq 0$$

Remark 2

We consider the following special choice $v(x) = e^{\alpha x}$ with $\alpha > 0$; in case there exists any degree of stop-loss dominance of Y over X one has

$$\frac{1}{\alpha} \ln E(e^{\alpha X}) \leq \frac{1}{\alpha} \ln E(e^{\alpha Y}),$$

in other words, the exponential premium for X is smaller than the one for Y.

Remark 3

By means of successive partial integrations the following representation is obtained

$$_k\overline{F}_X(x) = \int_x^\infty \frac{(t-x)^k}{k!} \, dF_X(t)$$

4.3. Some mathematical results

In this section we formulate and prove some results which will enable us to deduce some general results on ordering of risks which can be represented as compound processes, eventually weighted by some structure function.

Lemma 1

Let $\varphi(\nu)$ with $\nu = 0, \ldots, n$ denote a finite sequence such that

$$\Delta^r \varphi(\nu) \geq 0 \qquad \nu = 0, 1, n-r$$

then

$$\sum_{\nu=0}^{n} \varphi(\nu)\, a_\nu \geq 0$$

if and only if :

$$\sum_{\nu=0}^{n} \binom{\nu-k}{r-1} a_\nu \geq 0 \qquad k = 0, 1, \ldots, n-r+1$$

$$\sum_{\nu=0}^{n} a_\nu = 0$$

$$\sum_{\nu=0}^{n} \nu^{r-1} a_\nu = 0 \tag{4.3}$$

where we put $\binom{p}{m} = 0$ in case $p < m$.

Proof

The conditions (4.3) are certainly necessary because :

$$\varphi(\nu) = \binom{\nu-k}{r-1}$$

as well as $\varphi(\nu) = \pm 1, \pm\nu, \ldots, \pm\nu^{r-1}$ are such that $\Delta^r \varphi(\nu) \geqslant 0$.

Indeed :

$$\Delta\varphi(\nu) = \binom{\nu-k}{r-2}$$

and hence

$$\Delta^{r-1} \varphi(\nu) = \binom{\nu-k}{0} = 1.$$

The proof of the converse half of the theorem proceeds as follows Of course (3.10) can be cast into the form :

$$\varphi(t) = \varphi(0) + \Delta\varphi(0)t + \sum_{j=1}^{n-1} [\Delta^2 \varphi(j-1)] \binom{t-j}{1}$$

for $t = 0,1,\ldots, n$.

Consequently, application of this result to $\Delta\varphi(t)$ provides us with :

$$\Delta\varphi(t) = \Delta\varphi(0) + \Delta^2 \varphi(0)t + \sum_{j=1}^{n-2} [\Delta^3 \varphi(j-1)] \binom{t-j}{1}$$

for $t = 0,1,\ldots, n-1,n$.

Hence

$$\varphi(t) = \varphi(0) + \Delta\varphi(0)t + \Delta^2\varphi(0) \frac{t(t-1)}{2!}$$

$$+ \sum_{j=1}^{n-2} [\Delta^3 \varphi(j-1)] \sum_{t'=0}^{t-1} \binom{t'-j}{1} \quad t=0,1,\ldots, n$$

Use can be made of the well known formula :

$$\sum_{k=0}^{m} \binom{p+k}{p} = \binom{p+m+1}{p+1}$$

to give :

$$\varphi(t) = \sum_{j=1}^{n-2} [\Delta^3 \varphi(j-1)] \binom{t-j}{2} + \varphi(0) + \Delta\varphi(0)t + \Delta^2\varphi(0)\frac{t(t-1)}{2!}$$

Successive application of the same technique gives the following general result

$$\varphi(t) = \varphi(0) + \Delta\varphi(0)t + \Delta^2\varphi(0)\frac{t(t-1)}{2!} + \ldots + \Delta^{r-1}\varphi(0)\frac{t(t-1)\ldots(t-r+2)}{(r-1)!}$$
$$+ \sum_{j=1}^{n-r+1} [\Delta^r \varphi(j-1)] \binom{t-j}{r-1}$$

The result of the converse half of the theorem then follows immediately from the calculation of $\Sigma\, \varphi(\nu)a_\nu$ by means of the r.h.s. of this equation.

Lemma 2.

Let

$$\psi_{X_1\ldots X_n}(k) = \int_\alpha^\infty dx_1 \int_{x_1}^\infty dx_2 \ldots \int_{x_{k-2}}^\infty dx_{k-1}(1-F_{X_1} * F_{X_2} * \ldots * F_{X_n}(x_{k-1}))$$

where $X_1, X_2, \ldots X_n, X_{n+1}$, are independent, then :

$$\Delta\psi_{X_1\ldots X_n}(k) = \psi_{X_1\ldots X_{n+1}}(k) - \psi_{X_1\ldots X_n}(k)$$

$$= E(X)\psi_{Y_1 X_1 \ldots X_{n+1}}(k-1)$$

where Y_1 is independent of $X_1, \ldots, X_{n+1}$.

Proof

We get

$$\Delta\psi_{X_1\ldots X_n}(k) = \int_\alpha^\infty dx_1 \ldots \int_{x_{k-2}}^\infty dx_{k-1}(F_{X_1} * \ldots * F_{X_n}(x_{k-1})$$

$$- F_{X_1} * \ldots * F_{X_n} * F_{X_{n-1}}(x_{k-1}))$$

which evidently can be written as :

$$\int_0^\infty dF_{X_1} * F_{X_2} * \ldots * F_{X_n}(y)\{\int_\alpha^\infty dx_1 \ldots \int_{x_{k-2}}^\infty dx_{k-1}\, [H(x_{k-1}-y) - F_{X_{n+1}}(x_{k-1}-y)\,]\}$$

But

$$\int_{k_{k-2}}^\infty dx_{k-1}\, [H(x_{k-1}-y) - F_{X_{n+1}}(x_{k-1}-y)\,] = E(X)\, [\,1-F_Y(x_{k-2}-y)\,],$$

and hence :

$$\Delta\psi_{X_1\ldots X_n}(k) = E(X)\psi_{Y_1 X_1 \ldots X_n}(k-1) \qquad \text{q.e.d.}$$

Lemma 3.

Let $\psi_{X_1 \ldots X_n}(k)$ be defined as in lemma 2, then

$$\Delta^k \psi_{X_1 \ldots X_n}(k) \geq 0.$$

Proof

Repeated application of lemma 2 gives :

$$\Delta^{k-1}\psi_{X_1 \ldots X_n}(k) = E(X)^{k-1} \left[1-F_{Y_1} * \ldots * F_{Y_{k-1}} * F_{X_1} * \ldots * F_{X_n}(\alpha) \right]$$

and of course :

$$\Delta^k \psi_{X_1 \ldots X_n}(k) = E(X)^{k-1}\left[F_{Y_1} * \ldots * F_{X_n}(\alpha) - F_{Y_1} * \ldots * F_{X_n} * F_{X_{n+1}}(\alpha)\right] \geq 0$$

q.e.d.

4.4. Preservation of ordering under mixing and convolution

The following four theorems are formulated to indicate that stochastic dominance as well as stop-loss dominance is preserved under mixing and under convolution or addition.

Theorem 4

Stochastic dominance is preserved under mixing.

Proof

Trivial

Theorem 5

Stochastic dominance is preserved under addition.

Proof

Let $X_i \overset{n}{<} Y_i$ for i = 1,2 and let X_1 and X_2 as well as Y_1 and Y_2 be independent. Let further $X = X_1 + X_2$ and $Y = Y_1 + Y_2$, then we have to prove that the conditions of definition 1 hold.

$$E((2b-X)^k) \leq E((2b-Y)^k) \qquad k = 1,2,\ldots, n-2$$

These inequalities are true since

$$E((b-X_1+b-X_2)^k) = \sum_{j=0}^{k} \binom{k}{j} E((b-X_1)^j) E((b-X_2)^{k-j})$$
$$\leqslant \sum_{j=0}^{k} \binom{k}{j} E((b-Y_1)^j) E((b-Y_2)^{k-j})$$
$$= E((b-Y_1+b-Y_2)^k)$$

On the other hand we still have to prove that

$$_{n-1}F_X(x) \leqslant {}_{n-1}F_Y(x)$$

We have :

$$_{n-1}F_X(x) = \frac{1}{(n-1)!} \int_0^x (x-t)^{n-1}\, dt \int_0^t F_{X_2}(t-y)\, dF_{X_1}(y)$$

Fubinization gives the following result :

$$_{n-1}F_X(x) = \int_0^x dF_{X_1}(y) \frac{1}{(n-1)!} \int_0^{x-y} (x-y-\nu)^{n-1}\, dF_{X_2}(\nu)$$

Hence :

$$_{n-1}F_X(x) \leqslant \int_0^x dF_{X_1}(y) \frac{1}{(n-1)!} \int_0^{x-y} (x-y-\nu)^{n-1}\, dF_{Y_2}(\nu)$$

But, the r.h.s. of this inequality denotes $_{n-1}F_{X_1+Y_2}(x)$ which of course is equal to $_{n-1}F_{Y_2+X_1}(x)$, so that :

$$_{n-1}F_X(x) \leqslant \int_0^x dF_{Y_2}(y) \frac{1}{(n-1)!} \int_0^{x-y} (x-y-\nu)^{n-1}\, dF_{X_1}(\nu)$$

Hence :

$$_{n-1}F_X(x) \leqslant \int_0^x dF_{Y_2}(y) \frac{1}{(n-1)!} \int_0^{x-y} (x-y-\nu)^{n-1}\, dF_{Y_1}(\nu)$$

and finally

$$_{n-1}F_X(x) \leqslant {}_{n-1}F_Y(x)$$

Theorem 6

Stop loss dominance is preserved under mixing

Proof

Trivial

Theorem 7

Stop-loss dominance is preserved under addition

Proof

The proof proceeds along the same lines as the proof of the preservation of stochastic dominance under addition.

4.5.4.5. The notion of dominance for claim frequencies

In this section we consider a compound risk or random variable

$$S = X_1 + X_2 + \dots + X_N$$

where the X_i are i.i.d. and independent of N. In what follows we will compare S and $S' = X_1 + \dots + X_{N'}$ in the framework of stop-loss dominance.
Let again :

$$p(N=n) = p_n,$$
$$p(N'=n) = p'_n$$

Definition 3

The condition for first degree stop-loss dominance of N' over $N(N \overset{1}{\ll} N')$ is that :

$$\overline{F_N}(x) \leqslant \overline{F_{N'}}(x)$$

The condition for second degree stop-loss dominance of N' over $N(N \overset{2}{\ll} N')$ is that :

$$E((N-j)+) \leqslant E((N'-j)+) \qquad j = 0,1,\dots$$
$$E(N) = E(N')$$

The conditions for n^{th} degree stop-loss dominance of N' over $N(N^n \ll N')$ is that

$$E\binom{n-j}{n-1} \leqslant E\binom{N'-j}{n-1} \qquad j = 0,1,\dots$$
$$E(N^k) = E(N'^k) \qquad k = 1,\dots,n-1$$

We are now in the position to extend a result on stop-loss ordering significantly.

Theorem 8

Let $N^n \ll N'$ then $S^n \ll S'$.

Proof

The conditions of definition 3 hold. In order to able to prove the theorem we have to show that the following set of inequalities hold

$$E(S^k) \leqslant E(S'^k) \qquad \text{for } k = 1,2,n-2$$

$$E((S-\alpha)_+^{n-1}) \leqslant E((S'-\alpha)_+^{n-1}) \qquad \alpha \varepsilon \mathcal{R}_+$$

The condition for nthe degree stop-loss dominance of N' over $N(N^n < N')$ is that :

$$\sum_j (p_j - p'_j) \int_0^\infty x^k dF^{*j}(x) \leqslant 0$$

$$\sum_j (p_j - p'_j) \int_0^\infty \frac{(x-\alpha)^{n-1}}{(n-1)!} dF^{*j}(x) \leqslant 0$$

The set of inequalities follows immediately from the definition of stop-loss dominance for claim frequencies and because of lemma 1 and lemma 3 of this section.

Remark

Because in definition 3 we requite equality of the moments of a certain order, stop-loss dominance coincide with stochastic dominance for claim frequencies.

References

[1] Beard R.E., Pentikainen T., Personen E.: Risk Theory (eds) Methuen's monographs, London, (1969).

[2] Beekman J. : Two Stochastic Processes (eds). Halsted Press, New York (1974).

[3] Bühlmann H. : Mathematical Methods in Risk Theory. (eds) Springer, New York.

[4] Gerber H.U. : An Introduction to Mathematical Risk Theory (eds). S.S. Huebner Foundation, University of Pennsylvania (1979).

[5] Jewell W.S. Models in Insurance : Paradigms, Puzzles, Communications, and Revolutions. Transactions of the 22st International Congress of Actuaries (1980) pp.S87-S141.

[6] Seal H.L. : Stochastic Theory of a Risk Business. Wiley New York, (1969).

[7] Goovaerts M.J., De Vylder F., Haezendonck J., Insurance Premiums, North-Holland Publishing Company (1983).

[8] Bühlmann H., B. Gagliardi, H. Gerberand E. Straub, : "Some inequalities for stop-loss premiums" Astin Bulletin, Vol. IX, p. 75-83 (1977).

[9] R. De Groot, L. D'Hooge, M.J. Goovaerts : "On an extension of some stop-loss inequalities based on convex analysis"

Transaction of the 21th International Congres of Actuaries Zurich-Lausanne (1980), 169-177.

[10] B. Gagliardi and B. Straub, "Eine obere Grenze für Stop-Loss Prämien, M.V.S.V., 74, 215-221.

[11] H. Gerber and M.J. Goovaerts, "On the representation of additive principles of premium calculation" S.A.J.(1981),221-22

[12] H. Gerber, "An introduction to Mathematical Risk Theory" Heubner Foundation Monograph (1981)

[13] M.J. Goovaerts, "Some further results on ordering of risks Blätter, Vol. XV, 1, (1981), 1-6.

[14] M.J. Goovaerts, "On a partial ordering among risks", I.A.W. Report 7907.

[15] M.J. Goovaerts, "On ordering and danger of claim frequency distributions" Astin Bulletin, (1981), 72-76.

[16] M.J. Goovaerts and F. De Vylder, "Premium calculation principles some properties" Het Verzekeringsarchief,(1980),5-1

[17] G. Hanoch and H. Levy, "The Efficiency Analysis of Choices Involving Risk" Review of Economic Studies, Vol. 36, No. 2 (1969).

[18] Hardy, Littlewood and Polya, "Inequalities" (eds) Cambridge University Press (1959).

[19] J. Hoem, "Sammenhenger mellom riskoerog nullnyttepremier"

[20] W.H. Jean, "The Geometric Mean and Stochastic Dominance", The Journal of Finance, Vol. XXXV, No. 1, (1980), 151-158

[21] S. Karlin and W. Studden, "Tchebycheff systems : with application in analysis and statistics" (1966) Interscience publishers, John Wiley & Sons.

[22] G.C. Taylor, "Stop-loss premiums and danger of the claim frequency distribution" B.A.R.A.B. (1981), 71-74

[23] H.G. Verbeek, "A stop-loss inequality for compound Poisson processes with a unimodal claimsize distribution" Astin Bulletin 9 (1-2) (1977) pp. 247-256.

LIMIT THEOREMS FOR RISK PROCESSES

J. Haezendonck, K. Jansen, F. Delbaen
Universiteit Antwerpen and
Vrije Universiteit Brussel

0. Abstract :

We give a short review of limit theorems for the probability distribution of a general risk process.

1. Introduction :

Let us consider a risk process :

$$Z_t = \sum_{n=1}^{N_t} X_n \qquad (t \in \mathbb{R}_+) \qquad (1.1)$$

where $\{N_t : t \in \mathbb{R}_t\}$ is a counting process (i.e. an increasing process on $\mathbb{N}_o = \{0,1,2,\ldots\}$, starting at 0 and with jumps of size one) and where $\{X_n : n \in \mathbb{N}\}$ is a sequence of i.i.d. random variables, independent of the counting process.

In this note we shall give a review of limit theorems for $Z_t (t \to \infty)$. More details of these limit theorems can be found in Jansen, Haezendonck, Delbaen (1983).

F. de Vylder et al. (eds.), Premium Calculation in Insurance, 235–239.

As a general rule we suppose that $E[N_t]$ and $V[N_t]$ (= variance of N_t) exist and are finite for each t.Furthermore we suppose that $E[X_n] = \alpha$ exists and is finite.

We shall have to consider two cases : the variance $V[X_n]$ is finite or equal to infinity.

2. The case of claim amounts of finite variance.

In this section we assume that the variance $V[X_n] = \sigma^2$ is finite and positive.

First we introduce the following notations :

$$M_t = \alpha N_t, \; S_t^2 = \sigma^2 N_t \qquad (t \in \mathbb{R}_+) \tag{2.1}$$

$$N_t^\star = \frac{N_t - E[N_t]}{V[N_t]^{1/2}}, \quad Z_t^\star = \frac{Z_t - E[Z_t]}{V[Z_t]^{1/2}} \quad (t \in \mathbb{R}_+) \tag{2.2}$$

We also shall have to consider the following conditions :

(a) $\lim_{t \to \infty} N_t = \infty$ a.s. (or in probability)

(b) $\lim_{t \to \infty} E[N_t] = \infty$.

(c) $\lim_{t \to \infty} E[N_t] / V[N_t]$ exists in $\overline{\mathbb{R}}_+ = \mathbb{R}_+ \cup \{\infty\}$

(d) $\liminf_{t \to \infty} E[N_t]/V[N_t] < \limsup_{t \to \infty} E[N_t] / V[N_t]$

(e) $\lim_{t \to \infty} F_{N_t^\star} = N(0,1)$ ($F_{N_t^\star}$ is the c.d.f. of $N_t^\star$ and $N(0,1)$ represents the standard normal distribution).

(f) $F_{N_t^\star}$ has some limit c.d.f. Q on $\mathbb{R}$.

Here follow the main results.

Proposition 1.

If property (a) is fulfilled, then :

$$\lim_{t \to \infty} F_{(Z_t - M_t) / S_t} = N(0,1) \tag{2.3}$$

Proposition 2.

Suppose that the properties (a), (c) and (f) hold

Define

$$d = \lim_{t \to \infty} \left(1 + \frac{\sigma^2}{\alpha^2} \frac{E[N_t]}{V[N_t]}\right)^{-1/2} \tag{2.4}$$

Then :

$$\lim_{t \to \infty} F_{(Z_t - M_t) / V[Z_t]^{1/2}} = N(0,(1-d^2)^{1/2}) \tag{2.5}$$

Proposition 3.

Suppose that the properties (a), (c) and (f) are satisfied and let U be any random variable with c.d.f. Q, then

$$\lim_{t \to \infty} F_{Z_t^\star} = N(0,(1-d^2)^{1/2}) \star F_{dU} \tag{2.6}$$

where d is defined by (2.4)

Proposition 4.

Suppose that the properties (a), (d) and (f) hold. Then $Z_t^\star$ $(t \to \infty)$ has a limit distribution if and only if $F_{N_t^\star}$ converges to the standard normal distribution. In that case the limit distribution of $Z_t^\star$ is also the standard normal distribution.

For the proofs of these propositions as well as for illustrations, examples and counterexamples of these results one refers to J.H.D. (1983).

3. The case of claim amounts of infinite variance.

In this case we have to make a general assumption. We suppose the existence of a coefficient p with $1 < p < 2$ and of a real number $\tau > 0$ such that $\sum_{i=1}^{n} (X_i-\alpha)/\tau n^{1/p}$ has a limit distribution S if n tends to infinity.

This hypothesis is meaningful as is shown by the following example.

Let X_n have a Pareto distribution with characteristic exponent $p(1 < p < 2)$, i.e.

$$F_{X_n}(x) = 1-(\delta/x)^p \quad \text{if } x \geq \delta \qquad (3.1)$$
$$= 0 \quad \text{if } x < \delta$$

where δ is some positive real number

Then one shows that

$$\lim_{n\to\infty} F_{\sum_{i=1}^{n}(X_i-\alpha)/\delta n^{1/p}} = S$$

where S is determined by the characteristic function

$$\hat{S}(t) = \frac{\Gamma(2-p)}{p-1} \left(\cos\frac{1}{2}\pi p - i\frac{t}{|t|}\sin\frac{1}{2}\pi p\right)|t|^p. \qquad (3.3)$$

Let us write

$$V_p[Z_t] = \tau^p E[N_t] + \alpha^p V[N_t]^{p/2} \qquad (3.4)$$

We replace condition (c) of the preceding section by

(c') $\lim_{t\to\infty} E[N_t]/V[N_t]^{p/2}$ exists in $\overline{\mathbb{R}}_+ = \mathbb{R}_+ \cup \{\infty\}$

Proposition 5.

If condition (a) is fulfilled, then

$$\lim_{t\to\infty} F_{(Z_t-M_t)/\tau N_t^{1/p}} = S \qquad (3.5)$$

Proposition 6.

If the conditions (a), (c') and (f) hold, then

$$\lim_{t\to\infty} F_{(Z_t - M_t)\,/\,V_p[Z_t]^{1/p}} = F_{(1-b^p)^{1/p}X} \tag{3.6}$$

where

$$b = \lim_{t\to\infty} \left(1 + \frac{\tau^p}{\alpha^p}\,\frac{E[N_t]}{V[N_t]^{p/2}}\right)^{-1/p} \tag{3.7}$$

and where X is any random variable with c.d.f.S

Proposition 7.

If the conditions (a), (c') and (f) hold, and if U and X are two random variables such that $F_X = S$ and $F_U = Q$, then

$$\lim_{t\to\infty} F_{(Z_t - E[Z_t])\,/V_p[Z_t]^{1/p}} = F_{(1-b^p)^{1/p}X} \star F_{bU} \tag{3.8}$$

Again we refer to J.H.D. (1983) for proofs, examples and further comment on these results.

References :

Haezendonck, J. (1977). Une généralisation du théorème de la limite centrale de Lindeberg. C.R. Acad. Sci. Paris, Série A 285, 797-800.

Jansen, K. , Haezendonck, J., Delbaen, F. (1983). Limit distributions for risk processes in case of claim amounts of finite expectation. To appear in : Insurance : Mathematics and Economics.

Rychlik, Z. (1976). A central limit theorem for sums of a random number of independent random variables. Colloq. Mathematicum 35, 147-158.

SEMI-MARKOV MODELS IN ECONOMICS AND INSURANCE

Jacques JANSSEN

Brussels

This paper is subdivised into three parts :

(i) the first gives a short survey of the semi-Markov theory in view to construct the models used in the two following parts

(ii) the semi-Markov risk models : in this chapter, we develop the main ideas to build these risk models and we compare them with the classical ones

(iii) other semi-Markov models in insurance : here, we show how the consideration of non-homogeneous semi-Markov processes can be used to modelize some social security problems and how the concept of economics environment can lead to semi-Markov risk models.

CONTENTS

F. de Vylder et al. (eds.), Premium Calculation in Insurance, 241–261.

Chapter I. A survey of semi-Markov theory

§1. INTRODUCTION

The aim of this chapter is to present the semi-Markov modelization as an alternative to some classical approaches in risk theory and other fields to break down the basic assumption of independence of sequences of random variables (in short r.v.) and also to consider models with several types of risks.

§2. THE SEMI-MARKOV PROCESS AS A POINT PROCESS

2.1. The claim arrival process

Let us suppose we observe a random phenomenon such as claim arrivals. To do this, we introduce the following sequences of r.v.

(i) $(X_n, n \geq 1)$ where X_n is the interarrival time between claim numbers $(n-1)$ and n.
(At $t = 0$, the process starts just after the arrival of a claim, called claim number 0)

(ii) $(T_n, n \geq 0)$ where :

$$T_0 = 0,\ T_n = \Sigma_{k=1}^{n} X_k,\ n > 0. \tag{2.1}$$

T_n is the time of the occurence of claim number n.

(iii) $(J_n, n \geq 0)$ where the r.v. J_n represents the type of claim number n.
Here we introduce a new concept with regard to the classical risk models : the claims can be subdivised into m classes $(1 \leq m < \infty)$ noted by the set

$$I = \{1, \ldots, m\} \tag{2.2}$$

of claim types.
For example, we can subdivise between good, mean and bad claims or between claims of different insurance branches.

We will suppose that the successive observed claim types - J_o, J_1, ..., J_n, ... - form an homogeneous Markov chain with

$$P = (p_{ij}) \tag{2.3}$$

as transition probability matrix, that is :

$$\mathbb{P}[J_{n+1}=j \mid J_k, k \leq n, J_n=i] = p_{ij} . \tag{2.4}$$

Remark. Of course for J_o, we can

(i) fix $J_o = j_o$

or

(ii) give an initial distribution $(p_j, j \in I)$:

$$p_j = \mathbb{P}[J_o = j] \tag{2.5}$$

To be able to give a mathematical treatment of the model, we have to put on a suitable measurable space $(\Omega, \mathcal{A})$ a probability measure $\mathbb{P}$ defining completely the probabilistic evolution of the process. The semi-Markov assumption puts this measure by introducing the two-dimensional process

$$((J_n, X_n), n \geq 0) \tag{2.6}$$

as a particular Markov process on $I \times \mathbb{R}^+$ as state space in the sense that

$\forall i,j \in I$, $\forall x \in \mathbb{R}^+$

$$\mathbb{P}[J_{n+1}=j, X_{n+1} \leq x \mid (J_k, X_k), k \leq n, J_n=i] = Q_{ij}(x) \tag{2.7}$$

where :

(i) $\forall i,j \in I$, Q_{ij} is a mass function nul on $(-\infty, 0)$,

(ii) $\forall i \in I$, $\Sigma_j Q_{ij}(+\infty) = 1$. (2.8)

Equivalently, we can say that the two-dimensional process $((J_n, T_n), n \geq 0)$ satisfies :

$$\mathbb{P}[J_{n+1}=j, T_{n+1} \leq t \mid (J_k, T_k), k \leq n, J_n=i, T_n=s] = Q_{ij}(t-s) \tag{2.9}$$

so that the (J-T) process is also a particular Markov process with $I \times \mathbb{R}^+$ as state space with an "additive" property related to the second component.

This leads to the following definitions.

Def.2.1. Any m x m matrix $\mathbb{Q}$ satisfying (i) and (ii) above is called a positive semi-Markov matrix (or kernel).

Def.2.2. The process $((J_n, X_n), n \geq 0)$ is called a positive J-X process (JANSSEN (1983)).

Def.2.3. The process $((J_n, T_n), n \geq 0)$ is called a Markov additive process.

2.2. Some basic properties.

We list some basic properties presented in detail in JANSSEN (1983).

Property 2.1. The process of claim types $(J_n, n \geq 0)$ - called the J-process - is a Markov chain of matrix P given by

$$P = \mathbb{Q}(+\infty) \tag{2.10}$$

with

$$\mathbb{Q}(x) = (Q_{ij}(x)). \tag{2.11}$$

Property 2.2. The r.v. $(X_n, n \geq 1)$ are conditionally independent given the J-process, i.e., for all $n \geq 1$, for all $x_1, \ldots, x_n$ belonging to $\mathbb{R}^+$:

$$\mathbb{P}[X_1 \leq x_1, \ldots, X_n \leq x_n | J_k, k \leq n] = \Pi_{k=1}^{n} F_{J_{k-1}J_k}(x_k) \tag{2.12}$$

where

$$F_{ij}(x) = \mathbb{P}[X_n \leq x \mid J_{n-1} = i, J_n = j] \tag{2.13}$$

(of course if $p_{ij} > 0$; otherwise we can put for F_{ij} the value of a mass function with a unit mass at 1, say U_1).
The d.f. F_{ij} is called the conditional distribution function of X_n given $J_{n-1}=i$, $J_n=j$. Also the d.f. H_i defined by

$$H_i(x) = \Sigma_j \, p_{ij} \, F_{ij}(x) \tag{2.14}$$

is called the unconditional d.f. of X_n given $J_{n-1}=i$.
We introduce the corresponding means supposed to be finite :

$$b_{ij} = \int_{\mathbb{R}} x \, d \, F_{ij}(x) \tag{2.15}$$

(called the conditional mean of X_n given $J_{n-1}=i$, $J_n=j$)

$$\eta_i = \int_{\mathbb{R}} x \, d \, H_i(x) \tag{2.16}$$

(called the unconditional mean of X_n given $J_{n-1}=i$)
Of course, from (2.14), we get :

$$\eta_i = \Sigma_j \, p_{ij} \, b_{ij} \, . \tag{2.17}$$

Property 2.3.
If

$$Q^n_{ij}(t) = \mathbb{P}[J_n=j, T_n \leq t | J_o=i] \tag{2.18}$$

(with $n \geq 0$, $i,j \in I$, $t \geq 0$), we have :

$$Q^o_{ij}(t) = \delta_{ij} \, U_o(t) \tag{2.19}$$

($U_o(t)$ being the Heaviside d.f.)

$$Q^n_{ij}(t) = Q^{(n)}_{ij}(t) \tag{2.20}$$

where $Q_{ij}^{(n)}$ is the general term of the n-fold convoluted of the matrix $\mathbb{Q}$. In matrix notation :

$$\mathbb{Q}^{\underline{n}}(t) = \mathbb{Q}^{(n)}(t) \quad (2.21)$$

it is clear that

$$\mathbb{Q}^{(n)}(+\infty) = P^n \quad (2.22)$$

2.3. The associated counting processes.

Excluding the possibility of explosion, we can graphically represent a (J-X) process as in the fig.1, that is as a jump process. This also gives the possibility to associate counting processes.

Def.2.4. If the r.v. $N_i(t)$, $i \in I$, $t \in \mathbb{R}^+$ represents the total number of passages in state in on (0,t], the vector process

$$\bar{N}(t) = (N_1(t), \ldots, N_m(t)) \quad (2.23)$$

with $\mathbb{N}^m$ as state space, is called the Markov renewal counting process of kernel Q.

Def.2.5. If

$$N(t) = \Sigma_{j=1}^{m} N_j(t) \quad (2.24)$$

the process $(Z_t, t \geq 0)$ with

$$Z_t = J_{N(t)} \quad (2.25)$$

is called the semi-Markov process of kernel $\mathbb{Q}$.
The following interpretations are evident :

1) The r.v. $N_i(t)$ gives the total number of claims of type i in (0,t].
2) The r.v. N(t) gives the total number of claims on (0,t].
3) The r.v. Z_t represents the type of the last claim occured before or at time t.

§3. THE RENEWAL MATRIX

Having the preceding counting processes, we have to consider the mean numbers of the different possible claims. Also we do the following definitions.

Def.3.1. The renewal matrix $\mathcal{H}(t)$ is defined as

$$\mathcal{H}(t) = (H_{ij}(t)) \quad (3.1)$$

where

$$H_{ij}(t) = \mathbb{E}\,[N_j(t) \mid J_0 = i] \quad (3.2)$$

The semi-Markov theory developed by PYKE (1962) shows the follo-

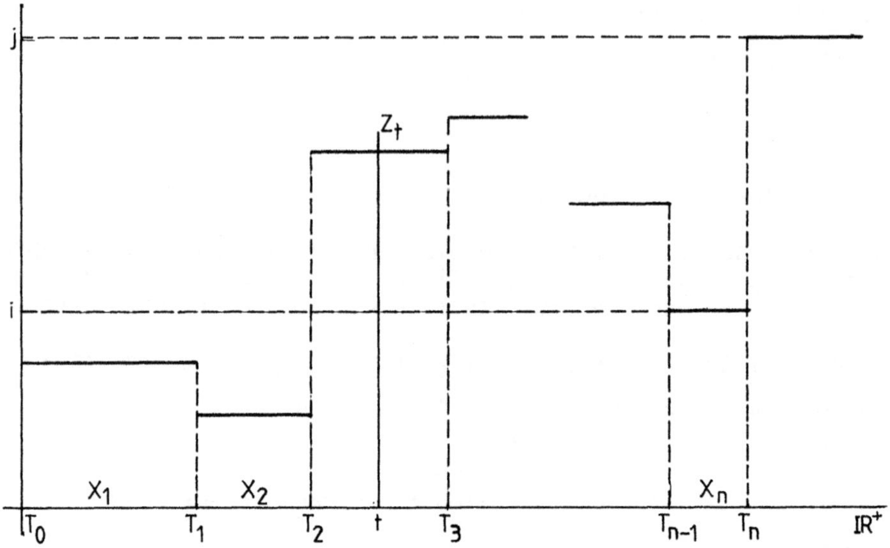

Fig. 1

A trajectory of N_t
(corresponding to fig.1)

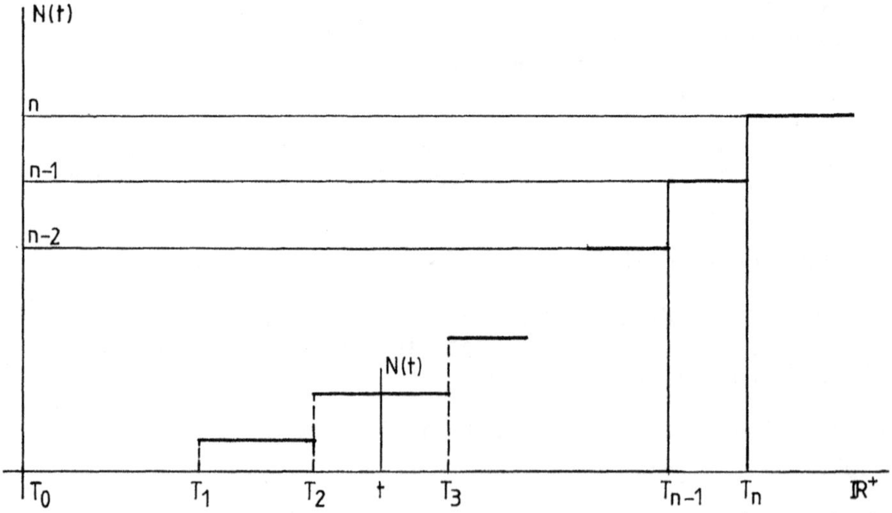

Fig. 2

wing results (see also JANSSEN (1982), (1983)).

Property 3.1. The matrix $\mathcal{H}(t)$ can be expressed as :

$$\mathcal{H}(t) = \Sigma_{n=1}^{\infty} \mathbb{Q}^{(n)}(t) \tag{3.3}$$

Let us suppose that the matrix P is ergodic and let us note by $\pi = \pi_1, \ldots, \pi_m)$ the unique stationary probability distribution related to P. We then have

Property 3.2.

$$\lim_{t} \frac{H_{ij}(t)}{t} = \frac{\pi_j}{\Sigma_k \pi_k \eta_k} . \tag{3.4}$$

It follows that for large t, we have the following approximation:

$$\mathbb{E}[N_j(t) | J_o = i] \simeq \frac{t}{\mu_{ij}} \tag{3.5}$$

$$\text{if } \mu_{jj} = \frac{1}{\pi j} \Sigma_k \pi_k \eta_k . \tag{3.6}$$

Now introduce the d.f. eventually defective G_{ij} of the first return time to the type j given that $J_o = i$. The relations between the matrices $\mathcal{H}$ and $G = (G_{ij})$ are :

$$H_{ij}(t) = G_{ij}(t) + G_{ij} * H_{jj}(t), \tag{3.7}$$

$$H_{jj}(t) = \Sigma_{n=1}^{\infty} G_{jj}^{(n)}(t) \tag{3.8}$$

because the process of the successive returns to the type j given that $J_o = i$ is a general renewal process eventually defective.
If P is ergodic, the mean return time from j to j is given by (3.6).
We thus have the following properties.

Property 3.3.

$$\mathbb{P}[N(t)=n, J_{N(t)}=j | J_o=i] = \int_0^t [1-H_j(t-\xi)] \, dQ_{ij}^{(n)}(\xi), \tag{3.9}$$

$$\mathbb{P}[N(t)=n | J_o=i] = \Sigma_j Q_{ij}^{(n)}(t) - \Sigma_j Q_{ij}^{(n+1)}(t). \tag{3.10}$$

The last result is a direct consequence by summation of (3.9) on j and from relation (2.16).

As a direct consequence of the renewal theory, we get the two last propositions

Property 3.4.

$$\mathbb{P}[N_j(t)=n \mid J_0=i] = \begin{cases} 1 - G_{ij}(t), & n = 0 , \\ G_{ij} * [G_{jj}^{(n-1)}(t) - G_{jj}^{(n)}(t)], & n > 0 . \end{cases} \tag{3.11}$$

Property 3.5.

For large t, the d.f. of $N_j(t)$ is approximated by a normal distribution of mean $\frac{t}{\mu_{jj}}$ and variance $\frac{t\,\sigma_j^2}{\mu_{jj}^3}$ in the sense that the corresponding reduced variable converges weakly to the reduced normal distribution.
In short :

$$N_j(t) \sim N\left(\frac{t}{\mu_{jj}} , \frac{t\,\sigma_j^2}{\mu_{jj}^3}\right) \tag{3.12}$$

(with σ_j^2 is the variance of the return time from j to j).

§4. ASYMPTOTICAL RESULTS. STATIONARITY.

Let us consider the transition probability related to the semi-Markov process $(Z_t, t \geq 0)$ starting at time 0 with $J_0 = i$:

$$P_{ij}(t) = \mathbb{P}[Z_t = j \mid Z_0 = i] \tag{4.1}$$

It can be shown that (JANSSEN (1983), p.51) :

$$= \Sigma_{n=0}^{\infty} \mathbb{Q}^{(n)} * (I - H) \tag{4.2}$$

where H is the diagonal matrix with $(H_1, \ldots, H_m)$ as diagonal elements. Moreover, we have

Proposition 4.1.

If P is ergodic, then

$$\lim_{t} P_{ij}(t) = \frac{\pi_j\, \eta_j}{\Sigma_k\, \pi_k\, \eta_k} \tag{4.3}$$

Instead of starting with $J_0 = i$, we can, as mentioning in §2 (see (2.5)), take an arbitrary distribution for J_0. However to obtain a stationary process for

$$(J_N(t), J_{N(t)+1}, T_{N(t)+1} - t), t \geq 0) \tag{4.4}$$

we have not only to choose a particular distribution for J_0 but also for (J_0, X_1, J_1). This gives the next proposition (PYKE (1961)).

Proposition 4.2.

If P is ergodic and if

$$\mathbb{P}[J_o = i] = \frac{\pi_i \eta_i}{\Sigma_k \pi_k \eta_k}, \quad (4.5)$$

$$\mathbb{P}[X_1 \leqslant x, J_1 = j | J_o = i] = \frac{p_{ij}}{\eta_i} \int_o^x (1 - F_{ij}(y))\, dy \quad (4.6)$$

then the process (4.4) is stationary and

$$\mathbb{P}[J_{N(t)} = j, J_{N(t)+1} = k, T_{N(t)+1} - t \leqslant x] = \frac{\pi_j p_{jk}}{\Sigma_i \pi_i \eta_i} \int_o^x (1 - F_{jk}(y))dy. \quad (4.7)$$

In particular, in this case the relations (3.5) are exact and for all t

$$\mathbb{P}[Z_t = j] = \frac{\pi_j \eta_j}{\Sigma_k \pi_k \eta_k}. \quad (4.8)$$

§5. PARTICULAR CASES.

We mention breefly three particular choices of the matrix $\mathbb{Q}$ to get back some classical risk models.

$$\text{(i) } m = 1 : Q_{11} = F \quad (5.1)$$

We have a renewal process for the $(X_n, n \geqslant 1)$ and we find back the framework of the ANDERSEN risk model

$$\text{(ii) } Q_{ij}(x) = (p_{ij} U_1(x) \quad (5.2)$$

In this case, all the X_n variables are a.s. equal to 1 and we only have a classical Markov chain model as those used in the bonus-malus theory.

$$\text{(iii) } Q_{ij}(x) = p_{ij}(1 - e^{-\lambda_i x}) \quad (5.3)$$

Then the semi-Markov process $(Z_t, t \geqslant 0)$ is in fact a Markov process in continuous time with I as state space.

Chapter II. Semi-Markov risk models

§1. MOTIVATION FOR CONSIDERING THE SEMI-MARKOV RISK MODEL

We can at least distinguish three basic motivations for considering the semi-Markov risk model :

(1) To break down some restrictive assumptions of classical models concerning the independence conditions, rarely fulfilled in practical situations

(2) To construct a plausible model giving possibilities of analytical developments (the influence of parameters must be studied)

(3) To have a model which is numerically treatable (computation of premium rates, algorithms for ruin probabilities, ...)

To do so, we suppose that $I = \{1,\dots,m\}$ is the set of possible claims and that the process $((J_n, X_n), n \geq 0)$ introduced in Chapter I is still a positive (J-X) process.

1.1. Basic assumption

To define a risk model, we have now to consider the process of claim sizes not still introduced. To do so, let $(Y_n, n \geq 1)$ the successive claim sizes, supposing that the claim amount Y_0 of the claim just arriving at time 0 is nul. Of course, the claim sizes must be correlated to the claim types but with a "sort" of independence of the claim arrivals. To combine these conditions, we have introduced (see JANSSEN (1982)) a basic assumption with a positive two-dimensional semi-Markov theory.

Basic assumption

The process $((J_n, X_n, Y_n), n \geq 0)$ satisfies the following two-dimensional semi-Markov property :

$$\mathbb{P}[J_n=j,\ X_n\leq x,\ Y_n\leq y \mid (J_k,X_k,Y_k),\ k\leq n-1,\ J_k=i]$$
$$= p_{ij}\ {}^{A}F_{ij}(x)\ .\ {}^{B}F_{ij}(y) \qquad (2.1)$$

with J_0 fixed, $X_0 = Y_0 = 0$ a.s. and where the matrices

$${}^{A}\mathbb{Q} = ({}^{A}Q_{ij}) \text{ and } {}^{B}\mathbb{Q} = ({}^{B}Q_{ij}) \qquad (2.2)$$

$$\text{with } {}^{A}Q_{ij} = p_{ij}\ {}^{A}F_{ij},\quad {}^{B}Q_{ij} = p_{ij}\ {}^{B}F_{ij} \qquad (2.3)$$

are two positive semi-Markov kernels (see def. 2.1, chapter I). In complement to the variables $(T_n, n \geq 0)$ defined by (2.1), chapter I, we introduce the total amount of the first n claims S_n:

$$S_n = \Sigma_{k=0}^{n} Y_k \tag{2.4}$$

Of course, we can write all the relations of chapter I replacing the kernel $\mathbb{Q}$ by ${}^A\mathbb{Q}$ or ${}^B\mathbb{Q}$ because, from (2.1), it is clear that both processes $((J_n, X_n), n \geq 0)$, $((J_n, Y_n), n \geq 0)$ are positive (J-X) processes respectively of kernels ${}^A\mathbb{Q}$ and ${}^B\mathbb{Q}$.
The relation (2.1) clearly shows that the process of arrivals times $(X_n, n \geq 0)$ and the process of successive claim sizes $(Y_n, n \geq 0)$ though not independent are nevertheless conditionally dependent given $(J_n \geq 0)$, so that :

$$\mathbb{P}\,[X_n \leq x, Y_n \leq y \mid J_{n-1} = i, J_n = j]$$
$$= {}^AF_{ij}(x)\ .\ {}^BF_{ij}(y), \tag{2.5}$$

$i, j \in I$, $x, y \in \mathbb{R}^+$

In risk theory, it seems sufficient to take :

$${}^AF_{ij} = {}^AF_j\ , \tag{2.6}$$

$${}^BF_{ij} = {}^BF_j \tag{2.7}$$

for all $i,j \in I$. This assumption means, in fact, that the inter-arrival distribution is only influenced by the future claim and not by the last occured claim and that the claim size distribution only depends of the observed claim.

Convention of notation

From now, we use the symbols of chapter I adding the left upper indice A or B following that we consider the process $((J_n, X_n))$ or $((J_n, Y_n))$.

1.2. The premium process

From now, we always suppose P ergodic.
Using property 3.2 (chapter I), we can approximate the mean cost at time t of the $N_j(t)$ claims of type j by

$${}^B\eta_j \cdot \frac{t}{{}^A\mu_{jj}} \tag{2.8}$$

so that, the expectation of the total claim amount at time t is approximated by :

$$\mathbb{E}\,[S_{N(t)}] \simeq t\ \Sigma_j \frac{{}^B\eta_j}{{}^A\mu_{jj}} \tag{2.9}$$

It follows that, if one wants a constant premium rate $\tilde{c}$ for an asymptotically fair game insurers-insurance company, we must take:

$$\tilde{c} = \Sigma_j \frac{B\eta_j}{A_{\mu_{jj}}} \tag{2.10}$$

or by relation (3.6) of chapter II

$$\tilde{c} = \frac{\Sigma_j \pi_j B\eta_j}{\Sigma_k \pi_k A\eta_k} \tag{2.11}$$

For the constant premium rate including a positive loading factor δ, we get

$$c = (1+\delta)\,\tilde{c} \tag{2.12}$$

Of course, for the stationary version for the process $((J_n, X_n), n \geq 0)$, the preceding relations hold exactly.

The justification of a positive loading factor δ can be done as in the classical Andersen model because JANSSEN (1982) showed that if $\delta = 0$, the ruin on $[0,\infty)$ is a certain event whatever the value of the initial reserve u is.

1.3. The total claim amount

This amount may be studied with the following mass functions :

$$M_{ij}(t,y) = \mathbb{P}[S_{N(t)} \leq y, J_{N(t)} = j \mid J_0 = i] \tag{2.13}$$

If $$Q_{ij}(x,y) = {}^{A}Q_{ij}(x) \,.\, {}^{B}F_{ij}(y) \tag{2.14}$$

we get

$$M_{ij}(t,y) = \sum_{n=0}^{\infty} Q_{ij}^{(n)}(.,y) * (1 - {}^{A}H_j(.))(t) \tag{2.15}$$

expressions leading to complicated numerical problems.

§2. RUIN PROBLEM OVER AN INFINITE HORIZON.

Giving u as initial reserve, we have non-ruin on $[0,\infty)$ if and only if :

$$\text{for all } t > 0 : S_{N(t)} \leq u + ct \tag{2.16}$$

We can then introduce the following non-ruin probabilities :

$R_i(u)$ = probability of non ruin given that $J_0 = i$,

$R_i^j(u)$ = probability of non ruin and $\lim_n J_n = j$ given that $J_0 = i$,

$R(u)$ = probability of non ruin with π as initial distribution for J_0

$R^j(u)$ = probability of non ruin and $\lim_n J_n = j$ with π as initial distribution for J_0

Supposing that $c = 1$ and that $\tilde{c}$ is strictly inferior to 1, we get (JANSSEN (1983)) :

$$\lim_{u\to\infty} R_i^j(u) = \pi_j \tag{2.17}$$

$$\lim_{u\to\infty} R_i(u) = 1 \tag{2.18}$$

In fact :

$$R_i^j(u) = \pi_j\, R_i(u),\ i,j \in I,\ u \in \mathbb{R}^+ \tag{2.19}$$

and the mass functions $(R_i(u),\ i \in I)$ are the unique solution satisfying (2.18) of the Wiener-Hopf system of integral equations :

$$R_i^+(u) = U_0(u) \sum_k R_k^+ * C_{ik}(u),\ i \in I \tag{2.20}$$

with

$$C_{ij}(z) = \int_{\mathbb{R}} {}^B F_{ij}(\xi+z)\ {}^A Q_{ij}(d\xi) \tag{2.21}$$

It is possible to have a theoretical study of (2.20) (JANSSEN (1970) and to treat this system from a numerical point of view using discretization.

§3. SOME PARTICULAR MODELS

The presentation of the semi-Markov risk model of §1 shows that such a model is characterized by a triple $(P, {}^A F, {}^B F)$ where :

$$P = (p_{ij}) \tag{3.1}$$

$${}^A F = ({}^A F_{ij}) \tag{3.2}$$

$${}^B F = ({}^B F_{ij}) \tag{3.3}$$

We will use the queueing notations ./. to characterize the next following particular models :

(1) The SM/G model

There is no restriction on P and ${}^A F$ but ${}^B F$ has all its elements identical :

$${}^B F_{ij} = B \tag{3.4}$$

This means that the r.v. $(Y_n,\ n \geq 1)$ are i.i.d. with B as d.f.

(2) The G/SM model

It is the symmetrical of the preceding one : no restriction on P and ${}^B F$ but ${}^A F$ has all its elements identical :

$${}^A F_{ij} = A \tag{3.5}$$

This means that the r.v. $(X_n,\ n \geq 1)$ are i.i.d. with A as d.f.

(3) The P/SM model ("P" for Poisson)

It is the particular G/SM model for which

$$A(x) = \begin{cases} 0 & x < 0, \\ 1-e^{-\lambda x} & x \geq 0 \end{cases} \tag{3.6}$$

So the claim arrivals form a Poisson process. This model has been studied by JANSSEN (1980) both for the transient (i.e. finite horizon) and infinite ruin problems to extend the well-known result of Benes-Prabhu.

(4) The G/G model
It is the intersection of (1) and (2) which gives the Andersen risk model with A and B respectively as d.f. for the interarrivals and the claim sizes.

(5) The M/SM model ("M" for Markov)
There is no restriction on P and ${}^{B}F$ but the d.f. ${}^{A}F_{ij}$ are choosen as :

$${}^{A}F_{ij}(x) = \begin{cases} 0 & , \ x < 0 \\ 1-e^{-\lambda_i x} & , \ x \geq 0 \end{cases} \tag{3.7}$$

This means that the claim arrival process is a Markov process continuous in time.

(6) The M'/SM model
It is a modification of the preceding one to obtain a more adequate model. There is no restriction on P and ${}^{B}F$ but for ${}^{A}F$, we have :

$${}^{A}F_{ij}(x) = \begin{cases} 0 & , \ x < 0 \\ 1 - e^{-\lambda_j x} & , \ x \geq 0 . \end{cases} \tag{3.8}$$

(7) The SM_0/SM_0 model
Here, the matrix P has all its rows identical to $= (\pi_1,\ldots,\pi_m)$ and moreover for all $i,j \in I$:

$$\begin{aligned} {}^{A}F_{ij} &= {}^{A}F_j \\ {}^{B}F_{ij} &= {}^{B}F_j \end{aligned} \tag{3.9}$$

In this case, the processes $(X_n, n \geq 1)$ and $(Y_n, n \geq 1)$ are two renewal processes depending each of other. ${}^{A}F$ and ${}^{B}F$ are the d.f. of these two renewal processes with

$${}^{A}F = \Sigma_i \ \pi_i \ {}^{A}F_i, \quad {}^{B}F = \Sigma_j \ \pi_j \ {}^{B}F_j \tag{3.10}$$

(8) The $S\bar{M}/S\bar{M}_0$ model

The matrix P is as in (7) but now

$${}^{A}F_{ij} = {}^{A}F_i \qquad {}^{B}F_{ij} = {}^{B}F_i \tag{3.11}$$

Here too we have two renewal processes but with another dependence that those in (7).

The last two particular cases are called semi-Markov of zero order respectively of first and second kind.

(9) The mixed models $SM_o/S\bar{M}_o$ and $S\bar{M}_o/SM_o$.
see JANSSEN (1983).

For all these models, there are particular results on the non-ruin probabilities (see JANSSEN (1983) for the details).
Here are two examples :

(1) SM_o/SM_o model
All the R_i^+ are equal to the common value R^+ solution of the Wiener-Hopf integral equation :

$$R^+(u) = U_o(u)\ R^+ * \Sigma_k\ C_k(u) \tag{3.12}$$

with

$$C_j(u) = \pi_j \int_{\mathbb{R}} {}^B F_j\ (\xi+u)\ {}^A F_j(d\xi) \tag{3.13}$$

(2) $S\bar{M}_o/S\bar{M}_o$ model
The probabilities R_i are given by :

$$R_i^+(u) = U_o(u)\ R^+ * F_i(u)\ ,\ F_i(u) = \int_{\mathbb{R}} {}^B F_i(\xi+u)\, {}^A F_i(d\xi) \tag{3.14}$$

where

$$R(u) = \Sigma_k\ \pi_k\ R_k^+(u) \tag{3.15}$$

is solution of the Wiener-Hopf integral equation

$$R^+(u) = U_o(u)\ R^+ * \Sigma_i\ \pi_i\ F_i(u)\ . \tag{3.16}$$

So, in both examples, we reduce the problem to the resolution of <u>one</u> Wiener-Hopf integral equation.

Chapter III. Other semi-Markov models in economics and insurance

There are many applications of semi-Markov processes in Operations Research (reliability, queueing, dynamic programming, ...) and in other fields (Medicine, Demography, ...). In this last chapter, we want to present two other applications in the field of insurance : (1) the first concerns the application of non-homogeneous semi-Markov processes to a social pension problem and (2) the second shows how the semi-Markov processes can be used to modelize the influence of the economical environment in a risk process. A complete exposition of these models will be find for (1) in JANSSEN and DEDOMINICIS (1983) and for (2) in REINHARD (1983).

§1. A SOCIAL PENSION PROBLEM

1.1. The problem

To compute premiums for insurance against professional diseases (as silicosis for example), it is necessary to compute the mean disability degree at pre-assigned epochs. Let us suppose we retain m invalidity degrees $S_1, \ldots, S_m$ the last being 100 % and including survival pension of the death.
An insurer can go from degree S_i to degree S_j with a probability p_{ij}. Without other assumption, the problem is easily treated with a Markov chain model (see YNTEMA (1965), JANSSEN (1967)) to get a limit distribution for the invalidity degree.
If moreover we suppose that the passage from degree S_i to degree S_j takes a random time depending both of the initial and final degrees, we can modelize with a semi-Markov model (JANSSEN (1967)).
The next step is to suppose that the transitions are influenced by the time at which they occur. The consideration of the time parameter leads to a non-homogeneous semi-Markov model (JANSSEN and DEDOMINICIS (1983)).

1.2. The non-homogeneous semi-Markov model

Keeping the notation of chapter I, we replace the basic assumption (2.9) by

$$\mathbb{P}[J_n=j,\ X_n \leqslant x \mid (J_k,X_k),\ k \leqslant n-1,\ J_{n-1} = i] \qquad (1.1)$$

$$= Q_{ij}(T_{n-1},\ T_{n-1} + x)$$

$i,j \in I,\ n \geqslant 1,\ x \in \mathbb{R}^+$

This defines a non-homogeneous J-X process. The kernel is now a matrix $\mathbf{Q}(s,t)$ of mass functions $Q_{ij}(s,t)$ satisfying :

$$\text{(i) } \forall\, i,j \in I,\ \forall\, t,s \in \mathbb{R}^+,\ t \leqslant s : Q_{ij}(s,t) = 0 \qquad (1.2)$$

(ii) $\forall\, i \in I,\ \forall\, s > 0$:

$$\Sigma_j Q_{ij} (s, + \infty) = 1 \tag{1.3}$$

In this case, the discrete time process $(J_n, n \geq 0)$ is a multiple Markov chain of matrix function $p_{ij}(s)$ where

$$p_{ij}(s) = Q_{ij}(s, +\infty) \tag{1.4}$$

For the invalidity problem, it is as a first approximation sufficient to take as in chapter I

$$p_{ij}(s) = p_{ij} \tag{1.5}$$

i.e. a classical homogeneous Markov chain.

For $p_{ij} > 0$, we have the relation defining the conditional d.f. of X_n given $J_{n-1} = i$, $J_n = j$; $T_{n-1} = s$.

$$F_{ij}(s,t) = p_{ij}^{-1} Q_{ij}(s,t) \tag{1.6}$$

From (1.6) we can also characterize a non-homogeneous (J-X) process by the couple $(P, (F_{ij}(s,t))$ instead of the kernel $Q(s,t)$. The counting process $(N(t), t \geq 0)$ is introduced as in (2.24) of chapter I so that we can define the non-homogeneous semi-Markov process $(Z_t, t \geq 0)$ as usual :

$$Z(t) = J_{N(t)} \tag{1.7}$$

(supposing that there is no explosion).
Define now the probability transitions of the Z-process

$$P_{ij}(s,t) = \mathbb{P}[Z(t)=j \mid Z(s) = i, N(s-) < N(s)] \tag{1.8}$$

$i,j \in I$, $0 \leq s < t$.

It can be shown that these probabilities satisfy the following system :

$$P_{ij}(s,t) = \delta_{ij}(1-H_i(s,t)) + \Sigma_k \int_s^t P_{kj}(u,t)Q_{ik}(s,du) \tag{1.9}$$

$i,j \in I$

where

$$H_i(,st) = \Sigma_j F_{ij}(s,t) \tag{1.10}$$

The system (1.9) can be solved numerically by a discretization method (see JANSSEN and DEDOMINICIS (1983). Then, to solve the problem it suffices to compute

$$\bar{S}_i(s,t) = \Sigma_j P_{ij}(s,t) S_j \tag{1.11}$$

which gives the mean disability degree at time t starting at time s with a degree S_i.
If we put an initial distribution $(p_1, \ldots, p_m)$ at the degree at

time s, we get

$$\bar{S}_i(s,t) = \Sigma_k\, p_k\, \bar{S}_k(s,t) \tag{1.12}$$

Some numerical results in JANSSEN and DE DOMINICIS show a remarkable stability.

§2. THE RISK MODEL WITH A SEMI-MARKOV ENVIRONMENT

The idea of modelizing the environment was introduced by PURDUE (1974) in a queueing model. This idea was introduced recently by REINHARD (1983) in the framework of risk theory.
The basic assumptions are the following.

(i) The environment process

The economical environment is represented by the sequence of random variables $(I_n,\ n \geq 0)$ giving the values at the jumps of the environment. The occurence times of the jumps are given by the sequence $(U_n,\ n \geq 0)$ with

$$U_n = \Sigma_{\nu=1}^{n} A_\nu$$
$$U_o = A_o = 0 \text{ a.s.}$$

where the r.v. A_n represent the time intervals between two jumps. We suppose that the process $((I_n,\ A_n),\ n \geq 0)$ is a positive (J-X) process of kernel $K = (K_{ij})$ with m possible states for the environment. The associated semi-Markov process will be noted by $(I(t),\ t \geq 0)$.

(ii) The risk process

Let, as usual, T_n be the occurence time of claim number n, (X_n) the successive interarrival times and define J_n by

$$J_n = I(T_n)$$

i.e. the environment state at time T_n, (Y_n) the successive claim sizes.
The premium rate is supposed to be 1 (see fig.3).
To modelize how the environment process influences the risk process, we do the following assumptions :

(H_1) : The processes (X_n), (Y_n) are conditionally independent given (J_n)

(H_2) : $\forall\ i \in I : \mathbb{P}[\,Y_n \leq y\ (J_k,\ X_k,\ Y_k),\ k \leq n-1,\ J_n=i] = F_i(y)$

(H_3) : The process of the total number of claims on $(0,t]$ - $(N(t),\ t \geq 0)$ satisfies the two next conditions :
(1) $\forall\ i \in I,\ \forall\ t \geq 0,\ \forall\ h > 0$:
$I(u) = i,\ u\ \ (t,\ t+h] \Rightarrow N(t+h)-N(t)$ has a Poisson distribution of parameter $\alpha_i h$

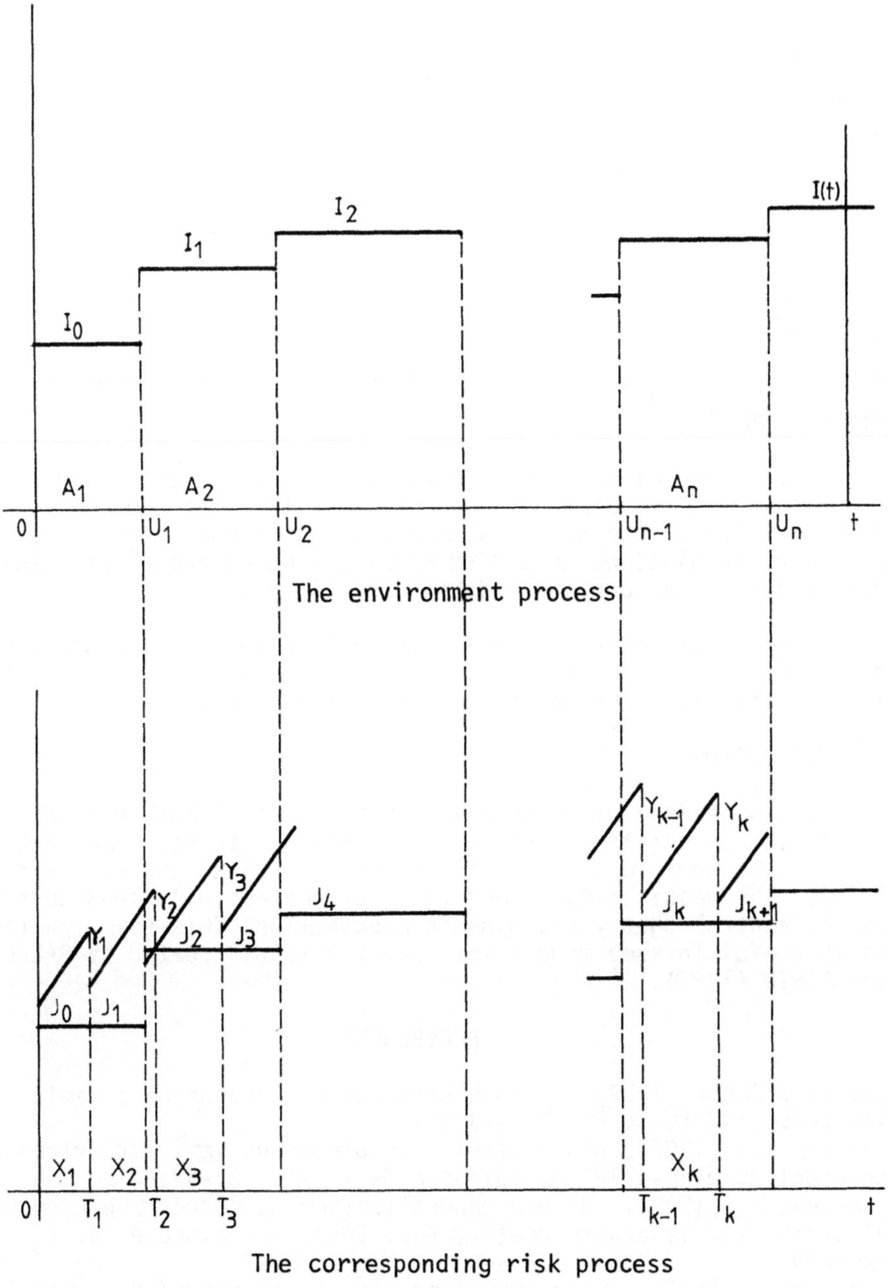
I(t)
I2
I1
I0
A1
A2
An
0
U1
U2
Un-1
Un
t
The environment process
Yk-1
Yk
Y3
Y2
Y1
J4
Jk
Jk+1
J2
J3
J0
J1
X1
X2
X3
Xk
0
T1
T2
T3
Tk-1
Tk
t
The corresponding risk process

Fig. 3

(2) Given $(I(t), t \geq 0)$, the process $(N(t), t \geq 0)$ has independent increments.

Then, the next proposition holds (Reinhard (1983))

Proposition

The process $((J_n, X_n, Y_n), n \geq 0)$ defines a semi-Markov risk model of kernel $\mathbf{Q}$ where (see (2.1), ch.II) :

$$Q_{ij}(x,y) = F_j(y)\, V_{ij}(x)$$

so that $V = (V_{ij})$ is the semi-Markov kernel of the (J-X) process $((J_n, X_n), n \geq 0)$

Corollary

If, for all j $F_j(y) = 1 - e^{-y/b_j}$ $(y \geq 0)$, the risk process is of type SM/M'.

The consequence of this last result is the possibility to get numerical results using the concept of duality introduced by JANSSEN (1983). Indeed, a risk model SM/M' is equivalent, from the numerical resolution, to a M/SM model for which numerical results are possible (see Janssen and Reinhard (1983)).

Let us say that it is also possible to introduce a premium rate depending of the environment : if the environment state is i at time t, the premium rate at this time will be c_i.

§3. CONCLUSION

As the possibility of numerical results for semi-Markov risk models is now existing, the next step will be to use some parameters and to compare the results of both classical and semi-Markov models. The possibility of modifying parameters will lead to the use of control theory via dynamic programming. This seems particularly useful in some models for pension funds treated by BALCER and SAHIN (1983).

REFERENCES

Balcer & Sahin (1983), A stochastic theory of pension dynamics, Insurance : Math. & Ec. 2, 179-197.

Janssen, J. (1966), Applications des processus semi-markoviens à un problème d'invalidité, Bulletin de l'ARAB, 35-52.

Janssen, J. (1970), Sur une généralisation du concept de promenade aléatoire sur la droite réelle, Ann. Inst. H. Poincaré, B, VI, 249-269.

Janssen, J. (1980), Some transient results on the M/SM special semi-Markov model in risk theory and queueing theory, ASTIN Bull.,

11, 41-51.
Janssen, J. (1982), Modèles de risque semi-markoviens, Cahiers du C.E.R.O., 24, 2-3-4,
Janssen, J. (1983), Semi-Markov processes and applications in risk and queueing theories, Cooperativa Universitaria editrice Napoletana, Napoli, 182 p.
Janssen, J. & DeDominicis, R., Finite non-homogeneous semi-Markov processes : Theoretical and computational aspects, Tech. report PS-02-83, Université Libre de Bruxelles.
Janssen, J. & Reinhard, J-M., Formes explicites de probabilités de ruine pour une classe de modèles de risque semi-markoviens, Tech. report PS-01-83, Université Libre de Bruxelles.
Purdue, P. (1974), The M/M/1 Queue in a Markovian environment, JORSA 22, 562-569.
Pyke, R. (1961), Markov renewal processes with finitely many states, Ann. Math. Stat., 32, 1243-1259.
Reinhard, J-M. (1983), On a class of semi-Markov risk models obtained as classical risk models in a markovian environment, to appear ASTIN Bulletin.
Yntena, L. (1962), A Markovian treatment of silicosis, Third Int. Conf. of Social Security Actuaries and Statisticians, Madrid.

LOSS DISTRIBUTIONS: ESTIMATION, LARGE SAMPLE THEORY, AND APPLICATIONS

Stuart A. Klugman

Department of Statistics and Actuarial Science
The University of Iowa

Given a sample of insurance losses, it is desired to fit a parametric probability distribution. A number of heavy (right) tailed models are presented, followed by five methods of parameter estimation. Numerical procedures are given with emphasis on the method of scoring for maximum likelihood estimation from grouped data. The latter also provides estimates of the asymptotic variance of the estimators. Several illustrations are presented using a lognormal model for automobile losses. The final example concerns multiple liability limits. An analysis of variance procedure is used to test for differences among the severities and then a James-Stein type credibility formula is presented to improve the estimator under a specific limit.

1. INTRODUCTION

Calculation of pure premiums usually involves separate estimation of frequency and severity (the expected individual loss given that one has occurred). If a sufficiently large sample of losses is available, the sample mean will provide a satisfactory estimate of severity. The more common situation is a relatively small sample with grouped data. In addition, the data are likely to be affected by truncation (deductible) or censoring (policy limit). Finally, we may desire to estimate the effects of inflation or of a coverage modification. In this paper we recommend the use of a fitted, continuous, parametric probability distribution. The rationale for this decision and an inventory of suitable models are given in Section 2.

In the next section five methods of parameter estimation are

F. de Vylder et al. (eds.), Premium Calculation in Insurance, 263–284.

discussed. Included are suggested numerical methods of implementing these procedures. An example of losses on automobile physical damage policies is introduced at this time.

In Section 4 large sample theory developed in conjunction with maximum likelihood estimation in the previous section is used to provide further analysis. In particular, it gives an approximation to the variance of an estimator of a function of the parameters. Of course, a special case is the severity.

The remaining sections cover the case of samples from slightly different populations. In Section 5 an analysis of variance procedure is developed which allows for comparison of the severities. If they are identical, the samples may be combined for the purpose of determining severity. If they are different, a credibility procedure may be used to provide an improved estimate. This is presented in Section 6. These two sections are accompanied by an illustration involving automobile bodily injury (liability) policies with five different policy limits. In Section 7 the differences (and similarities) of the procedure in Section 6 and that of Hachemeister (6) are presented. A suggestion for combining these two approaches concludes this section.

The material presented in Sections 2-4 is taken from Loss Distributions by R. V. Hogg and S. A. Klugman to be published by John Wiley and Sons in 1984. Valuable assistance in the preparation of that material was provided by C. C. Hewitt and G. S. Patrik. The research was conducted under a grant from the Actuarial Education and Research Fund. The author is indebted to all of the above for their generous help. Thanks are also due to the Insurance Services Office which supplied the data used in the examples. Finally, the relations given in Section 7 were motivated by the presentations given by B. Zehnwirth at this NATO Advanced Study Institute.

2. CONTINUOUS, PARAMETRIC MODELS

In most all situations, the available information consists of a sample of losses. What is desired is the distribution of the random variable which represents the loss process. We assume that the observations are independent and identically distributed and that future losses will be described by an identical random variable. Three reasonable choices for a model are 1) the empirical distribution (which assigns mass $1/n$ to each sample point), 2) a nonparametric adjustment of the empirical distribution, and 3) a member of a parametric family of continuous distributions. In this section we present arguments in favor of the third alternative. The second part of this section gives a number of distributions which provide reasonable models for insurance losses.

The empirical distribution and nonparametric methods based on it have two major drawbacks. The first is that they do not model the right tail of the distribution. In particular, the empirical distribution places no mass beyond the largest observation. The second is that we are often (as will be illustrated in subsequent sections) interested in evaluating the d.f. at specific values which are not likely to be data points. This requires interpolation and a natural method of doing this is not readily apparent for the empirical distribution. This problem is increased when the data are compiled in grouped (frequency table) form. Then the interpolation must be done between the class boundaries.

Both of the above problems are handled by a continuous distribution. An additional benefit is that the estimated parameters help to characterize the loss process and to enable comparisons of distributions. Finally, large sample theory for parameter estimates and estimates of functions of parameters is well developed. This allows for the construction of confidence intervals and for the creation of hypothesis tests.

Most insurance loss distributions are skewed to the right with an especially heavy right tail. There are two "natural" distributions from which a number of heavy-tailed distributions may be created. Two methods of creating such distributions are by transformation and by mixing. Two common transformations which shift probability to the right are $Y = \exp(X)$ and $Y = X^{1/\tau}$ $(\tau < 1)$. Mixing occurs when one of the parameters is considered to be random with a specified parametric distribution. The new random variable is the marginal distribution of the original variable. That is, let $f(x;\theta_1,\ldots,\theta_p)$ be the continuous p.d.f. of X given that $\Theta_p = \theta_p$ and let $g(\theta_p;\tau_1,\ldots,\tau_q)$ be the continuous p.d.f. of Θ_p. Then the new, mixed distribution will have p.d.f.

$$h(x;\theta_1,\ldots,\theta_{p-1},\tau_1,\ldots,\tau_q) = \int f(x;\theta_1,\ldots,\theta_p)g(\theta_p;\tau_1,\ldots,\tau_q)d\theta_p.$$

The "natural" distributions are (where $f(\cdot)$ is the p.d.f. and $F(\cdot)$ is the d.f.):

1. Normal: $f(x;\mu,\sigma) = (\sigma\sqrt{2\pi})^{-1}\exp(-\frac{1}{2}(\frac{x-\mu}{\sigma})^2)$.
2. Gamma: $f(x;\alpha,\lambda) = \frac{\lambda^\alpha}{\Gamma(\alpha)} x^{\alpha-1} \exp(-\lambda x)$.

A special case with $\alpha = 1$ is the exponential distribution.

Derived distributions:

3. Lognormal: $X \sim \text{Normal}(\mu,\sigma)$, $Y = \exp(X)$.

$$f(y;\mu,\sigma) = (y\sigma\sqrt{2\pi})^{-1}\exp(-\frac{1}{2}(\frac{\ln y-\mu}{\sigma})^2).$$

4. Weibull: $X \sim \text{Exponential}(c)$, $Y = X^{1/\tau}$.

$$f(y;c,\tau) = c\tau y^{\tau-1}\exp(-cy^{\tau}), \quad F(y;c,\tau) = 1-\exp(-cy^{\tau}).$$

5. Loggamma: $X \sim \text{Gamma}(\alpha,\lambda)$, $Y = \exp(X)$.

$$f(y;\alpha,\lambda) = \frac{\lambda^{\alpha}}{\Gamma(\alpha)} y^{-\lambda-1}(\ln y)^{\alpha-1}.$$

6. Transformed gamma: $X \sim \text{Gamma}(\alpha,\lambda)$, $Y = X^{1/\tau}$.

$$f(y;\alpha,\lambda,\tau) = \frac{\tau\lambda^{\alpha}}{\Gamma(\alpha)} y^{\alpha\tau-1}\exp(-\lambda y^{\tau}).$$

7. Pareto: $X \sim \text{Exponential}(\Theta)$, $\Theta \sim \text{Gamma}(\alpha,\lambda)$.

$$f(x;\alpha,\lambda) = \alpha\lambda^{\alpha}(\lambda+x)^{-\alpha-1}, \quad F(x;\alpha,\lambda) = 1-\lambda^{\alpha}(\lambda+x)^{-\alpha}.$$

8. Burr: $X \sim \text{Pareto}(\alpha,\lambda)$, $Y = X^{1/\tau}$.

$$f(y;\alpha,\lambda,\tau) = \alpha\tau\lambda^{\alpha}y^{\tau-1}(\lambda+y^{\tau})^{-\alpha-1}, \quad F(y;\alpha,\lambda,\tau) = 1-\lambda^{\alpha}(\lambda+y^{\tau})^{-\alpha}.$$

It can also be obtained by mixture with $X \sim \text{Weibull}(\Theta,\tau)$ and $\Theta \sim \text{Gamma}(\alpha,\lambda)$.

9. Generalized Pareto: $X \sim \text{Gamma}(k,\Theta)$, $\Theta \sim \text{Gamma}(\alpha,\lambda)$.

$$f(x;\alpha,k,\lambda) = \frac{\Gamma(\alpha+k)\lambda^{\alpha}x^{k-1}}{\Gamma(\alpha)\Gamma(k)(\lambda+x)^{\alpha+k}}.$$

10. Transformed beta: $X \sim \text{transformed gamma}(k,\Theta,\tau)$, $\Theta \sim \text{transformed gamma}(\alpha,\lambda^{1/\tau},\tau)$.

$$f(x;\alpha,k,\lambda,\tau) = \frac{\Gamma(\alpha+k)\tau\lambda^{\alpha}x^{\tau k-1}}{\Gamma(\alpha)\Gamma(k)(\lambda+x^{\tau})^{\alpha+k}}.$$

Special cases are Burr $(k=1)$ and generalized Pareto $(\tau=1)$. It can also be obtained by $X \sim \text{generalized Pareto}(\alpha,k,\lambda)$ and $Y = X^{1/\tau}$. For an application of this distribution see Venter (18).

All of the above distributions have support $(0,\infty)$ except the loggamma which has support $(1,\infty)$.

3. PARAMETER ESTIMATION

Having selected a family of distributions, the next step is to estimate the parameters from the available data. In this section we present five estimation techniques. Data from losses on automobile physical damage insurance will be used to illustrate the methods.

Before giving the details of the methods, we must mention two factors which complicate the estimation. The first is coverage modification. The model we seek describes losses as suffered by the insured. However, depending on the terms of the policy, the insurance may not pay the full amount of the loss. The most common modifications are the deductible and the policy limit. The data usually are recorded in terms of the amount paid, not the actual loss. More formally, if $F(x;\underset{\sim}{\theta})$ ($\underset{\sim}{\theta}$ is a $p\times 1$ vector of parameters) is the d.f. for actual losses and $G(x;\underset{\sim}{\theta})$ is the d.f. for the amount paid, with a deductible of d and a limit of u we have

$$
\begin{aligned}
G(x;\underset{\sim}{\theta}) &= 0 && x < d\\
&= (F(x;\underset{\sim}{\theta})-F(d;\underset{\sim}{\theta}))/(1-F(d;\underset{\sim}{\theta})) && d \le x < u\\
&= 1 && x \ge u.
\end{aligned}
$$

We define the empirical d.f., $G_n(x)$, as

$$G_n(x) = \frac{1}{n}\sum_{i=1}^{n} I_{[0,x]}(x_i),$$

that is, the proportion of observations less than or equal to x.

A second problem is that the data may be grouped. That is, there are $k+1$ class boundaries, $c_0 < \ldots < c_k$ where c_0 is usually equal to d (the deductible) and $c_k = \infty$. In this situation we do not have the complete empirical distribution. Instead, for $c_{i-1} < x \le c_i$,

$$G_n^* = G_n(c_i).$$

In all applications we will only require values of $G_n^*(x)$ at class boundaries and so the * will be dropped from the notation. The context will make it clear if the data are individual or grouped.

Of the five estimation methods, the first two are "crude" methods. They are (relatively) easy to compute, but are not likely to produce accurate results. Their major purpose is to provide starting values for the more refined methods which follow. The first method is percentile matching and is especially useful if the d.f. can be written in closed form (e.g., Burr). It requires the selection of p values from the empirical distribution, say, $z_1,\ldots,z_p$ where each z is an x_i for individual data or a c_i for grouped data. The method is completed by solving the following p equations for $\underset{\sim}{\theta}$:

$$G(z_i;\underset{\sim}{\theta}) = G_n(z_i),\quad i=1,\ldots,p.$$

On the left hand side, the function $G(x;\underset{\sim}{\theta})$ must be replaced by the appropriate expression involving $F(x;\underset{\sim}{\theta})$. If there is a deductible the left hand side will be a complicated function of $\underset{\sim}{\theta}$ making the equations difficult to solve. As our intention is merely to provide a rough estimate, it is usually more convenient to simply use $F(x;\underset{\sim}{\theta})$ on the left hand side.

The second method is the method of moments. It involves matching the sample moments to the population moments. Let the ith population moment be

$$\mu_i(\underset{\sim}{\theta}) = \int x^i dG(x;\underset{\sim}{\theta})$$

and let the ith sample moment (for individual data) be

$$m_i = \frac{1}{n} \sum_{j=1}^{n} x_j^i .$$

For grouped data, one solution would be to replace each observation by the midpoint of the class. In any event, the estimates are found by solving the p equations:

$$\mu_i(\underset{\sim}{\theta}) = m_i, \quad i = 1,\ldots,p.$$

Once again, it may be more convenient (but, of course, less accurate) to replace $G(x;\underset{\sim}{\theta})$ by $F(x;\underset{\sim}{\theta})$ in the calculation of $\mu_i(\underset{\sim}{\theta})$.

At this point we introduce a numerical example. Losses on 27,618 claims under automobile physical damage were recorded by the Insurance Services Office. The data were grouped and there was no deductible or limit (so $F(x;\underset{\sim}{\theta}) = G(x;\underset{\sim}{\theta})$. The empirical d.f. is given in Table 1.

c_i	$F_n(c_i)$
100	.15092
250	.47730
500	.72174
750	.84695
1,000	.90162
2,000	.97697
3,000	.99218
4,000	.99623
5,000	.99841
6,000	.99909
7,000	.99935
8,000	.99960
9,000	.99967
15,000	.99996
20,000	1.00000

Table 1. Automobile Physical Damage. Empirical D.F. (n = 27,618).

We elect to fit a lognormal model. This may not be the best choice, but it is sufficiently tractable to provide a good illustration. Using percentile matching with $z_1 = 250$ and $z_2 = 1{,}000$, we have

$$\Phi(\log 250 - \mu)/\sigma) = .47730$$

$$\Phi(\log 1{,}000 - \mu)/\sigma) = .90162$$

where $\Phi(\cdot)$ is the standard normal d.f. The inverse is readily available reducing the equations to

$$5.5215 - \mu = -.05693\sigma$$

$$6.9078 - \mu = 1.29084\sigma.$$

The solution is $\hat{\mu} = 5.5801$ and $\hat{\sigma} = 1.0286$. The method of moments estimates are a bit more difficult to obtain. We note here that given a specific model there may be a clever method of solving the estimation equations. For example, in the lognormal case we could take logarithms and then proceed as if we were dealing with a normal distribution. However, it is not our intention to give procedures which are applicable only to special situations, but rather to give procedures which can be used in all situations.

The three more "refined" methods are all based on the minimization of an objective function. Each in some manner measures the closeness of the fitted model to the observed data. The first is called minimum distance (md). For individual data the objective function is

$$K = \sum_{i=1}^{n} w(x_i)[G(x_i;\underset{\sim}{\theta}) - \frac{n}{n+1}G_n(x_i)]^2.$$

The function $w(x)$ is a weight function which allows for greater emphasis to be placed at those points where good fit is most important. One obvious choice for the weight function is

$$w(x) = n/G(x;\underset{\sim}{\theta})(1-G(x;\underset{\sim}{\theta}))$$

or

$$w(x) = n/G_n(x)(1-G_n(x)).$$

This gives each term in the sum an approximate chi-square(1) distribution and thus each term makes an equal contribution to the total. While the first version of $w(x)$ more accurately meets this objective, the second is much easier to use. The factor of $n/(n+1)$ in the expression for K is introduced to shift the empirical d.f. down. Fitting a continuous model to the left hand endpoints of the step function $G_n(x)$ does not capture the essence of the empirical d.f. With a perfect fit, the model d.f. would lie entirely to the left and above $G_n(x)$. For grouped data there is not such a problem; the objective function is

$$K = \sum_{i=1}^{n} w(c_i)[G(c_i;\underset{\sim}{\theta}) - G_n(c_i)]^2 .$$

The second method is only applicable to grouped data. It is based on the familiar chi-square goodness-of-fit test and has as the objective function

$$Q = \sum_{i=1}^{k} \frac{(f_i - E_i)^2}{E_i}$$

where f_i is the number of observations in the ith class and

$$E_i = n[G(c_i;\underset{\sim}{\theta}) - G(c_{i-1};\underset{\sim}{\theta})].$$

We call this the minimum chi-square (mc) method. For computational purposes it is easier to use f_i in place of E_i in the denominator. The two estimators are asymptotically equivalent (Moore (13)).

The final method is that of maximum likelihood (ml). To be consistent with the above two methods, we minimize the negative of the logarithm of the likelihood function. For individual losses it is

$$L = -\sum_{i=1}^{n} \ln\, g(x_i;\underset{\sim}{\theta})$$

and for grouped data it is

$$L = -\sum_{i=1}^{k} f_i \ \ln[G(c_i;\underset{\sim}{\theta}) - G(c_{i-1};\underset{\sim}{\theta})].$$

When the groupings are identical the mc and ml estimators are asymptotically equal (Moore (13)).

All three of the minimizations must be done by numerical means. A general method is based on minimizing an arbitrary sum-of-squares. That is, minimize

$$\sum_{i=1}^{n} r(x_i;\underset{\sim}{\theta})^2 .$$

Since all three objective functions are summations, all that need be done is use the square root of the summand as $r(x_i;\underset{\sim}{\theta})$. There are a number of numerical procedures available which perform the minimization. One popular method is by Marquardt (10). His method requires the derivatives of $r(x_i;\underset{\sim}{\theta})$ with respect to $\underset{\sim}{\theta}$. Derivative-free approaches (e.g., Brown and Dennis (2) and Ralston and Jennrich (15)) are much easier to use and are available in many computer packages. The values in this paper were computed by the IMSL subroutine ZXSSQ. In general, good starting values are required. Usually the ones provided by the first two methods presented in this section will be close enough to produce convergence.

For the automobile example, the md estimates are $\hat{\mu} = 5.6049$ and $\hat{\sigma} = .99953$. The mc estimates are $\hat{\mu} = 5.6111$ and $\hat{\sigma} = .99345$.

For maximum likelihood estimates from grouped data an alternate method is available. It is called the method of scoring (Rao (16), Chapter 5) and is based on derivatives of the logarithm of the likelihood function. For convenience, let $p_i(\underset{\sim}{\theta}) = G(c_i;\underset{\sim}{\theta}) - G(c_{i-1};\underset{\sim}{\theta})$, $i = 1,\ldots,k$. Then the logarithm of the likelihood function is

$$L(\underset{\sim}{\theta}) = \sum_{i=1}^{k} f_i \ln p_i(\underset{\sim}{\theta})$$

and the derivatives are

$$S_j(\underset{\sim}{\theta}) = \partial L(\underset{\sim}{\theta})/\partial\theta_j = \sum_{i=1}^{k} \frac{f_i}{p_i(\underset{\sim}{\theta})} \frac{\partial p_i(\underset{\sim}{\theta})}{\partial\theta_j}, \quad j = 1,\ldots,p.$$

Instead of setting each $S_j(\underset{\sim}{\theta})$ (the scores) equal to zero, we use the Taylor series expansion about an initial estimate of $\underset{\sim}{\theta}$. The expected value of the Taylor expansion is set equal to zero and then solved for an improved estimate of $\underset{\sim}{\theta}$. The procedure continues until there is little change in the value of $\underset{\sim}{\theta}$. The formula is based on the $p \times p$ information matrix $I(\underset{\sim}{\theta})$ with rsth element

$$I(\underset{\sim}{\theta})_{rs} = \sum_{i=1}^{k} \frac{1}{p_i(\underset{\sim}{\theta})} \frac{\partial p_i(\underset{\sim}{\theta})}{\partial\theta_r} \frac{\partial p_i(\underset{\sim}{\theta})}{\partial\theta_s}.$$

The iterative step is then

$$\hat{\underset{\sim}{\theta}}_{m+1} = \hat{\underset{\sim}{\theta}}_m + I(\hat{\underset{\sim}{\theta}}_m)^{-1}\underset{\sim}{S}(\hat{\underset{\sim}{\theta}}_m)/n,$$

where $\underset{\sim}{S}(\hat{\underset{\sim}{\theta}}_m) = S_1(\hat{\underset{\sim}{\theta}}_m),\ldots,S_p(\hat{\underset{\sim}{\theta}}_m))'$.

A side benefit is that $I(\theta)^{-1}$ is the asymptotic covariance matrix of $\sqrt{n}(\hat{\underset{\sim}{\theta}}^*-\underset{\sim}{\theta}^*)$ where $\underset{\sim}{\theta}^*$ is the true parameter and $\hat{\underset{\sim}{\theta}}$ is the ml estimator. A reasonable approximation is given by $I(\hat{\underset{\sim}{\theta}})^{-1}$.

For some of the distributions given in the previous section, it is easy to obtain the needed derivatives. It is more difficult for the gamma, loggamma, transformed gamma, generalized Pareto, and transformed beta. In these cases the d.f. is not available in closed form. The derivatives can be taken inside the integral but then the integral must be evaluated by numerical means.

One major advantage of the method of scoring is that the

algorithm is relatively simple and can be implemented on a personal computer. For the automobile example the results are (and were obtained in just three iterations):

$$\hat{\mu} = 5.6101 \qquad \hat{\sigma} = .99305$$

$$\widehat{\text{Var}}(\hat{\mu}) = .000038656 \qquad \widehat{\text{Var}}(\hat{\sigma}) = .000023949$$

$$\widehat{\text{Correlation}}(\hat{\mu},\hat{\sigma}) = -.098005.$$

In Table 2, values of the d.f. for this model are presented at the same amounts used in Table 1.

c_i	$F(c_i)$
100	.15578
250	.46444
500	.72865
750	.84543
1,000	.90435
2,000	.97750
3,000	.99209
4,000	.99656
5,000	.99829
6,000	.99907
7,000	.99946
8,000	.99966
9,000	.99978
15,000	.99997
20,000	.99999

Table 2. Automobile Physical Damage. Lognormal Model D.F.

4. LARGE SAMPLE THEORY

Having found maximum likelihood estimates and an estimate of the associated information matrix, we are prepared to obtain estimates of quantities of actuarial interest. In addition, we would like an estimate of the variance of these estimators.

The key results are contained in two theorems. The first is from Rao (16, Theorem 5e.2(iv)) and relates to $\tilde{\hat{\theta}}$, the ml estimate found by the method of scoring. The requirements are (i) $\tilde{\theta} \neq \tilde{\beta} \Rightarrow p_i(\tilde{\theta}) \neq p_i(\tilde{\beta})$ for at least one i, (ii) $p_i(\tilde{\theta})$ admits all first-order partial derivatives and they are continuous at $\tilde{\theta}^*$, and (iii) $I(\tilde{\theta}^*)$ is nonsingular. If so, then $\sqrt{n}(\tilde{\hat{\theta}}-\tilde{\theta}^*)$ converges in distribution to a multivariate normal random variable with mean 0 and covariance matrix $I(\tilde{\theta}^*)^{-1}$.

Now let $\gamma = h(\underset{\sim}{\theta})$ be any single-valued function of $\underset{\sim}{\theta}$ which has all first and second order derivatives in a neighborhood of $\underset{\sim}{\theta}^*$. Then (Anderson (1, Theorem 4.2.5)), if $\hat{\gamma} = h(\hat{\underset{\sim}{\theta}})$ and $\gamma^* = h(\underset{\sim}{\theta}^*)$, $\sqrt{n}(\hat{\gamma}-\gamma^*)$ converges in distribution to $N(0,\sigma_h^2)$ where

$$\sigma_h^2 = \underset{\sim}{\eta}(\underset{\sim}{\theta}^*)' I(\underset{\sim}{\theta}^*)^{-1}\underset{\sim}{\eta}(\underset{\sim}{\theta}^*).$$

The vector $\underset{\sim}{\eta}(\underset{\sim}{\theta}^*)$ has as its ith element $\eta_i(\underset{\sim}{\theta}^*) = \partial h(\underset{\sim}{\theta}^*)/\partial\theta_i$.

An approximate level $1-\alpha$ confidence interval for * is given by

$$\hat{\gamma} \pm z_{\alpha/2}\, \hat{\sigma}_h/\sqrt{n}$$

where $\hat{\sigma}_h$ is computed using $\hat{\underset{\sim}{\theta}}$ in place of $\underset{\sim}{\theta}^*$ and $z_{\alpha/2}$ is the appropriate percentile from the standard normal distribution.

Before presenting some examples, we introduce a new symbol. The limited expected value of a random variable X at u is defined to be the expected value of X censored at u. That is, all probability beyond u is concentrated at u. More formally:

$$E[X;u] = \int_0^u x dF(x;\underset{\sim}{\theta}) + u(1-F(u;\underset{\sim}{\theta})).$$

This, of course, is just the severity for a policy with limit u.

If p is the frequency of loss (to the individual, regardless of any payment by the insurance), we have the following expressions for the pure (net) premium.

No modification	$pE[X]$
Limit of u	$pE[X;u]$
Deductible of d	$p(E[X]-E[X;d])$.

If there is inflation at rate r and a deductible of d, the new pure premium will be

$$(1+r)p(E[X]-E[X;\frac{d}{1+r}]).$$

It is easy to see that this is greater than $1+r$ times the pure premium prior to inflation.

For the automobile example we might desire a 90% confidence interval for the pure premium with a deductible of 100 assuming p is known to be .2 and r is fixed at .05. For a lognormal loss variable we have

$$E[X;u] = \exp(\mu+\sigma^2/2)\Phi(\frac{\ln u-\mu-\sigma^2}{\sigma}) + u(1-\Phi(\frac{\ln u-\mu}{\sigma})).$$

Therefore, the pure premium is (with $z = (\ln \frac{100}{1.05} - \mu)/\sigma$),

$$\gamma = h(\mu,\sigma) = 1.05(.2)[\exp(\mu+\sigma^2/2)(1-\Phi(z-\sigma)) - \frac{100}{1.05}(1-\Phi(z))].$$

At $\hat{\mu} = 5.6101$ and $\hat{\sigma} = .99305$ we have $\hat{\gamma} = 74.938$. The derivatives with respect to the parameters are,

$$\eta_1(\mu,\sigma) = 1.05(.2)[\exp(\mu+\sigma^2/2)(1-\Phi(z-\sigma)+\phi(z-\sigma)/\sigma) - \frac{100}{1.05}\phi(z)/\sigma]$$

and

$$\eta_2(\mu,\sigma) = 1.05(.2)[\exp(\mu+\sigma^2/2)\{\sigma(1-\Phi(z-\sigma)) + (z+\sigma)\phi(z-\sigma)/\sigma\} - \frac{100}{1.05}z\phi(z)/\sigma]$$

where $\phi(z)$ is the standard normal p.d.f. For the automobile data they are estimated as $\eta_1 = 92.051$ and $\eta_2 = 95.955$. The inverse of the estimated information matrix is

$$\begin{bmatrix} 1.0676 & -.082357 \\ -.082357 & .66142 \end{bmatrix}$$

Combining all the above produces $\hat{\sigma}_h^2 = 13{,}681$ and a 90% confidence interval for the pure premium is

$$74.938 \pm 1.645\sqrt{13{,}681}/\sqrt{27{,}618}, \quad \text{or}$$
$$74.938 \pm 1.158.$$

Now suppose that the frequency p also had to be estimated. Assume (all models and numbers connected with frequency were created for this illustration and have no basis in fact) a Poisson distribution with mean λ and suppose that $\hat{\lambda} = .2$ had been estimated on the basis of 5,000 losses on 25,000 policies. Further assume that $\hat{\lambda}$ and $(\hat{\mu},\hat{\sigma})$ are independent. The pure premium can be written as $t(\lambda,\mu,\sigma)$ and the large sample theory still applies. In the Poisson case we have $t(\lambda,\mu,\sigma) = \lambda h(\mu,\sigma)$ and so the extra derivatives are easy to obtain. The function $h(\mu,\sigma)$ as used here is the same as above but with the factor $p = .2$ removed. The enhanced information matrix will be

$$\begin{bmatrix} 1/\lambda^* & \underset{\sim}{0}' \\ \underset{\sim}{0} & I(\underset{\sim}{\theta}^*) \end{bmatrix}$$

and the asymptotic variance of $\hat{\lambda}h(\hat{\mu},\hat{\sigma}) - \lambda^* h(\mu^*,\sigma^*)$ will be given by

$$[h(\mu^*,\sigma^*), \lambda^*\underset{\sim}{\eta}(\mu^*,\sigma^*)'] \begin{bmatrix} n_1/\lambda^* & \underset{\sim}{0}' \\ \underset{\sim}{0} & n_2 I(\underset{\sim}{\theta}^*) \end{bmatrix} \begin{bmatrix} h(\mu^*,\sigma^*) \\ \lambda^*\underset{\sim}{\eta}(\mu^*,\sigma^*) \end{bmatrix}$$

where n_1 and n_2 are the sample sizes for frequency and severity respectively. For the above example we have a variance of

$$.2(374.69)^2/25{,}000 + .2^2(342{,}025)/27{,}618 = 1.6185$$

and the revised confidence interval is

$$74.938 \pm 1.645\sqrt{1.6185}, \quad \text{or}$$
$$74.938 \pm 2.093.$$

Finally, suppose there is uncertainty about the inflation rate r. This could be incorporated by giving a prior distribution for r which reflects our opinion of the inflation rate. It could then be mixed to produce the marginal loss distribution. This would then change the expected value used in determining the pure premium. When the model distribution has a scale parameter the effect of inflation is easily modeled. An example of this is given in Venter (18). Further simplification results if the distribution for r is conjugate to the loss distribution. Examples of this nature are given in Meyers and Schenker (11).

5. AN ANALYSIS OF VARIANCE

Suppose that we have several populations with similar policies. We may believe that the severities are sufficiently similar so that the samples may be combined for ratemaking purposes. In this section an analysis of variance procedure is constructed to provide a test of this hypothesis.

The procedure will be illustrated with samples of automobile bodily injury (liability) losses under different policy limits. In computing premiums for increased limits, it would be beneficial to be able to combine observations from policies with different limits (Lange (7)). In addition, by fitting a parametric model, information on losses under one limit can be used to obtain severities under higher limits. This procedure has been recommended by Finger (4) and Miccolis (12). The data for the example are presented in Table 3.

The analysis of variance is based on the variance components model (Graybill (5), Chapters 16 and 17). Let X_{ij} be the jth $(j = 1,\ldots,n_i)$ loss observed in the sample from policies with the ith $(i = 1,\ldots,k)$ limit. The model is

$$X_{ij} = \mu + a_i + b_{ij}$$

where μ is an unknown constant, $a_1,\ldots,a_k$ are a random sample from a population with mean 0 and variance σ_a^2, and the b_{ij}

Class limit*	Policy limit 5,000	10,000	25,000	50,000	100,000
250	.31857	.32524	.30137	.30488	.25407
500	.47708	.47179	.42374	.44095	.38034
1,000	.61849	.57837	.56575	.56194	.50561
2,000	.74463	.70611	.69452	.68549	.63509
3,000	.81430	.77900	.77215	.75610	.71780
4,000	.85366	.82367	.81507	.80745	.77073
6,000	.87283	.89028	.86986	.86425	.83531
7,000		.90674	.88402	.87644	.85245
8,000		.92006	.90502	.89730	.87244
9,000		.93025	.91735	.90725	.88348
11,000		.93966	.93196	.92908	.90371
12,000			.93516	.93261	.90970
14,000			.94703	.93806	.91944
16,000			.95525	.94961	.92981
19,000			.95982	.95764	.93782
21,000			.96484	.96502	.94436
23,000			.96849	.96727	.94640
30,000			.97123	.97593	.96120
35,000				.97818	.96558
40,000				.98042	.96928
45,000				.98267	.97212
55,000				.98556	.97952
60,000					.98168
70,000					.98409
80,000					.98723
100,000					.99143
n	3,861	1,276	2,190	3,116	16,212

*For a given policy limit, the last entry in that column is for a class limit equal to the policy limit, not the class limit given in this column. All losses above that amount have been censored at the policy limit.

Table 3. Empirical Distribution Functions. Automobile Bodily Injury Losses.

are a random sample from a population with mean 0 and variance σ_b^2. If $\hat{\mu}_1,\ldots,\hat{\mu}_k$ are unbiased estimators of the class means (that is, given a_i, $E[\hat{\mu}_i] = \mu + a_i$) and if $\hat{\sigma}_1^2,\ldots,\hat{\sigma}_k^2$ are unbiased estimators of $\mathrm{Var}(\hat{\mu}_1),\ldots,\mathrm{Var}(\hat{\mu}_k)$ respectively, then unbiased estimators of the three model parameters are:

$$\hat{\mu} = \frac{1}{N}\sum_{i=1}^{k} n_i\hat{\mu}_i$$

$$\hat{\sigma}_b^2 = \frac{1}{N-k}\sum_{i=1}^{k}(n_i-1)n_i\hat{\sigma}_i^2 ,$$

and

$$\hat{\sigma}_a^2 = \{\frac{1}{k-1}\sum_{i=1}^{k} n_i(\hat{\mu}_i-\hat{\mu})^2 - \hat{\sigma}_b^2\}/n_0$$

where $N = n_1 + \cdots + n_k$ and $n_0 = (N^2 - \sum_{i=1}^{k} n_i^2)/N(k-1)$.

The null hypothesis is that the k populations have the same mean. Under the above model this is equivalent to $\sigma_a^2 = 0$. If the $\hat{\mu}_i$ have a normal distribution and the $\hat{\sigma}_i^2$ have a chi-square distribution, then the test statistic

$$F = \frac{\hat{\sigma}_b^2 + n_0\hat{\sigma}_a^2}{\hat{\sigma}_b^2}$$

will, under the null hypothesis, have an F distribution with $k-1$ and $N-k$ degrees of freedom. The grouped data estimates of severity are, at least asymptotically, unbiased and normal. Assuming a chi-square distribution for the $\hat{\sigma}_i^2$ must be done out of necessity.

It is appropriate to question the use of ml estimators rather than the data itself. The example provides the answer. Suppose we want to obtain the severity for a limit of 25,000. Sample means may be available from the three high limit populations but are not available from the two low limit ones. However, we can obtain $\hat{\mu}_i = E[X_i; 25{,}000]$ from all five populations from the fitted lognormal models. For the automobile data the important quantities are:

Limit	5,000	10,000	25,000	50,000	100,000
$\hat{\mu}_i$	2378.2	2627.4	2890.5	3017.1	3677.1
$n_i\hat{\sigma}_i^2$	26,276,000	27,594,000	28,505,000	29,945,000	36,619,000

$$\hat{\mu} = 3296.9$$

$$\hat{\sigma}_b^2 = 33{,}243{,}000$$

$$\hat{\sigma}_a^2 = 426,140$$
$$n_0 = 3907.5$$
$$F = 51.090.$$

With 4 and 26,650 degrees of freedom, the cutoff value for a test at significance level .001 is 4.62. So there is clear evidence that if the five populations were all sold policies with a 25,000 limit, at least two would produce different severities (assuming that upon purchasing a different limit their driving habits and the settlement process remain unchanged). In view of the extraordinarily large sample sizes, this result is hardly surprising.

6. A CREDIBILITY PROCEDURE

In the previous section we were able to determine that the five populations were different. It appears that, in setting the premium for a 25,000 limit, we may only use the data from policies with that limit. If the sample sizes had been much smaller, this may not be a reasonable solution. Credibility procedures have been designed for just this purpose. They allow for the incorporation of data from similar (but not necessarily identical), experience. In this section we develop such a procedure which uses the ml estimates. We then note its similarity to the model used in the previous section. Finally, we show how the estimator may be improved by following an empirical Bayes approach.

The basic credibility model as given in Buhlmann (3) may be formulated using the notation of the previous section. Let $\underset{\sim}{\theta}_1,\ldots,\underset{\sim}{\theta}_k$ be vector parameters which define the k populations. They are considered to be a random sample from some multivariate probability distribution. We are interested in estimating $\mu_i = h(\underset{\sim}{\theta}_i)$, a single-valued function of the parameters. Now let $\mu = E[h(\underset{\sim}{\theta})]$ and $a_i = \mu_i - \mu$, $i = 1,\ldots,k$. Finally, let X_{ij}, $j = 1,\ldots,n_i$ be a random sample from the ith population. Let $b_{ij} = X_{ij} - \mu - a_i$. So, the model is unchanged, but the objective is different. We now fix i and desire an estimate of μ_i. The best (posterior least-squares) estimate is

$$\tilde{\mu}_i = Z_i\bar{X}_i + (1-Z_i)\mu$$

where $\bar{X}_i = \sum_{j=1}^{n_i} X_{ij}/n_i$, the ith sample mean, and $Z_i = \sigma_a^2/(\sigma_a^2+\sigma_b^2/n_i)$ is the credibility factor. From the previous section we have estimates of $\mu, \sigma_a^2,$ and σ_b^2 and since the sample means are unavailable, we use $\hat{\mu}_i = h(\underset{\sim}{\hat{\theta}}_i)$ in place of $\bar{X}_i$. It is possible to improve the estimation. Straub (17) showed that the least-squares estimate of μ should be credibility weighted and not weighted by the sample size. That is,

$$\hat{\mu} = \sum_{j=1}^{k} Z_j \hat{\mu}_j / \sum_{j=1}^{k} Z_j .$$

(Straub's result was for individual data but the results follow for any unbiased estimator of μ_i.) This result is appealing. Each $\hat{\mu}_j$ contributes on the basis of credibility, not the sample size. Of course, this version depends on the Z_j which in turn depends on $\hat{\sigma}_b^2$ which depends on $\hat{\mu}$. The obvious solution is to begin with a sample size weighted $\hat{\mu}$ and then iterate until the estimates stabilize.

For the automobile data the results are $\hat{\mu} = 2920.2$, $\hat{\sigma}_a^2 = 666{,}210$, $\hat{\sigma}_b^2 = 33{,}243{,}000$, and $Z_3 = .97772$. So, the credibility estimate of severity with a 25,000 limit is

$$\tilde{\mu}_3 = .97772(2890.5) + .02228(2920.2) = 2891.2.$$

A similar problem has been studied by statisticians under the heading of "James-Stein estimation." Numerical results are usually obtained via an empirical Bayes approach. One advantage is that adjustments are made so that the estimates of the credibility factors Z_i are less biased. The most recent contribution is by Morris (14) and references to earlier work may be found in that paper.

Morris allows for some generalization in that the null hypothesis in the corresponding analysis of variance need not be one of equal means. We again begin with $\mu_i = h(\underset{\sim}{\theta}_i)$ as the quantity of interest and assume it is a realization of a random variable with mean $\underset{\sim}{r}_i'\underset{\sim}{\beta}$ and variance σ_a^2 (Morris does not require equal variances). Here r_i is a $q \times 1$ vector of known values and $\underset{\sim}{\beta}$ is a $q \times 1$ vector of unknown regression coefficients. Setting $\underset{\sim}{r}_i = 1$ and $\underset{\sim}{\beta} = \mu$ yields the model used in the first part of this section. The rest of the model gives $\hat{\mu}_i$ as an unbiased estimator of μ_i with variance σ_i^2. In most applications $\hat{\mu}_i$ is the sample mean and σ_i^2 is estimated by the sample variance divided by the sample size. In our case, with grouped ml estimates, we use $\hat{\mu}_i$ and $\hat{\sigma}_i^2$ as constructed in Section 5.

The first step is to estimate the regression coefficients $\underset{\sim}{\beta}$. The weighted least-squares solution is

$$\hat{\underset{\sim}{\beta}} = (R'DR)^{-1}R'D\hat{\underset{\sim}{\mu}}$$

where $\hat{\underset{\sim}{\mu}}' = (\hat{\mu}_1,\ldots,\hat{\mu}_k)$, $D^{-1} = \text{diagonal}(\hat{\sigma}_1^2,\ldots,\hat{\sigma}_k^2)$, and R is $k \times q$ with ith row equal to $\underset{\sim}{r}_i'$. For the case $\underset{\sim}{r}_i' = 1$ and $\underset{\sim}{\beta} = \mu$ we have $\hat{\mu} = \Sigma\hat{\sigma}_i^2\hat{\mu}_i/\Sigma\hat{\sigma}_i^2$. This is, if the variances of the observations are assumed equal, a sample size weighted grand mean. In the general setting, the empirical Bayes estimate as given by

Morris is

$$\tilde{\mu}_i = \hat{Z}_i \hat{\mu}_i + (1-\hat{Z}_i) \underset{\sim}{r}'_i \hat{\underset{\sim}{\beta}}$$

with

$$\hat{Z}_i = 1 - \frac{k-q-2}{k-q} \frac{\hat{\sigma}_i^2}{\hat{\sigma}_i^2 + \hat{\sigma}_a^2}$$

and

$$\hat{\sigma}_a^2 = \sum_{i=1}^{k} \bar{W}_i \{ \frac{k}{k-q} (\hat{\mu}_i - \underset{\sim}{r}'_i \hat{\underset{\sim}{\beta}})^2 - \hat{\sigma}_i^2 \}$$

where

$$\bar{W}_i = (\hat{\sigma}_i^2 + \hat{\sigma}_a^2)^{-1} / \sum_{i=1}^{k} (\hat{\sigma}_i^2 + \hat{\sigma}_a^2)^{-1} .$$

Should $\hat{\sigma}_a^2$ be negative, replace it with zero. Since $\bar{W}_i$ depends on $\hat{\sigma}_a^2$, it is necessary to iterate for $\hat{\sigma}_a^2$. The estimator given in Section 5 provides a good starting value. It is surprising, as noted in the discussion of Lindley (8) that Morris uses least-squares to estimate $\underset{\sim}{\beta}$, a procedure which the rest of the analysis discredits. The result, in the one-dimensional case, was a sample size weighted mean instead of the preferred credibility weighted mean. A subject for future research would be an investigation of the improvement (both theoretical and empirical) in replacing D by a diagonal matrix of credibilities. This was not done in the examples which follow.

Morris concludes by providing an estimate of the variance of $\tilde{\mu}_i$. It is

$$s_i^2 = \hat{\sigma}_i^2 \{ \hat{Z}_i + \frac{\hat{q}_i}{k} (1-\hat{Z}_i) \} + \frac{2}{k-q-2} (1-\hat{Z}_i)^2 \frac{\Sigma \bar{W}_i \hat{\sigma}_i^2 + \hat{\sigma}_a^2}{\hat{\sigma}_i^2 + \hat{\sigma}_a^2} (\hat{\mu}_i - \underset{\sim}{r}'_i \hat{\underset{\sim}{\beta}})^2$$

where $\hat{q}_i$ is the ith diagonal element of

$$k \bar{W}_i R (R' D R)^{-1} R' \sum_{i=1}^{k} (\hat{\sigma}_i^2 + \hat{\sigma}_a^2)^{-1} .$$

We now reanalyze the automobile data using the above technique. We begin with $\underset{\sim}{r}'_i = 1$ and $\underset{\sim}{\beta} = \mu$. The starting values are taken from Section 5. The regression coefficient is $\hat{\mu} = 3225.9$. The iteration proceeds as follows:

Iteration		$\bar{W}_1$	$\bar{W}_2$	$\bar{W}_3$	$\bar{W}_4$	$\bar{W}_5$
1	426,140	.20174	.19506	.19888	.20044	.20388
2	348,770	.20210	.19402	.19863	.20052	.20472
3	348,840	.20210	.19402	.19863	.20052	.20472

At the fourth iteration all six values were unchanged. The results for the 25,000 limit are $\hat{Z}_3 = .98201$ and

$$\tilde{\mu}_3 = .98201(2890.5) + .01799(3225.9) = 2986.5.$$

The estimated standard deviation of $\tilde{\mu}_3$ is 113.06.

We observe that the 5 estimates $\hat{\mu}_1,\ldots,\hat{\mu}_5$ are increasing. This is consistent with a prior belief that those who purchase higher limits tend to have larger settlements. This is despite the fact that the five estimates are all based on censoring at 25,000. A better estimation may be possible if a straight line is fitted to the means. One choice of design matrix is (dropping the 000)

$$\begin{bmatrix} 1 & 5 \\ 1 & 10 \\ 1 & 25 \\ 1 & 50 \\ 1 & 100 \end{bmatrix}$$

The least squares regression coefficients are $\hat{\beta}_1 = 2403.1$ and $\hat{\beta}_2 = 12.779$. The resulting estimates are

i	$\hat{\mu}_i$	$\underset{\sim}{r}_i'\underset{\sim}{\beta}$
1	2378.2	2467.0
2	2627.4	2530.9
3	2890.5	2722.6
4	3017.1	3042.0
5	3677.1	3681.0

After nine iterations, we have $\hat{\sigma}_a^2$ and the following credibility estimates.

i	$\hat{Z}_i$	$\tilde{\mu}_i$	s_i
1	.76812	2398.8	76.45
2	.70700	2599.1	127.14
3	.72872	2844.9	100.04
4	.74552	3023.4	86.21
5	.85622	3677.7	45.48

The new estimate of $\tilde{\mu}_3$ is 2844.9 and the standard deviation has been reduced by 12% from the previous model.

7. A DIFFERENT REGRESSION MODEL

The regression model introduced in Section 6 is appropriate when there is a linear structure relating the population parameters. Another form of linear structure was introduced by Hachemeister (6). Here the structure is placed on the observations within each sample. In this section we conpare the two models in order to clarify their differences. We close by noting that the two models can be combined into a single general model.

Hachemeister's model, in our notation, requires a fixed number of observations from each population. That is, let $\hat{\mu}_i,\ldots,\hat{\mu}_{im}$ be ordered estimates from the ith population. They may be single observations or estimates. The vector of estimates $\hat{\underset{\sim}{\mu}}_i$ has mean $Y_i\underset{\sim}{\tau}_i$ where Y_i is an $m\times r$ design matrix and $\underset{\sim}{\tau}_i$ is an $r\times 1$ vector of regression coefficients. The covariance matrix of $\hat{\underset{\sim}{\mu}}_i$ is $\sigma_i^2 P_i$ where P_i is diagonal and known (usually the inverses of the sample sizes) and σ_i^2 is unknown. The next step is to consider $\underset{\sim}{\tau}_i$ and σ_i^2 as realizations of random variables with means $\underset{\sim}{\tau}$ and σ^2 and covariances Γ and γ^2.

To see how this relates to the variance components model of Section 5, let $\hat{\mu}_{ij} = X_{ij}$ be a single observation from population i. Let $\underset{\sim}{Y}_i$ be an $m\times 1$ vector of 1's and $\underset{\sim}{\tau}_i$ be the scalar $\mu+a_i$. Then $E[X_{ij}] = \mu+a_i$ and $\mathrm{Var}(\underset{\sim}{X}_{ij}) = \sigma_i^2$. In addition, $\mu+a_i$ has mean μ and variance $\Gamma = \sigma_a^2$. So $X_{ij} = \mu+a_i+b_{ij}$ where a_i has mean 0 and variance σ_a^2. Since

$$\sigma_i^2 = \mathrm{Var}(X_{ij}) = \sigma_a^2 + \mathrm{Var}(b_{ij}),$$

we have

$$\mathrm{Var}(b_{ij}) = \sigma_i^2 - \sigma_a^2 = \sigma_b^2$$

if we assume that all populations have the same variance. The one difference is that σ_i^2 is treated in the variance components model as a constant while in Hachemeister's model it is considered to be a random variable. It is shown in the paper by Zehnwirth in these proceedings that the resulting estimator is not affected by this consideration.

We now have two regression models at our disposal. One (Morris) provides a structure between the populations. The other (Hachemeister) provides a structure within each population. The following model incorporates both features. Let X_{ij} be the jth $(j=1,\ldots,m)$ observation from the ith $(i=1,\ldots,k)$ population. The observations from the ith population are $\underset{\sim}{X}_i' = (X_i,\ldots,X_{im})$ and given $\underset{\sim}{\beta}_i$ have mean $Y_i\underset{\sim}{\beta}_i$ and covariance matrix Σ_1. We next assume that each $\underset{\sim}{\beta}_i$ is random with mean $H\underset{\sim}{\beta}$ and covariance matrix Σ_2. All the above models assume Σ_1

and Σ_2 to be diagonal. Setting $H = I$ gives Hachemeister's model and setting each $\underset{\sim}{Y}_i = (1,\ldots,1)'$ gives Morris' model. Placing prior distributions on Σ_1, Σ_2, and $\underset{\sim}{\beta}$ gives a Bayesian model. This is similar to the random coefficient model developed in Lindley and Smith (9). An empirical Bayes version of this model has yet to be constructed.

REFERENCES

1. Anderson, T.W. 1958, An Introduction to Multivariate Statistical Analysis, New York: Wiley.

2. Brown, K.M. and Dennis, J.E. 1972, "Derivative Free Analogues of the Levenberg-Marquardt and Gauss Algorithms for Nonlinear Least Squares Approximations," Numerische Mathematik, 18, pp. 289-297.

3. Buhlmann, H. 1967, "Experience Rating and Credibility," ASTIN Bull., 4, pp. 199-207.

4. Finger, R.J. 1976, "Estimating Pure Premiums by Layers - An Approach," Proc. Casualty Act. Soc., 63, pp. 34-56.

5. Graybill, F.A. 1961, An Introduction to Linear Statistical Models, Vol. I, New York: McGraw-Hill.

6. Hachemeister, C. 1975, "Credibility for Regression Models with Applications to Trend," in Credibility, Theory and Applications (P.M. Kahn, ed.), New York: Academic Press, pp. 129-163.

7. Lange, J.T. 1969, "The Interpretation of Liability Increased Limits Statistics," Proc. Casualty Act. Soc., 56, pp. 163-173.

8. Lindley, D. 1983, Comment on "Parametric Empirical Bayes Inference: Theory and Applications," J. Amer. Statist. Assoc., 78, pp. 61-62.

9. Lindley, D. and Smith, A. 1972, "Bayes Estimates for the Linear Model," J. Royal Statist. Soc. (B), 34, pp. 1-41.

10. Marquardt, D.W. 1963, "An Algorithm for Least Squares Estimation of Nonlinear Parameters," J. Soc. Industrial and Appl. Math., 11, pp. 431-441.

11. Meyers, G. and Schenker, N. 1983, "Parameter Uncertainty in the Collective Risk Model," Proc. Casualty Act. Soc., 70 (to appear).

12. Miccolis, R.S. 1977, "On the Theory of Increased Limits and

Excess loss Pricing," Proc. Casualty Act. Soc., 64, pp. 27-73.

13. Moore, D.S. 1978, "Chi-square Tests," in Studies in Statistics (R.V. Hogg, ed.), Mathematical Association of America, pp. 66-106.

14. Morris, C.N. 1983, "Parametric Empirical Bayes Inference: Theory and Applications," J. Amer. Statist. Assoc., 78, pp. 47-55.

15. Ralston, M.L. and Jennrich, R.I. 1978, "DUD, A Derivative-Free Algorithm for Nonlinear Least Squares," Technometrics, 20, pp. 7-14.

16. Rao, C.R. 1965, Linear Statistical Inference and Its Applications, New York: Wiley.

17. Straub, E. 1975, "Credibility in Practice," in Credibility, Theory and Applications (P.M. Kahn, ed.), New York: Academic Press, pp. 347-359.

18. Venter, G.G. 1983, "Transformed Beta and Gamma Distributions and Aggregate Losses," Proc. Casualty Act. Soc., 70 (to appear).

RATING OF NON PROPORTIONAL REINSURANCE TREATIES BASED ON ORDERED CLAIMS

Erhard Kremer

University of Hamburg

ABSTRACT

The purpose of the present paper is to give a survey of mathematical models and methods for rating special nonproportional reinsurance treaties, so-called non-proportional reinsurance treaties based on ordered claims. The emphasis lies on describing the different main approaches with a thorough formulation of the basic stochastic assumptions, whereas details dealing with the practical implementation are mostly avoided. The rating of the nonproportional reinsurance treaties in the view of the following presentation consists mainly in applying or modifying concepts and results of Mathematical Statistics.

I. INTRODUCTION

Various reinsurance forms exist nowadays, each having its own special field of efficient applicability. The requirement on the solvability of an insurance company often is the reason for concluding a so-called proportional reinsurance treaty, i.e. a quota or surplus share. Whereas the need for protection against large claims explains the importance of the so-called non-proportional reinsurance treaties, i.e. the famous <u>collective treaty</u>:

the stop-loss cover

F. de Vylder et al. (eds.), Premium Calculation in Insurance, 285–314.

and the various forms of individual reinsurance treaties:

the excess-of-loss cover,
the largest claims treaty
and the ECOMOR-treaty.

The most frequently used nonproportional treaty is still the excess-of-loss cover, whereas the stop-loss treaty is offered in practice only in a few branches and the largest claims and the ECOMOR-treaties are very seldomly applied. When concluding such a nonproportional reinsurance treaty, one of the main problems for the reinsurer is the calculation of a risk-adequate premium. A vast body of literature exists for the stop-loss treaty dealing with the problem of estimating the premium (see e.g. Vajda (1951), Conolly (1955), Bohman & Esscher (1963), Ammeter (1964), Bowers (1969), Bühlmann et al. (1974), Taylor (1977), Goovaerts et al. (1980), Panjer (1980), Gerber (1980),(1982), Heilmann (1981), De Vylder (1982), De Vylder & Goovaerts (1982)). The mathematical theory of that treaty is very well-developed and fairly general statements have been derived. For details we refer the interested reader to the above literature.

In the present paper we want to discuss the second class of individual reinsurance treaties, i.e. the excess-of-loss cover, the largest claims treaty and the ECOMOR cover. We will present basic stochastic concepts for the premium calculation of these important treaties. The aim is to give a unifying survey of older and newer mathematical approaches, emphasis lying more on the theoretical than practical aspects. It becomes apparent that the rating of our nonproportional reinsurance treaties is strongly related to the field of Mathematical Statistics, most models and methods are based on (modified) results of that science. The paper starts with defining a general form of a reinsurance treaty including all above individual nonproportional reinsurance treaties. When using asymptotic arguments simple premium formulas can be given for the general treaty. In the following chapters the general result is applied to the special nonproportional reinsurance treaties, i.e. to the excess-of-loss, the largest claims and the ECOMOR cover. Using more restricted model assumptions more specialized rating methods can be given. We present the underlying simple stochastic models and review the resulting premium calculation methods, some of them being used quite frequently in practice.

II. THE GENERAL NONPROPORTIONAL REINSURANCE TREATY BASED ON ORDERED CLAIMS

Let us consider a collective K of risks and let N denote the random variable of the number of claims. The corresponding claim amounts are described by the random variables $X_1,\dots,X_N$. Denote by:

$$X_{N:1} \leq \dots \leq X_{N:N}$$

the claims ordered in increasing size. Furthermore let h be a measurable function:

$$h : [0,\infty) \to [0,\infty)$$

and for each natural number n let b_{ni}, $i=1,\dots,n$ be real constants with

$$b_{ni} \in [0,1], \text{ for all } i,n.$$

For fixed nonnegative integer l and (in case $l \geq 1$) given constants c_j, $j=1,\dots,l$, natural numbers q_j, $j=1,\dots,l$ define:

$$R_n(y_1,\dots,y_n) :=$$

$$\sum_{i=1}^{n} b_{ni} h(y_i) + \sum_{j=1}^{l} c_j h(y_{n-q_j}) \cdot 1_{[0,n)}(q_j) \qquad (2.1)$$

(1_M denoting the indicator function of the set M and defining in case l=0 the second summation to be equal to zero) and assume:

$$R_n(y_1,\dots,y_n) \in [0, \sum_{i=1}^{n} y_i]$$

$$\text{for all } 0 \leq y_1 \leq \dots \leq y_n \text{ and } n.$$

Interpreting:

$$R_N(X_{N:1},\dots,X_{N:N}) \cdot 1_{(0,\infty)}(N)$$

as the part of the total claim amount $\sum_{i=1}^{N} X_i$ taken by the reinsurer, the family $R = (R_n, n=1,2,\dots)$ of measurable mappings defines a reinsurance treaty. Obviously the stop-loss treaty is not contained in the model, nor the surplus share (since the factors b_{ni} do not depend on the special risk producing the i-th smallest claim $X_{N:i}$). Finally assume that for each n the:

b_{ni}, i=1,...,n are not identical,

excluding the quota share cover. We get:

Definition 1.

The family $R = (R_n, n=1,2,..)$ satisfying the above conditions, is called nonproportional reinsurance treaty based on ordered claims.

Functionals of type (2.1), applied to an ordered sample, are already studied intensively in the field of Mathematical Statistics. They are very important in the field of robust statistics and are well-known under the heading "linear combinations of functions of order statistics" (see e.g. Huber (1981), Moore (1968), Shorack (1972), Stigler (1974), chapter 8 in Serfling (1980)).

As special cases of the general treaty we get:

1. the excess-of-loss treaty with priority P,

covering all claims excess P, by the choice:

$$l=0,\ h(y) = \max(y-P,0)$$
$$b_{ni} = 1, \quad \text{for all } i.$$

2. the largest claims treaty,

covering the p largest claims, by the choice:

$$l=0,\ h(y) = y$$
$$b_{ni} = \begin{cases} 1, & \text{for } i>n-p \\ 0, & \text{for } i\le n-p. \end{cases}$$

3. the largest claims treaty with priority P,

covering the p largest claims excess the priority P, by the choice l=0, h as in the first example, the b_{ni} as in the second example.

4. <u>the ECOMOR-treaty,</u>

covering all claims excess the p-th largest claim, by the choice:

$$l=1,\ c_1=-p,\ q_1=p-1,\ h \text{ and } b_{ni} \text{ as in the example 2.}$$

In the examples 1, 2 and 4 the random variables describing the claim amount retained by the insurer, i.e.:

$$S_N = [\sum_{i=1}^{N} X_i - R_N(X_{N:1},\dots,X_{N:N})] \cdot 1_{(0,\infty)}(N),$$

are related to well-known procedures for treating outliers in statistical data. The S_N in example 1 corresponds to the classical <u>truncation method</u>, the S_N of example 2 to the <u>method of winzorizing</u> and the S_N in example 4 to the <u>method of data-trimming</u>, all used in Mathematical Statistics for robustifying an estimator (see Barnett and Lewis (1978), David (1980), Huber (1981)).

In the present section we want to state a simple premium formula for the general nonproportional reinsurance treaty based on ordered claims. In the last ten years various <u>principles for premium calculation</u> were defined and analyzed mathematically (see e.g. Benktander (1971), Berliner (1976), Goovaerts, De Vylder & Haezendonck (1981)). In the author's point of view the <u>standard deviation principle</u> is most appropriate for rating nonproportional reinsurance treaties, i.e. the <u>risk premium</u> for the treaty R is defined according:

$$P(R) = \mu(R) + \Lambda \cdot \sigma(R) \tag{2.2}$$

where:

$$\mu(R) = E(R_N(X_{N:1},\dots,X_{N:N}) \cdot 1_{(0,\infty)}(N))$$

$$\sigma(R) = (Var(R_N(X_{N:1},\dots,X_{N:N}) \cdot 1_{(0,\infty)}(N)))^{1/2}$$

and Λ is a suitable positive constant. In the following we do not care about the value of Λ, we regard it as being fixed. From the theoretical point of view it could be calculated by <u>ruin-theoretical considerations</u> (see e.g. Ammeter (1969) and the lectures of

Prof. Gerber), asymptotically it might be chosen as a suitable quantile of the standard normal distribution. Now for our treaty R the problem of premium calculation is reduced to giving handy expressions for $\mu(R)$ and $\sigma^2(R)$. Simple formulas for both parameters can be derived by asymptotic arguments.

We consider a sequence of growing collectives K_k, $k=1,2,\dots$, i.e. the claim number N_k of collective K_k satisfies:

$$\lim_{k\to\infty} E(N_k) = \infty .$$

For the collective K_k we define a nonproportional reinsurance treaty $R_k = (R_{kn}, n=1,2,\dots)$ based on ordered claims by (2.1) with:

$$b_{ni} = b_{ni}^{(k)}, \quad c_j = c_j^{(k)}, \quad q_j = q_j^{(k)},$$

depending on k and assume:

(A.1) the claim amounts $X_i, i=1,2,\dots$ are independent and identically distributed with the continuous distribution function F.

(A.2) N_k is stochastically independent of the claim amounts X_i, $i=1,2,\dots$

(A.3) there exist constants $s_j \in (0,1)$, $j=1,\dots,l$ such that:

$$\lim_{k\to\infty} \frac{q_j^{(k)}}{E(N_k)} = s_j, \quad j=1,\dots,l.$$

(A.4) there exist (in case $l \geq 1$) constants c_j, $j=1,\dots,l$ such that:

$$\lim_{k\to\infty} \frac{c_j^{(k)}}{E(N_k)} = c_j, \quad j=1,\dots,l$$

(A.5) there exists a function $b/[0,1]$ and numbers

$$0 = t_0 < t_1 < \dots < t_{m+1} = 1,$$

such that the sequence of functions $b_n^{(k)}$, defined according:

$$b_n^{(k)}(o) = b_{n1}^{(k)}$$

$$b_n^{(k)}(u) = b_{ni}^{(k)}, \text{ for } u \in ((i-1)/n, i/n], \; i=1,\ldots,n,$$

satisfies

$\lim_{k\to\infty} b_{n_k}^{(k)} = b$ uniformly on closed subintervals of $[0,1] - \{t_1,\ldots,t_m\}$ for each sequence $\{n_k\}$ satisfying $n_k \to \infty$, $n_k/E(N_k) \to 1$.

(A.6) $b = b_s + b_d$, where b_s is of bounded variation and continuously differentiable and b_d is a step funktion with steps at $t_1,\ldots,t_m$.

(A.7) h is nondecreasing on $(0,\infty)$, continuous at the points $F^{-1}(1-s_j)$, $j=1,\ldots,l$ and F is strictly increasing (from the left and right hand side) at the points $F^{-1}(1-s_j)$, $j=1,\ldots,l$, $F^{-1}(1-t_i)$, $i=1,\ldots,m$, with the convention:

$$F^{-1}(u) := \inf\{x : F(x) \geq u\}.$$

In case $l \geq 1$ we assume in addition that $\overline{F} = h \circ F^{-1}$ is differentiable at the points $1-s_j$, $j=1,\ldots,l$. Denote the derivative by $\overline{F}'$.

Let us define:

$$\mu = \int b(F(x))h(x)F(dx) + \sum_{j=1}^{l} c_j \cdot \overline{F}(1-s_j)$$

$$\begin{aligned}\sigma^2 = \iint &(\min(F(s),F(t)) - F(s)\cdot F(t)) \cdot \\ &\cdot b(F(s))\cdot b(F(t))h(ds)h(dt) + 2\cdot \sum_{j=1}^{l} c_j \cdot \overline{F}'(1-s_j)\cdot \\ &\cdot \int [\min(F(t),1-s_j) - F(t)(1-s_j)]b(F(t))h(dt) + \\ &+ \sum_{i=1}^{l}\sum_{j=1}^{l} c_i\cdot c_j\cdot \overline{F}'(1-s_i)\overline{F}'(1-s_j)\cdot \\ &\cdot [\min(1-s_i,1-s_j)-(1-s_i)(1-s_j)].\end{aligned}$$

Results from the theory of linear combinations of functions of order statistics (see Shorack (1972), Stigler (1974)) suggest to use:

$$\mu_k = E(N_k) \cdot \mu$$

$$\sigma_k^2 = E(N_k) \cdot \sigma^2 + Var(N_k) \cdot \mu^2$$

as approximations for $\mu(R_k)$ and $\sigma^2(R_k)$ when k, i.e. $E(N_k)$ is large. In fact we have (under some minor additional regularity conditions):

Theorem 1

(a) Under the additional assumption:

$$\lim_{k\to\infty} \frac{\sqrt{Var(N_k)}}{E(N_k)} = 0$$

holds:

$$\lim_{k\to\infty} \frac{\mu(R_k)}{E(N_k)} = \mu \text{ resp.} \lim_{k\to\infty} \frac{\mu(R_k)}{\mu_k} = 1$$

(b) Consider case l=0 and assume in addition:

$$\lim_{k\to\infty} \frac{Var(N_k)}{E(N_k)} = c \in [0,\infty),$$

then:

$$\lim_{k\to\infty} \frac{\sigma^2(R_k)}{E(N_k)} = \sigma^2 + c \cdot \mu^2 \text{ resp. } \lim_{k\to\infty} \frac{\sigma^2(R_k)}{\sigma_k^2} = 1.$$

For nonrandom N_k similar statements can be found in the papers of Jung (1954), Bickel (1967), Stigler (1974). The proof of (a) is given in Kremer (1983d), and similarly part (b) can be shown by carrying over the results from nonrandom (see Stigler (1974)) to random N_k. We conjecture that the statement of part (b) is correct also in case $l \geq 1$. □

According to the above considerations one concludes that the risk premium for a general nonproportional reinsurance treaty $R(= R_k)$ based on ordered claims can be calculated for a large portfolio by approximating $\mu(R), \sigma^2(R)$ by:

$$\hat{\mu}(R) = E(N) \cdot \mu \tag{2.3}$$

$$\hat{\sigma}^2(R) = E(N) \cdot \sigma^2 + Var(N) \cdot \mu^2 \tag{2.4}$$

Formula (2.2) with the approximations (2.3), (2.4) is our desired <u>general premium formula</u>. In the following chapters this formula is applied to the special non-proportional reinsurance treaties and we derive from it simple rating methods.

III. THE EXCESS OF LOSS TREATY

According to example 1 the excess of loss treaty with priority P (in short: XL(P)) is defined as that non-proportional reinsurance treaty based on ordered claims, implying the reinsurer's claim amount:

$$R_N^{XL} := \sum_{i=1}^{N} (X_{N:i} - P) \cdot 1_{\{X_{N:i} \geq P\}} = \sum_{i=1}^{N} (X_i - P) \cdot 1_{\{X_i \geq P\}}.$$

Obviously Theorem 1 is applicable with:

$$b(u) = 1, \text{ for all } u \in [0,1],$$

yielding the simple formulas:

Corollary 1

In case of the excess-of-loss treaty with priority P one has:

$$\hat{\mu}(XL(P)) = E(N) \cdot \int_{[P,\infty)} (x-P)F(dx) \tag{3.1}$$

$$\hat{\sigma}^2(XL(P)) = E(N) \cdot \left[\int_{[P,\infty)} (x-P)^2 F(dx)\right] + [Var(N) - E(N)]\left(\int_{[P,\infty)} (x-P)F(dx)\right)^2. \tag{3.2}$$

According to Theorem 1, these formulas are asymptotically correct. Nevertheless it is well-known that they are exact already for each finite collective (see Ammeter (1955), Helbig (1953)). Now the premium can be calculated when one has estimates for the mean claim number $E(N)$, $Var(N)$ and the claim size distribution function F. Suppose we know claim numbers of past years $i=1,\dots,I$:

$$N_i,\ i=1,\dots,I$$

and claim amounts

$$X_{ij},\ j=1,\dots,N_i,\quad i=1,\dots,I,$$

corresponding variables assumed to be i.i.d.. Then we get the following two <u>rating methods</u>:

(1.) The Nonparametric Method:

We estimate $E(N)$ and $Var(N)$ by:

$$\hat{\mu}_N = \frac{1}{I} \cdot \sum_{i=1}^{I} N_i,\quad \hat{\sigma}_N^2 = \frac{1}{I-1} \cdot \sum_{i=1}^{I} (N_i - \hat{\mu}_N)^2$$

and F by the empirical distribution function:

$$\hat{F}(x) = \frac{1}{\sum_{i=1}^{I} N_i} \cdot \sum_{i=1}^{I} \sum_{j=1}^{N_i} 1_{(-\infty,x]}(X_{ij}). \ \square$$

(2.) The Parametric Method:

We might estimate $E(N)$, $Var(N)$ as in (1.) by $\hat{\mu}_N$, $\hat{\sigma}_N^2$. Nevertheless, assuming N being <u>Poisson</u>-distributed, i.e.:

$$\text{Prob}(N=n) = \lambda^n \cdot \frac{\exp(-\lambda)}{n!}\ ,\ n = 0,1,2,\dots (\lambda \in (0,\infty)) \tag{3.3}$$

implying $Var(N) = E(N)$, one only has to estimate $\lambda = E(N)$ by $\hat{\mu}_N$. Clearly the Poisson claim number assumption might be inappropriate (which has to be checked by a goodness-of-fit test). Alternatively one can try the <u>negative-binomial</u> assumption, i.e.:

$$\text{Prob}(N=n) = \binom{\alpha+n-1}{n} \cdot p^\alpha \cdot (1-p)^n,\ n = 0,1,2,\dots \tag{3.4}$$

$$(\alpha \in (0,\infty),\ p \in (0,1)),$$

and estimate the parameters α, p from the known past claim numbers by the moment estimators. As optimistic parametric model assumption for the claim size distribution one might choose the <u>exponential distribution function</u>, i.e.:

$$F(x) = 1 - \exp(-\beta x), \quad x \geq 0 \qquad (3.5)$$

and estimate β by the best estimator:

$$\hat{\beta} = \frac{\sum_{i=1}^{I} N_i - 1}{\sum_{i=1}^{I} \sum_{j=1}^{N_i} X_{ij}} . \qquad (3.6)$$

It is well-known (see Benktander (1970), Benktander & Segerdahl (1960)) that for large claims, e.g. in reinsurance, the exponential distribution is inappropriate and a better fit to the claims data can be reached by choosing the <u>Pareto distribution function</u>, i.e.:

$$F(x) = 1 - a^{\alpha} \cdot x^{-\alpha}, \; x \geq a, \; (a > 0, \; \alpha > 0) \qquad (3.7)$$

which in some sense is a pessimistic model assumption (see Benktander (1962), Berliner (1980)). The parameter a is usually given in practice, α has to be estimated, i.e. by the Maximum-likelihood-estimator:

$$\hat{\alpha} = \frac{\sum_{i=1}^{I} N_i}{\sum_{i=1}^{I} \sum_{j=1}^{N_i} \ln(\frac{X_{ij}}{a})} . \qquad (3.8)$$

Suppose the claim numbers are Poisson-distributed. Then we get under the claim-size distribution assumption (3.5) for (3.1), (3.2):

$$\hat{\mu}(XL(P)) = E(N) \cdot \frac{\exp(-\beta \cdot P)}{\beta}$$

$$\hat{\sigma}^2(XL(P)) = E(N) \cdot \frac{2}{\beta^2} \cdot \exp(-\beta \cdot P)$$

and under the more realistic claim-size distribution assumption (3.7):

$$\hat{\mu}(XL(P)) = E(N) \cdot a^{\alpha} \cdot \frac{P^{-\alpha+1}}{\alpha-1} \quad (\alpha > 1)$$

$$\hat{\sigma}^2(XL(P)) = E(N) \cdot a^{\alpha} \cdot P^{-\alpha+2} \cdot \frac{2}{(\alpha-1)(\alpha-2)} \quad (\alpha > 2)$$

(compare Helbig (1953), Harding (1968), Berliner (1980)), implying in case $\alpha=3$, $a=P$ the simple relation:

$$\hat{\sigma}(XL(P)) = \frac{2\cdot\hat{\mu}(XL(P))}{\sqrt{E(N)}} \tag{3.9}$$

(see Benktander (1970),(1977)) which is very helpful for practical rating in Motor Liability Insurance.

The above statistical rating methods were derived from the general Theorem 1. Clearly under additional assumptions more special methods can be deduced. In the remaining part of this chapter we scetch such methods, two of them being often used in reinsurance practice.

(3.) The Burning Cost Method:

Denote by Y_i the random variable describing the total claim amount taken by the reinsurer in year i, i.e.:

$$Y_i = \sum_{j=1}^{N_i} (X_{ij}-P)\cdot 1_{\{X_{ij} \geq P\}}, \quad i=1,\ldots,I+1$$

and let the random variable P_i denote the total (net-) premium income of the collective covered by the reinsurance treaty. Assume that:

$$\frac{E(Y_i)}{E(P_i)} = \rho \quad \text{is independent of } i=1,\ldots,I+1. \tag{3.10}$$

Having an estimate $\hat{\rho}$ for ρ based on the past values $X_i,P_i,i=1,\ldots,I$ and knowing $\mu_P = E(P_{I+1})$ (resp. having a good prediction of μ_P from the time series $(P_1,\ldots,P_I)$) the net premium $\mu(XL(P)) = E(Y_{I+1})$ of year I+1 can be predicted by,

$$\hat{\mu}(XL(P)) = \hat{\rho}\cdot\mu_P. \tag{3.11}$$

This simple forecasting method is well-known to the statistician under the heading <u>ratio estimation</u> (see Cochran (1963), chapter 6) and very suitable for situations, where X_i,P_i are highly correlated. An estimate $\hat{\rho}$ for ρ is called <u>ratio-estimate</u>.
For example assumption (3.10) is satisfied if:

$$Y_i = \rho\cdot P_i + e_i \tag{3.12}$$

where e_i, $i=1,\ldots,I+1$ are random variables with $E(e_i) = 0$, $\forall i = 1,\ldots,I+1$. One has (see Cochran (1963), chapter 6.9):

Theorem 2

Assume in (3.12) that conditionally given $(P_1,\ldots,P_I)$ the $e_i, i=1,\ldots,I$ are uncorrelated with:

$$E(e_i \mid P_1,\ldots,P_I) = 0, \quad Var(e_i \mid P_1,\ldots,P_I) = c \cdot P_i, \tag{3.13}$$

then conditionally given $(P_1,\ldots,P_I)$:

$$\hat{\rho} = \frac{\sum_{i=1}^{I} Y_i}{\sum_{i=1}^{I} P_i} \tag{3.14}$$

is the best linear unbiased estimator of ρ in model (3.12).

A statement similar to this theorem was also given in the actuarial literature by Franckx (1960). In reinsurance practice the method (3.11) with the estimator (3.14) is well-known under the name <u>burning cost method</u> (or <u>loss-ratio method</u>) (see Gerathewohl (1976), Grossmann (1977), Kremer (1983e)).

Often $\hat{\sigma}(XL(P))$ can be calculated by formula (3.9) which was based on the parametric assumption of Poisson-distributed claim numbers and Pareto-distributed claim sizes. A nonparametric choice of $\hat{\sigma}(XL(P))$ would be more suitable. Now, knowing $\sigma_P^2 = Var(P_{I+1})$ (resp. having a good estimate), under the model (3.12), (3.13) the variance $\sigma^2(XL(P))$ can be estimated by:

$$\hat{\sigma}^2(XL(P)) = \hat{\rho}^2 \cdot \sigma_P^2 + \hat{c} \cdot \mu_P,$$

where $\hat{c}$ is an estimator for c. An unbiased estimator for c is given by:

$$\hat{c} = \frac{1}{I-1} \cdot \sum_{i=1}^{I} P_i \cdot \left(\frac{Y_i}{P_i} - \hat{\rho}\right)^2. \quad \square$$

The following procedure is the most appropriate rating-method for some insurance branches:

(4.) The Exposure-Method:

Consider an insurance form with fixed sums insured, denote by $V(R_j)$ the sum insured of the risk R_j from the collective $K = \{R_j, j \in J\}$ and let I_j be a random variable, taking values in $[0,1]$ and describing the intensity of the claim amount produced by the risk R_j. We assume:

(a) the random variables $I_j, j \in J$ are independent, $V(R_i) = V(R_j)$ implies that I_i, I_j are identically distributed.

(b) the claim amount S_j of risk R_j is defined according:

$$S_j = g(I_j) \cdot V(R_j)$$

where g is a strictly increasing function:

$$g : [0,1] \to [0,\infty).$$

We get:

Theorem 3

Let $V_\ell, \ell = 1,\dots,L$ be the different sums insured in the collective such that:

$$V_\ell > \frac{P}{g(1)}$$

and denote by A_ℓ the number of risks R_j such that:

$$V(R_j) = V_\ell.$$

Then:

(a) $$\mu(XL(P)) = \sum_{\ell=1}^{L} A_\ell \cdot p_\ell \cdot V_\ell \cdot W_\ell$$

(b) $$\sigma^2(XL(P)) \leq \sum_{\ell=1}^{L} A_\ell \cdot p_\ell \cdot V_\ell \cdot W_\ell \cdot [V_\ell \cdot g(1) - P] - \sum_{\ell=1}^{L} A_\ell \cdot p_\ell^2 \cdot V_\ell^2 \cdot W_\ell^2$$

with

$$p_\ell = \int_{[0,1]} g(i)G_\ell(di)$$

$$W_\ell = \frac{\int_{[i_\ell,1]} g(i)G_\ell(di) - \frac{P}{V_\ell} \cdot (1-G_\ell(i_\ell))}{\int_{[0,1]} g(i)G_\ell(di)}$$

where G_ℓ is the distribution funktion of I_j with $V(R_j) = V_\ell$ and $i_\ell = g^{-1}(P/V_\ell)$. □

We refer to Kremer (1981) for the simple proof. In practice the reinsurer can ask the insurer for the V_ℓ, A_ℓ, p_ℓ, g and only has to calculate the weights W_ℓ, i.e. has to estimate the distribution functions G_ℓ from market data and to evaluate the formula above. It is trivial to show that in case:

$$g(i) = c \cdot i$$

$$G_\ell(i) = (1-\exp(-a \cdot i)), \quad i \geq 0$$

W_ℓ can be approximated according:

$$W_\ell \approx 1 - G_\ell\left(\frac{P}{V_\ell \cdot c}\right) \tag{3.15}$$

This suggests to replace an arbitrary g by a function:

$$\hat{g}(i) = \hat{c} \cdot i$$

and to calculate W_ℓ by the approximation (3.15) with $c = \hat{c}$ (see Kremer (1983a)). The rating method given by Theorem 3 is usually called Exposure-method (see Benktander (1954), Gerathewohl (1976), Kremer (1981)).□

Finally let me mention an interesting approach which was subject of an Astin colloquium at Trieste (1963):

(5.) The extreme value method:

This method is based on the following result. Let K_k k=1,2,... be a sequence of collectives such that the claim number N_k of collective K_k satisfies:

$$\lim_{k\to\infty} E(N_k) = \infty, \quad \lim_{k\to\infty} \frac{\sqrt{Var(N_k)}}{E(N_k)} = 0$$

Assume (A.1), (A.2) of section II and that for some finite a:

$$\int_a^\infty (1-F(y))dy < \infty,$$

and:

$$\lim_{t\uparrow F^{-1}(1)} \frac{1-F(t+x\cdot R(t))}{1-F(t)} = \exp(-x), \ \forall x \qquad (3.16)$$

where

$$R(t) = (1-F(t))^{-1} \cdot \int_{(t,F^{-1}(1))} (1-F(y))dy.$$

We get from the Theorems 2.1.3, 6.2.1 of Galambos (1978):

Theorem 4

Under the above assumptions there exist constants a_n and $b_n > 0$ such that with $n_k = [E(N_k)]$ ([x] denoting the integer part of x):

$$\lim_{k\to\infty} P(X_{N_k:N_k} \leq a_{n_k} + b_{n_k} \cdot x) = \exp(-\exp(-x)).$$

The constants a_n and b_n can be chosen as:

$$a_n = F^{-1}(1-\frac{1}{n}), \ b_n = R(a_n). \ \square$$

In reinsurance the conditions of Theorem 4, i.e. (3.16), will in most cases not be justifiable. It is more realistic to apply the Theorem to the transformed claims $\tilde{X}_i = \ln(X_i)$ with distribution function $\tilde{F}(x) = F(\exp(x))$ (see Ramachandran (1974)). With corresponding parameters $\tilde{a}_n$, $\tilde{b}_n$ the net risk premium $\mu(XL(P))$ can then be calculated (see Ramachandran (1976)) by the formula:

$$\frac{P\cdot\exp(-\tilde{a}_{[E(N)]}\cdot(\ln(P)-\tilde{b}_{[E(N)]}))}{(\tilde{a}_{[E(N)]}-1)}$$

so that the problem of calculating $\mu(XL(P))$ reduces to

estimating $\tilde{a}_{[E(N)]}$, $\tilde{b}_{[E(N)]}$ from the logarithms of known past extreme values $X^{(i)}_{N_i:N_i}$, $i=1,\ldots,I$. Suitable estimation methods can be found in the papers of Jung (1963) and Ramachandran (1974). For further details and extensions we refer the reader to the literature (Beard (1963), D'Hooge (1963), Galambos (1978), Gumbel (1958), Jung (1963), Pickants (1975), Ramachandran (1974), (1976), Oliveira (1977)). □

The question arises which method to apply in practice. This can only be answered by checking the validity of the model assumptions and judging the practicability of the resulting method. In some few insurance branches, i.e. fire and accident insurance, the exposure method is most appropriate, whereas the statistical methods (1.), (2.) seem to be very suitable for rating excess-of-loss treaties in motor insurance. The estimation methods were based on the assumptions of i.i.d. claim numbers and claim amounts of the different past years. This implies that the known past claim numbers have to be IBNR-adjusted and the known past claims IBNER- and inflation-adjusted (for definition of IBNR and IBNER see Bühlmann et.al. (1980)). Also for the application of the burning cost method these adjustments have to be carried out in advance, furthermore the past premiums P_i, $i=1,\ldots,I$ have to be revaluated (see Benktander (1968a), (1968b), (1969)). Nevertheless the basic model assumption (3.10) (resp. (3.12)) will hardly be justifiable in practice. In most cases

$$\rho_i = \frac{E(Y_i)}{E(P_i)}$$

will depend on the year i. Perhaps instead of (3.10) an autoregressive model:

$$\rho_{I+1} = \sum_{i=1}^{L} \alpha_i \cdot \rho_{I+1-i}$$

will be more realistic, yielding moving averages of the known loss ratios Y_i/P_i, $i=1,\ldots,L$ as estimator for ρ_{I+1}. The statistical method (2.) seems to be more appropriate in the light of these remarks than the method (3.) when rating excess-of-loss treaties in motor insurance. Finally the extreme value method can be helpful, e.g. in fire insurance, when only information on the largest claims is available. For sake of completeness it has to be mentioned that practitioners often calculate premium rates by a very elementary method, the so-called pay-back method, which is based on estimating (or better: guessing) return periods for the claims (some remarks can be found in the book of Gerathewohl (1976) and in Ramachandran (1980)).

Some additional problems occur in practice (e.g. the problem of calculating the influence of the stability clause (see Kremer (1982b)), the problem of rating high layers (see Benktander (1963), Schmitter (1978)), the special problem of rating cumule-excess-of-loss treaties (see Strickler (1960), Feilmeier et al. (1980), Kremer (1983b)), the problem of considering interest rates (see Kahane (1979)), the problem of calculating variable premiums by bounding (see Ammeter (1961), Clemeur (1983))), which we do not like to discuss here.

The above rating methods make use only of the individual claim experience of the special collective. Often the reinsurer has rather scarce claim experience, implying then that the credibility of the calculated premium is not sufficiently high. Then the premium estimate can be improved by applying more refined statistical methods, so-called credibility techniques (see e.g. Bühlmann & Straub (1970), for a survey Norberg (1980)). For details I refer to the lectures of Prof. De Vylder and Zehnwirth.

IV. The largest claims reinsurance treaty

More than 15 years ago an alternative to the classical excess of-loss treaty was developed, called largest claims treaty covering the p largest claims (in short: LC(p)), defining the reinsurers claim amount according:

$$R_N^{LC} = \sum_{i=1}^{p} X_{N:N-i+1}$$

(compare example 2). Theorem 1 can be applied with:

$$b_{ni}^{(k)} = \begin{cases} 1, & i=n-p_k+1,\dots,n \\ 0, & \text{otherwise} \end{cases}$$

$$b(u) = \begin{cases} 1, & u \geq 1-s \\ 0, & \text{otherwise} \end{cases}$$

where $s := \lim_{k\to\infty} (p_k/E(N_k))$, implying:

Corollary 2

In case of the largest claims treaty covering the p largest claims one can choose (with $s=p/E(N)$):

$$\hat{\mu}(LC(p)) = E(N) \cdot \int_{[F^{-1}(1-s),\infty)} x\, F(dx) \tag{4.1}$$

$$= \hat{\mu}(XL(F^{-1}(1-s))) + p \cdot F^{-1}(1-s)$$

$$\hat{\sigma}^2(LC(p)) = \hat{\sigma}^2(XL(F^{-1}(1-s))) +$$

$$+ \frac{Var(N)}{E(N)} \cdot s \cdot F^{-1}(1-s) \cdot [p \cdot F^{-1}(1-s) + \tag{4.2}$$

$$+ 2 \cdot \hat{\mu}(XL(F^{-1}(1-s)))]$$

Contrary to Corollary 1 these formulas for $\mu(LC(p))$ and $\sigma^2(LC(p))$ are only asymptotically exact, i.e. should be used only for large collectives. We conclude that the problem of calculating the premium of the largest claims cover can be led back to the problem of calculating the premium for an excess-of-loss treaty with priority

$$P = F^{-1}\left(1 - \frac{p}{E(N)}\right)$$

(compare also Benktander (1978), Kremer (1982b)). Nevertheless in practice $F^{-1}(1-p/E(N))$ is unknown and has to be estimated. Assume one can get information about the largest claims of the past years $i=1,\dots,I$ and that in the i-th year the claim number has been equal to N_i. If for the quotation year the expected claim number is estimated as being ν, then set:

$$t_i = \left[\frac{N_i \cdot p}{\nu}\right]$$

and estimate P by the consistent estimator:

$$\hat{P} = \frac{1}{I} \cdot \sum_{i=1}^{I} X^{(i)}_{N_i:N_i-t_i+1},$$

where $X^{(i)}_{N_i:N_i-j+1}$ denotes the j-th largest claim of the past year i (assuming as above all claims being inflation and IBNER-corrected). Now we can apply the methods of section III for rating the largest claims cover.

For small collectives the above approximations (4.1), (4.2) might be inappropriate. Then it is preferable to apply finite formulae, given in the following

Theorem 5

Assume the claim sizes are i.i.d. with continuous distribution function F, and independent of the claim number N. Denoting by M the moment-generating function of N, i.e.:

$$M(s) = \sum_{n=0}^{\infty} P(N=n) \cdot s^n$$

and by Γ the (complete) gamma-function, i.e.

$$\Gamma(n) = \int_0^{\infty} \exp(-t) \cdot t^{n-1} dt,$$

we have:

$$\text{(a)} \quad \mu(LC(p)) = \sum_{i=1}^{p} \frac{1}{\Gamma(i)} \cdot \int_0^1 F^{-1}(t)(1-t)^{i-1} M^{(i)}(t)dt$$

$$\text{(b)} \quad \sigma^2(LC(p)) = \sum_{i=1}^{p} \frac{1}{\Gamma(i)} \cdot \int_0^1 [F^{-1}(t)]^2 (1-t)^{i-1} M^{(i)}(t)dt + 2 \cdot \sum_{i=1}^{p-1} \sum_{j=i+1}^{p} \frac{1}{\Gamma(i)} \frac{1}{\Gamma(j-i)} \cdot \int_0^1 \int_0^t F^{-1}(s) F^{-1}(t)(t-s)^{j-i-1}(1-t)^{i-1} M^{(j)}(s)\, ds\, dt - [\mu(LC(p))]^2$$

where $M^{(i)}$ is the i-th derivative of M.

The simple proof consists in applying results from the theory of order statistics (see David (1980), chapter 2) conditionally given N and using the property of iterativity of the expectation operator (compare also Ciminelli (1976)). □

For applying the theorem one has to estimate the functions M and F, similarly as in the previous section. Under additional parametric assumptions the theorem simplifies considerably:

Corollary 3

Let N be Poisson-distributed with mean λ (see (3.3)) and X_i be Pareto-distributed with parameter α (see (3.7)). Then

(a) $$\mu(LC(p)) = \lambda^{1/\alpha} \cdot \sum_{i=1}^{p} \frac{\Gamma_\lambda(i-1/\alpha)}{\Gamma(i)} \cdot a \quad (\alpha > 1)$$

(b) $$\sigma^2(LC(p)) = \lambda^{2/\alpha} \cdot \Big[\sum_{i=1}^{p} \frac{\Gamma_\lambda(i-2/\alpha)}{\Gamma(i)} - \Big(\sum_{i=1}^{p} \frac{\Gamma_\lambda(i-1/\alpha)}{\Gamma(i)} \Big)^2 \Big] \cdot a^2 \quad (\alpha > 2)$$

where Γ_λ is the incomplete gamma-function, i.e.:

$$\Gamma_\lambda(n) = \int_0^\lambda \exp(-t)t^{n-1}dt.$$

Now the problem of premium calculation reduces to the estimation of λ and α (compare section III). For large λ the incomplete gamma-function may be replaced by the (complete) gamma-function, yielding the handy approximations:

$$\mu(LC(p)) \approx \lambda^{1/\alpha} \cdot \frac{\alpha}{\alpha-1} \cdot \frac{\Gamma(p+1-1/\alpha)}{\Gamma(p)} \cdot a \tag{4.3}$$

$$\sigma^2(LC(p)) \approx \lambda^{2/\alpha} \cdot \Big[\frac{\alpha}{\alpha-2} \cdot \frac{\Gamma(p+1-2/\alpha)}{\Gamma(p)} - \frac{\alpha^2}{(\alpha-1)^2} \cdot \frac{\Gamma^2(p+1-1/\alpha)}{\Gamma^2(p)}\Big] \cdot a^2 \tag{4.4}$$

(compare Ammeter (1964), Berliner (1972), Ciminelli (1976)). For exponential-distributed claim sizes some results can be found in Kupper (1971), and formulae for negative binomial-distributed claim numbers are given in Ciminelli (1976).

Clearly the <u>burning-cost method</u> can also directly be applied to the largest-claims cover by setting:

$$Y_i := \sum_{j=1}^{p} X^{(i)}_{N_i:N_i-j+1}$$

(assuming the claims being IBNER- and inflation-adjusted) and also the extreme-value method can easily be adapted (compare Theorems 2.8.1, 6.6.2 in Galambos (1978)).
The section will be closed by giving simple asymptotic bounds on the net premium $\mu(LC(p))$:

Theorem 6

Consider a sequence K_k, $k=1,2,\ldots$ of growing collectives, N_k denoting the claim number of collective K_k. Assume:

$$\lim_{k\to\infty} E(N_k) = \infty, \quad \lim_{k\to\infty} \frac{\sqrt{Var(N_k)}}{E(N_k)} = 0$$

and (A.1),(A.2) of section 2. Then one has for a largest claim treaty $LC_k(p)$ covering the p largest claims in collective K_k:

$$\limsup_{k\to\infty} \frac{\mu(LC_k(p))-p\cdot\tilde{\mu}}{\sqrt{E(N_k)}\cdot\tilde{\sigma}} \leq b_1(p)$$

with

$$b_1(p) := \sum_{j=1}^{p} \sqrt{\binom{2j-2}{j-1}/2^{2j-1}}$$

$$\tilde{\mu} = E(X_i), \quad \tilde{\sigma}^2 = Var(X_i).$$

The proof consists in applying bounds on the expectation of order-statistics (see David (1980), chapter 3) conditionally given N and then using the iterativity of the expectation (see Kremer (1983c)). □

According to this result we have for large collectives a conservative estimate of $\mu(LC(p))$:

$$p\cdot\tilde{\mu} + \sqrt{E(N)}\cdot\tilde{\sigma}\cdot b_1(p), \tag{4.5}$$

where the reinsurer only has to ask the insurer for estimates of $\tilde{\mu}$, $\tilde{\sigma}^2$ and $E(N)$.

V. The ECOMOR-treaty

In 1950 the French actuary Thépaut introduced an alternative to the excess-of-loss cover, developed for coping with strongly inflationary conditions. Under this so-called ECOMOR-cover (in short: ECOMOR(p)) the reinsurer takes the claim amount:

$$\sum_{i=1}^{p} (X_{N:N-i+1} - X_{N:N-p+1})$$

(see example 4 of section II), where p is a fixed positive integer. Since for this treaty the rating methods are similar to those for the largest claims cover, we restrict on giving some of the main formulas. We get from (2.3), the obvious approximation:

$$\hat{\mu}(\mathrm{ECOMOR}(p)) = \hat{\mu}(\mathrm{XL}(F^{-1}(1-p/E(N))))$$

Stating the corresponding more complicated approximation (see (2.4)) for $\sigma^2(\mathrm{ECOMOR}(p))$ is left to the interested reader.

Clearly under the conditions of theorem 5:

$$\mu(\mathrm{ECOMOR}(p)) = \mu(\mathrm{LC}(p)) - \frac{p}{\Gamma(p)} \cdot \int_0^1 F^{-1}(t)(1-t)^{p-1} M^{(p)}(t)\,dt$$

implying in case of the Poisson-claim number and Pareto claim size distribution:

$$\mu(\mathrm{ECOMOR}(p)) = \mu(\mathrm{LC}(p)) - p \cdot \lambda^{1/\alpha} \cdot \frac{\Gamma_\lambda(p-1/\alpha)}{\Gamma(p)} \cdot a$$

which can be approximated by the counterpart to (4.3):

$$\mu(\mathrm{ECOMOR}(p)) \approx \lambda^{1/\alpha} \cdot \frac{1}{\alpha-1} \frac{\Gamma(p-1/\alpha)}{\Gamma(p-1)} \cdot a$$

(compare Ammeter (1964), Seal (1969)). The analogous formula for the variance is given by (see Seal (1969), p.151):

$$\sigma^2(\mathrm{ECOMOR}(p)) \approx \lambda^{2/\alpha} \cdot \Big[\Big(\frac{\alpha}{\alpha-2} + p-1\Big) \cdot \frac{\Gamma(p-2/\alpha)}{\Gamma(p-1)} - \frac{1}{(\alpha-1)^2} \cdot \frac{\Gamma^2(p-1/\alpha)}{\Gamma^2(p-1)}\Big] \cdot a^2$$

Finally the bound (4.5) carries over to (see Kremer (1983c)):

$$\sqrt{E(N)} \cdot \tilde{\sigma} \cdot b_2(p)$$

with

$$b_2(p) = \sum_{j=1}^{p-1} [\binom{2j-2}{j-1} / 2^{2j-1} + \binom{2p-2}{p-1} / 2^{2p-1} -$$

$$- 2 \cdot \binom{p+j-2}{p-1} / 2^{p+j-1}]^{1/2}$$

Some remarks on the <u>extreme value method</u> can be found in the papers of Ramachandran (1976), Weissman (1978) and clearly the <u>burning cost method</u> is also applicable by setting

$$Y_i = \sum_{j=1}^{p} [X^{(i)}_{N_i:N_i-j+1} - X^{(i)}_{N_i:N_i-p+1}] ;$$

where $X^{(i)}_{N_i:N_i-j+1}$ denotes the j-th largest claim of the past year i (all claims being IBNER- and inflation corrected).

References

Ammeter, H. (1955): The calculation of premium rates for excess of loss and stop loss reinsurance treaties. *Nonproportional reinsurance.* Bruxelles 1955, p. 79.

Ammeter, H. (1961): Risikotheoretische Grundlagen der Erfahrungstarifierung. *Mitteilungen der Vereinigung Schweizerischer Versicherungsmathematiker* 61, p. 183.

Ammeter, H. (1964): The Rating of "Largest Claim" Reinsurance Covers. *Quarterly Letter from the Allgemeene Reinsurance Companies*, Jubilee Number 2, 1964, p. 5.

Ammeter, H. (1965): Practical Applications of the collective risk theory. *Scandinavian Actuarial Journal* 1969, p. 99.

Barnett, V. and Lewis, T. (1978): *Outliers in statistical data.* John Wiley & Sons, New York.

Beard, R.E. (1963): Some notes on the statistical theory of extreme values. *Astin Bulletin* 3, p. 6

Benktander, G. (1954): A method of fixing the premium of excess-of-loss in fire. *Transactions of the International Congress of Actuaries,* 1954, p. 823.

Benktander, G. (1962): A note on the most "dangerous" and skewest class of distributions. *Astin Bulletin* 2, p. 387.

Benktander, G. (1963): Notes sur la distribution conditionée du montant d'un sinistre par rapport a l'hypothese qu'il y a un sinistre dans l'assurance automobile. *Astin Bulletin* 2, p. 387.

Benktander, G. (1968 a) : The effect of inflation on excess layers. *The Review,* 1968, p. 5.

Benktander, G. (1968 b) : The dilemma of a motor excess underwriter. *The Review,* 1968, p. 9.

Benktander, G. (1969): The calculation of a motor excess rate. *The Review,* 1969, p. 68.

Benktander, G. (1970): Schadenverteilung nach Größe in der Nicht-Leben-Versicherung. *Mitteilungen der Vereinigung schweizerischer Versicherungsmathematiker* 1970, p. 263.

Benktander, G. (1971): Some aspects on reinsurance profits and loadings. *Astin Bulletin* 5, p. 314.

Benktander, G. (1977): The calculation of a fluctuation loading for an excess-of-loss cover. *Astin Bulletin* 9, p. 272.

Benktander, G. (1978): Largest claims reinsurance (LCR): A quick method to calculate LCR-risk rates from excess of loss risk rates. *Astin Bulletin* 10, p. 54.

Benktander, G. & Segerdahl, C.O. (1960): On the analytical representation of claim distributions with special reference to excess of loss reinsurance. *Transactions of the international congress of actuaries,* Brussels 1960, p. 626.

Berliner, B. (1972): Correlations between excess of loss reinsurance covers and reinsurance of the n largest claims. *Astin Bulletin* 6, p. 260.

Berliner, B. (1976): On the choice of risk loadings. *Transactions of the international congress of actuaries*, Tokyo 1976, p. 413.

Berliner, B. (1980): On the rating of excess of loss covers under inflationary conditions. *Internal report of the Swiss reinsurance company*, Zürich 1980.

Bickel, P.J. (1967): Some contributions to the theory of order statistics. *Proc. Fifth. Berkeley Symp. Math. Statist. Prob.*, p. 575.

Bowers Jr., N.L. (1969): An upper bound on the stop-loss net premium. *Transactions of the society of actuaries* 21, p. 211.

Bohman, H. & Esscher, F. (1963): Studies in risk theory with numerical illustrations concerning distribution functions and stop-loss premiums. Part 1 and 2. *Scandinavian Actuarial Journal* 1963, p. 173.

Bühlmann, H., Gagliardi, D., Gerber, H.U. and Straub, E. (1977): Some inequalities for stop-loss premiums. *Astin Bulletin* 9, p. 75.

Bühlmann, H. & Straub, E. (1970): Glaubwürdigkeit für Schadensätze. *Mitteilungen der Vereinigung Schweizerischer Versicherungsmathematiker* 1970, p. 111.

Bühlmann, H., Schnieper, R. and Straub, E. (1980): Claims reserves in casualty insurance based on a probabilistic model. *Mitteilungen der Vereinigung Schweizerischer Versicherungsmathematiker* 80, p. 21.

Ciminelli, E. (1976): On the distribution of the highest claims and its application to the automobile insurance liability. *Transactions of the international congress of actuaries*, p. 501.

Clemeur, H. (1983): Practical rating of variable accident excess-loss premiums. *Short communication given at the Nato-advanced study institute at the University of Leuven*, Belgium, July 1983.

Cochran, W.G. (1963): *Sampling Techniques.* John Wiley & Sons, New York.

Conolly, B.W. (1955): Unbiased premiums for stop-loss reinsurance. *Scandinavian Actuarial Journal* 1955, p. 127.

David, H. (1980): *Order statistics.* John Wiley & Sons, New York.

D'Hooge, L. (1963): Théorie des valeurs extrêmes et la tarification de l'excess of loss. *Astin Bulletin* 3, p. 163.

De Vylder, F. (1982): Best upper bounds for integrals with respect to measures allowed to vary under conical and integral constraints. *Insurance: Mathematics and Economics* 1, p. 109.

De Vylder, F. and Goovaerts, M.J. (1982): Analytical best upper bounds on stop-loss premiums. *Insurance: Mathematics and Economics* 1, p. 197.

Feilmeier, M. and Segerer, G. (1980): Einige Anmerkungen zur Rückversicherung von Kumulrisiken nach dem "Verfahren Strickler". *Blätter der deutschen Gesellschaft für Versicherungsmathematik* 1980, p. 611.

Franckx, E. (1960): Sur la tarification de l'Excess of Loss. *Transactions of the international congress of actuaries* 1960, p. 536.

Galambos, J. (1978): *The asymptotic theory of extreme order statistics.* John Wiley & Sons, New York.

Gerathewohl, K. (1976): *Rückversicherung - Grundlagen und Praxis. Band 1,* Verlag Versicherungswirtschaft, Karlsruhe.

Gerber, H.U. (1980): *An introduction to mathematical risk theory.* Huebner foundation.

Gerber, H.U. (1982): On the numerical evaluation of the distribution of aggregate claims and its stop-loss premiums. *Insurance: Mathematics and Economics* 1.

Goovaerts, M.J. and Declerq, M. (1980): On an application of a smoothing inequality to the estimation of stop-loss premiums. *Scandinavian Actuarial Journal* 1980, p. 33.

Goovaerts, M.J., De Vylder, F. and Haezendonck, H. (1981): *The Mathematics of Insurance Premiums, Part 1 & 2*. Universität Leuven.

Grossmann, M. (1977): *Rückversicherung - Eine Einführung*. Peter Lang Verlag.

Gumbel, E.J. (1958): *Statistics of Extremes*. Columbia University Press, New York.

Harding, V. (1968): The calculation of premiums for excess of loss reinsurances of motor business. *Transactions of the international congress of actuaries* 1968, p. 445.

Heilmann, W.-R. (1981): Improved methods for calculating and estimating maximal stop-loss premiums. *Blätter der deutschen Gesellschaft für Versicherungsmathematik*, p. 29.

Helbig, M. (1953): Mathematische Grundlagen der Schadenexzedentenrückversicherung. *Festschrift für Emil Bebler*, p. 183, Berlin.

Huber, P.J. (1981): *Robust Statistics*. John Wiley & Sons, New York.

Jung, J. (1954): On linear estimates defined by a continuous weight function. *Arkiv för Matematik Band 3*, Nr. 15.

Jung, J. (1963): On the use of extreme values to estimate the premium for an excess of loss reinsurance. *Astin Bulletin* 3, p. 178.

Kahane, Y. (1979): The theory of insurance risk premiums. A re-examination in the light of recent developments in capital market theory. *Astin Bulletin* 10, p. 223.

Kremer, E. (1981): Das allgemeine Exposure-Verfahren zur Tarifierung von Unfall-Schadenexzedenten-Verträgen. *Blätter der deutschen Gesellschaft für Versicherungsmathematik*, p. 195.

Kremer, E. (1982 a): Bemerkung zur Kalkulation der Wirkung der Stabilisierungsklausel in Schadenexzedentenrückversicherungsverträgen. *Blätter der deutschen Gesellschaft für Versicherungsmathematik*, p. 259.

Kremer, E. (1982 b): Rating of largest claims and ECOMOR reinsurance treaties for large portfolios. *Astin Bulletin* 13, p. 47.

Kremer, E. (1983 a): Ein Näherungsverfahren zur Kalkulation der progressiven Invalidität in Unfall-Schadenexzedenten-Rückversicherungsverträgen. *Blätter der deutschen Gesellschaft für Versicherungsmathematik*, p. 31.

Kremer, E. (1983 b): Ein Modell zur Tarifierung von Kumul-Schadenexzedentenverträgen in der Unfallversicherung. *Mitteilungen der Vereinigung schweizerischer Versicherungsmathematiker* 1983, p. 53.

Kremer, E. (1983 c): Distribution-free upper bounds on the premiums of the LCR and ECOMOR treaties. *Insurance: Mathematics and Economics* 1983.

Kremer, E. (1983 d): An asymptotic formula for the net premium of some reinsurance treaties. To appear in *Scandinavian Actuarial Journal*.

Kremer, E. (1983 e): The burning cost method and ratio estimation. *Unpublished manuscript*.

Kupper, J. (1971): Contributions to the theory of the largest claim cover. *Astin Bulletin* 6, p. 134.

Moore, D.S. (1968): An elementary proof of asymptotic normality of linear functions of order statistics. *Annals of Mathematical Statistics* 1968, p. 263.

Norberg, R. (1979): The credibility approach to experience rating. *Scandinavian Actuarial Journal* 1979, p. 181.

Oliveira, J.T. (1977): Statistical methodology for large claims. *Astin Bulletin* 9, p. 1.

Panjer, H. (1980): The aggregate claims distribution and stop-loss reinsurance. *Transactions of the Society of Actuaries* 32, p. 523.

Pickands, J. III (1975): Statistical inference using extreme order statistics. *Annals of Statistics* 3, p. 119.

Ramachandran, G. (1974): Extreme value theory and large fire losses. *Astin Bulletin* 7, p. 293.

Ramachandran, G. (1976): Extreme value theory and fire insurance. *Transactions of the international congress of actuaries* 1976, p. 695.

Ramachandran, G. (1980): Extreme value theory and Earthquake Insurance. *Transactions of the international congress of actuaries* 1980, p. 337.

Seal, H.L. (1969): *Stochastic Theory of a Risk Business.* Wiley & Sons, New York.

Serfling, R.J. (1980): *Approximation Theorems of Mathematical Statistics.* Wiley & Sons, New York.

Shorack, G.R. (1972): Functions of order statistics. *Annals of Mathematical Statistics* 43, p. 412.

Schmitter, H. (1978): Quotierung von Sach-Schadenexzedenten mit Hilfe des Paretomodells. ***SR-Fokus.***

Stigler, S.M. (1968): Linear functions of order statistics. *Annals of Mathematical Statistics* 40, p. 770.

Strickler, P. (1960): Rückversicherung des Kumulrisikos in der Lebensversicherung. *Transactions of the international congress of actuaries* 1960, p. 666.

Taylor, G.C. (1977): Upper bounds on stop-loss premiums under constraints on claim size distribution. *Scandinavian Actuarial Journal* 1977, p. 94.

Thépaut, A. (1950): Le traité d'excédent du coût moyen relativ. *Bulletin Trimestriel de l'Institut des Actuaires Francais*, No. 192.

Vajda, S. (1951): Analytical studies in stop-loss reinsurance. *Scandinavian Actuarial Journal* 1951, p. 158.

Weissman, I. (1978): Estimation of parameters and large quantiles based on the k largest observations. *Journal of the American Statistical Association* 73, p. 812.

RESISTANT LINE FITTING IN ACTUARIAL SCIENCE

P. Rousseeuw, A. Leroy, B. Daniels

Vrije Universiteit Brussel

An example on employers' liability insurance shows that the ordinary least squares method for fitting a straight line to data is very sensitive to outliers. As an alternative, the least trimmed squares (LTS) technique is discussed, which can resist the effect of a large fraction of contaminated data. The LTS is compared to other proposals, using both theoretical arguments and a numerical experiment. A complete portable FORTRAN program is given to enable the reader to apply the new method himself.

1. THE LEAST TRIMMED SQUARES METHOD

In order to see why one would ever need something called "resistant line fitting", let us look at an example on employers' liability insurance from the 1981-1982 report of the Belgian Association of Insurance Companies [22]. Table 1 lists the nation-wide number of reported accidents of people going to or coming from their work, from 1975 till 1981. Only the accidents leading to at least one day of absence were counted. When looking at these figures, we notice a slight downward trend over the years.

However, the number for 1979 is extraordinarily high. The reason lies in the fact that in the first months of 1979 it was extremely cold, with snow and ice on most days (one has to go back decades to find records of a similar winter). Therefore, the number of

F. de Vylder et al. (eds.), Premium Calculation in Insurance, 315–332.

year (x)	number of claims (y)
75	18031
76	18273
77	16660
78	15688
79	23805
80	15241
81	13295

Table 1. Number of accidents in 1975-1981.

fractures, sprainings and dislocations in January and February was unheard of. We call this observation an outlier.

Another way to look at such bivariate data is to make a scatterplot as in Figure 1, on which the points (x_i, y_i) are drawn (here $i=1,...,n$ where in this case the sample size n equals 7). The eye tells us that most of these points follow a decreasing trend, except for the 1979 point which is way off. Often one wants to fit a line to such points, that is, to find coefficients $\hat{a}$ and $\hat{b}$ such that the line $y=\hat{a}x+\hat{b}$ lies close to the plotted data. The most well-known and generally used technique for this is least squares (LS), due to Gauss, which determines $\hat{a}$ and $\hat{b}$ by

$$\underset{\hat{a},\hat{b}}{\text{minimize}} \; \Sigma_{i=1}^{n} r_i^2 \qquad (1)$$

where the residual r_i is defined as $y_i-(\hat{a}x_i+\hat{b})$. In the example, one obtains $y = -468.821x + 53852.8$ which is pictured as the dotted line in Figure 1. This dotted line is certainly not what we expected to see: it lies too high because it is affected very much by the fifth point. If we would take this outlier and move it up even higher, the dotted line would follow. Now this kind of behaviour is not what we really want. Indeed, outliers often occur in practice, usually as transcription errors or other mistakes. The LS line is too easily affected by them, because the square in (1) increases very fast in the case of an outlying value of y_i . Therefore, we say that LS is not resistant, because it cannot resist the effect of even a single outlier. We would actually prefer a line fitting procedure which discards such points, in an objective way (contrary to the common subjective deletion of points).

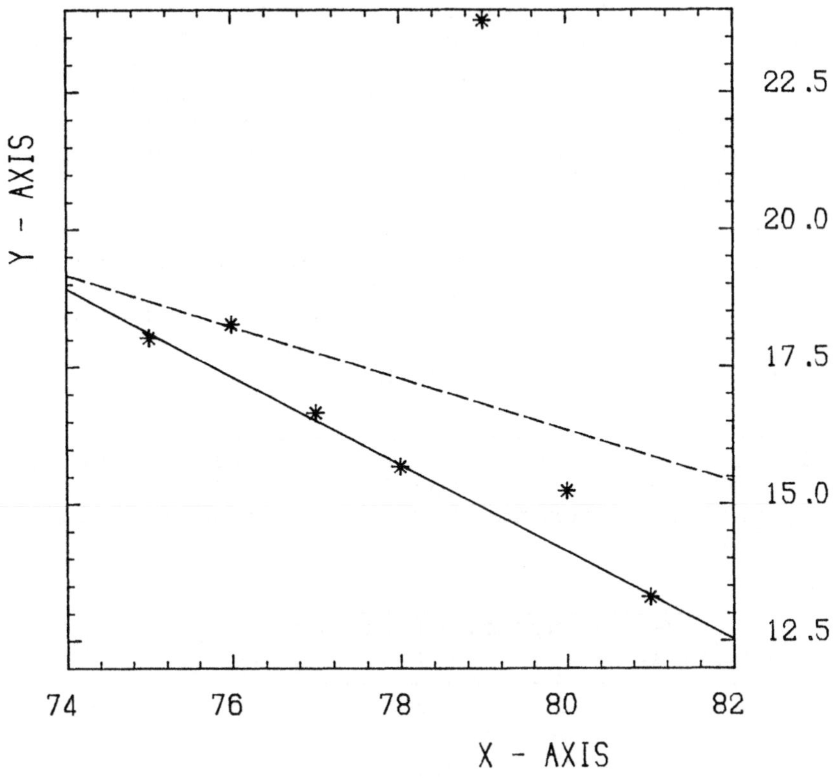

Figure 1. Scatterplot of Table 1.

The technique to be discussed in this paper is due to Rousseeuw [17]. Not only can it cope with one very extreme outlier, but it can even handle a large fraction of them. In fact, the method always follows the majority of the data. To achieve this, the squared residuals $r_1^2,\dots,r_n^2$ are ordered:

$$(r^2)_{1:n} \leq (r^2)_{2:n} \leq \dots \leq (r^2)_{n:n} \quad .$$

The largest ones aren't taken into account, because if some point does not follow the majority of the data, then its squared residual will be large at the "good" $\hat{a}$ and $\hat{b}$. Therefore, (1) is replaced by a sum over the first 50% of the squared residuals:

$$\underset{\hat{a},\hat{b}}{\text{minimize}} \; \Sigma_{i=1}^{h} \; (r^2)_{i:n} \tag{2}$$

where $h=[\frac{n}{2}]+1$ (the notation $[\frac{n}{2}]$ stands for the largest integer less than or equal to $\frac{n}{2}$). This method is called <u>least trimmed squares</u> (and abbreviated LTS), because the largest squared residuals are trimmed away. In the next section we shall discuss the resistance of this method, and compare it to other procedures that have been proposed.

In our example, the LTS line is given by $y = -797.9x + 77958.3$. It has been plotted as the solid line in Figure 1. This line has not been affected by the 1979 outlier, but fits the majority of the data. Even if the 1979 point would be moved further up, the LTS line would not change any more.

Let us now look at a more extreme example. In order to have the outlier on the border of the range of x , we consider the total number of claims of Belgian fire-insurance companies, from 1976 till 1980 (see [22]).

year (x)	number of claims (y)
76	16694
77	12271
78	12904
79	14036
80	13874

Table 2. Number of fires in 1976-1980.

The high frequency in 1976 was caused by the fact that in that year the summer was extremely hot and dry (compared to Belgian standards), causing even trees and bushes to catch fire spontaneously. Figure 2 shows both LS (dotted line) and LTS (solid line). The lines are very different, as one is decreasing whereas the other increases. In this example the outlier lies on the outside of the x-range, causing LS to misbehave even more than in the first example.

2. COMPARISON WITH OTHER METHODS

The quantity which describes the resistance of the LTS estimator most clearly is the <u>breakdown point</u> ε^* , invented by Hampel [5] and formulated in a more

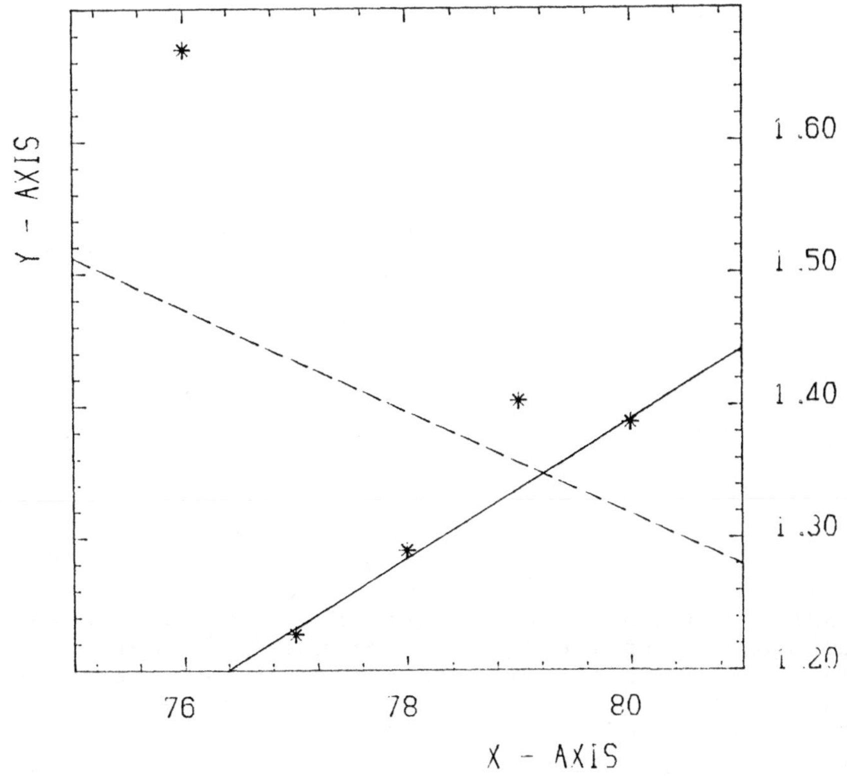

Figure 2. Scatterplot of Table 2.

simple setting by Donoho and Huber [3]. In words, it is the smallest fraction of outliers that can carry $\hat{a}$ and $\hat{b}$ over all bounds. For LS it is easily seen that $\varepsilon^* = 0\%$, because in any sample even a single outlier can pull the line as much as it wants to, provided it is bad enough.

There have been many proposals to improve on the resistance of LS. Already Edgeworth [4] proposed to replace the square in (1) by the absolute value, giving less weight to extreme observations. This so-called least absolute values method is fairly insensitive to outliers in the y-direction, but still cannot cope with grossly aberrant values of x_i . The latter points are called "leverage points" because one of them is enough to tilt the line, so again $\varepsilon^* = 0\%$. A generalization of this method was introduced by Huber [9], who proposed what is now called a Huber-type

M-estimator:

$$\underset{\hat{a},\hat{b}}{\text{minimize}}\ \Sigma_{i=1}^{n}\ \rho(r_i) \tag{3}$$

where ρ is quadratic in the middle and linear elsewhere. This estimator is more efficient in a statistical sense than least absolute values, but still $\varepsilon^* = 0\%$ because of the damage that can be done by leverage points. Therefore, general M-estimators were proposed by Mallows [13] and Schweppe (see Hill [8]), followed by refinements of Hampel [7], Krasker [11] and Krasker and Welsch [12]. However, Maronna, Bustos and Yohai [15] showed that all M-estimators must satisfy $\varepsilon^* \leq 33.3\%$. Various other estimators were introduced by Theil [21], Brown and Mood [2] , Sen [19], Jaeckel [10] and Andrews [1], but none of them reach up to $\varepsilon^* = 30\%$.

Therefore, one wonders whether it is possible to construct procedures with a breakdown point of 50%, which is the best we can hope to achieve (for still larger amounts of contamination, it may be impossible to distinguish the "good" from the "bad" part of the sample). The first such techniques were sketched by Hampel [6], one of them leading to the so-called least median of squares method (Rousseeuw [16]). Also Siegel's repeated median algorithm [20] can resist up to 50% of outliers. For the LTS, it is shown in [18] that $\varepsilon^* = 50\%$.

In order to investigate such breakdown properties empirically, we conducted an artificial experiment. To begin with, we generated 100 "good" observations according to the linear relation

$$y_i = ax_i + b + e_i$$

where a=1 , b=2 , x_i is uniformly distributed on the interval (1,4) and the error e_i is normally distributed with mean zero and standard deviation 0.2 . To these data, we applied least squares, the Huber-type M-estimator, the Mallows estimator with weights given by Maronna, the Schweppe estimator and the Mallows estimator with weights given by Hampel. For the calculation of these M-estimators, the ROBETH library was used [14]. Moreover, the repeated median algorithm and the LTS were also carried out. Because the data were well-behaved, all seven estimators yielded values of $\hat{a}$ and $\hat{b}$ which were very close to the original a

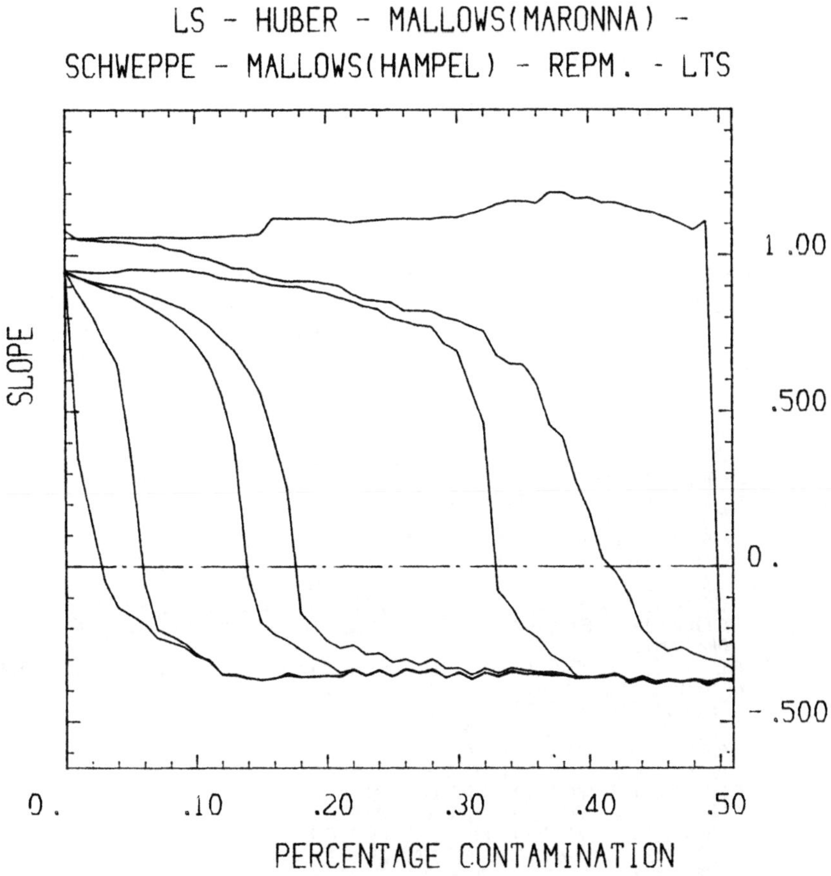

Figure 3. Breakdown plot of 7 estimators.

and b. However, we then started to contaminate the data. At each step we deleted one "good" point, and replaced it by a "bad" data point generated according to a bivariate normal distribution with mean (7,2) and standard deviation 0.5 . We repeated this until only 50 "good" points remained. The LS was immediately affected by these leverage points, so the estimated slope $\hat{a}$ became negative, moving away from the ideal value a=1 . In Figure 3, the value of $\hat{a}$ is plotted as a function of the percentage of outliers.

We see that LS breaks down first, and is then followed by the Huber-type estimator and the other M-estimators. (Note that only the Mallows estimator with Hampel weights comes close to the theoretical upper bound of 33.3% on ε^* !) The repeated median goes down gradually and becomes negative at about 40% , whereas the LTS holds on till the very end before breaking down.

3. FORTRAN PROGRAM

The FORTRAN program below calculates $\hat{a}$ and $\hat{b}$ from the data. It has been made portable by means of the PFORT verifier, so it runs without difficulty on any FORTRAN IV or FORTRAN 77 compiler (only the first line of the program depends on the computer system).

```
      PROGRAM LTSSUB(TAPE1,TAPE2)
CC  DIMENSION SOL1D(N- INT(N/2) +1) , X(N), Y(N), W(N)
CC  N = NUMBER OF OBSERVATIONS
      DIMENSION RLCL(3,3)
      REAL MINMT2,MINRC,MINCT,MADX,MADY
      COMMON SOL1D(500),ITEL,X(1000),Y(1000),W(1000),
     V RESID,RESRT,RCLS,CTLS,VARX,VARY
      COMMON /SLTS/ MINMT2,MINRC,MINCT,BSTSSQ
      COMMON /SUBR/ XMED,YMED,MADX,MADY,S2RES
      N=1000
CC  READ THE NUMBER OF OBSERVATIONS (X(I),Y(I)) AND
CC  THE OBSERVATIONS (X(I),Y(I))
      WRITE(2,8000)
      READ(1,8100) LENGTH
      CALL READXY(LENGTH)
      CALL CALCUL(LENGTH)
      WRITE(2,9000) XMED,YMED
      WRITE(2,9005) MADX,MADY
      WRITE(2,9020) MINRC,MINCT
      WRITE (2,9030) MINMT2
      WRITE(2,9040) RCLS,CTLS
      WRITE(2,9050) S2RES
 8000 FORMAT(/29H L E A S T   T R I M M E D
     1 23H S Q U A R E S  L I N E/1X,52(1H*)/)
 8100 FORMAT(I4)
 9000 FORMAT(13H MEDIAN X  = ,F15.6/13H MEDIAN Y  = ,
     1 F15.6)
 9005 FORMAT(13H MAD X      = ,F15.6/13H MAD Y      = ,
     1 F15.6///)
 9020 FORMAT(11H LTS    Y = ,
     1 F15.6,5H X + ,F15.6)
 9030 FORMAT(20H SCALE ESTIMATE    = ,
     1 F15.6///)
 9040 FORMAT(11H LS     Y = ,
     1 F15.6,5H X + ,F15.6)
 9050 FORMAT(20H SCALE ESTIMATE    = ,
     1 F15.6///)
      STOP
      END
CC
CC
      SUBROUTINE CALCUL(LENGTH)
CC  DIMENSION SOL1D(N- INT(N/2) +1) , X(N), Y(N), W(N)
CC  N = NUMBER OF OBSERVATIONS
```

```
      DIMENSION RLCL(3,3)
      REAL MINMT2,MINRC,MINCT,MADX,MADY
      COMMON SOL1D(500),ITEL,X(1000),Y(1000),W(1000),
     V RESID,RESRT,RCLS,CTLS,VARX,VARY
      COMMON /SLTS/ MINMT2,MINRC,MINCT,BSTSSQ
      COMMON /SUBR/ XMED,YMED,MADX,MADY,S2RES
      N=1000
      TEMIN= -1.55
      TEMAX = 1.55
      EPS1 = 0.020
      EPS2 = 0.001
      MINMT2 = 1.0E320
      BMT = 1.0E320
      RLCL(1,1)=1.0E320
      RLCL(1,2)=1.0E320
      RLCL(2,1)=1.0E320
      RLCL(2,2)=1.0E320
      IP=1
CC   CALCULATION OF THE LEAST SQUARES REGRESSION LINE
CC
      CALL LSTSQR(LENGTH)
CC   SCALING OF THE OBSERVATIONS SO THAT MEDIAN(X)=0,
CC   MEDIAN(Y)=0,MEDIAN DEVIATION(X)=1 AND
CC   MEDIAN DEVIATION(Y)=1
      XMED=AMDIAN(X,LENGTH)
      YMED=AMDIAN(Y,LENGTH)
      MADX=AMAD(X,LENGTH,XMED)
      MADY=AMAD(Y,LENGTH,YMED)
      IF (MADX.EQ.0.0) MADX=0.68*SQRT(VARX)
      IF (MADY.EQ.0.0) MADY=0.68*SQRT(VARY)
      DO 5000 I=1,LENGTH
      X(I)=(X(I)-XMED)/MADX
      Y(I)=(Y(I)-YMED)/MADY
 5000 CONTINUE
CC   SEARCHING OF THE TWO LEAST LOCAL MINIMA BY
CC   VARYING THE SLOPE A FOR ALL ANGLES BETWEEN
CC   TEMIN AND TEMAX WITH A GIVEN PRECISION
CC   (STEPSIZE) EQUAL TO EPS1
      TETA=TEMIN
      IC=0
 5010 IF(TETA.GT.TEMAX) GOTO 5030
 5020 CALL ALGO(TETA,LENGTH,IP)
      IF(MINMT2.GT.BMT) GOTO 5050
 5040 BMT=MINMT2
      BEFRC=MINRC
      BEFCT=MINCT
      IC=0
      GOTO 5090
 5050 IF(.NOT.(((IC.EQ.0).AND.(BMT.LE.RLCL(1,1)))
     1 .OR.(IP.EQ.1))) GOTO 5070
 5060 IC=1
      RLCL(1,2)=RLCL(1,1)
      RLCL(2,2)=RLCL(2,1)
      RLCL(1,1)=BMT
      RLCL(2,1)=BEFRC
      BMT=MINMT2
      BEFRC=MINRC
      BEFCT=MINCT
      GOTO 5090
 5070 IF(.NOT.((IC.EQ.0) AND.(BMT.LE.RLCL(1,2))))
     1 GOTO 5095
 5080 RLCL(1,2)=BMT
      RLCL(2,2)=BEFRC
      IC=1
 5095 BMT=MINMT2
```

```
      BEFRC=MINRC
      BEFCT=MINCT
 5090 TETA=TETA+EPS1
      IP=IP+1
      GOTO 5010
CC  SCAN WITH A GREATER PRECISION (EPS2) IN THE
CC  NEIGHBOURHOOD OF THE TWO LOCAL MINIMA
CC  FOUND ABOVE
CC
 5030  TETA=ATAN(RLCL(2,1))-EPS1
      TEMAX=ATAN(RLCL(2,1))+EPS1
 5100 IF(TETA.GT.TEMAX) GOTO 5120
 5110 CALL ALGO(TETA,LENGTH,IP)
      IF(MINMT2.GT.RLCL(1,1)) GOTO 5140
 5130 RLCL(1,1)=MINMT2
      RLCL(2,1)=MINRC
      RLCL(3,1)=MINCT
 5140 TETA=TETA+EPS2
      GOTO 5100
 5120 IF (RLCL(2,1).EQ.RLCL(2,2)) GOTO 5200
      TETA=ATAN(RLCL(2,2))-EPS1
      TEMAX=ATAN(RLCL(2,2))+EPS1
 5150 IF(TETA.GT.TEMAX) GOTO 5170
 5160 CALL ALGO(TETA,LENGTH,IP)
      IF (MINMT2.GT.RLCL(1,2)) GOTO 5190
 5180 RLCL(1,2)=MINMT2
      RLCL(2,2)=MINRC
      RLCL(3,2)=MINCT
 5190 TETA=TETA+EPS2
      GOTO 5150
 5170 IF (RLCL(1,1).GE.RLCL(1,2)) GOTO 5210
 5200 MINMT2=RLCL(1,1)
      MINRC=RLCL(2,1)
      MINCT=RLCL(3,1)
      GOTO 5220
 5210 MINMT2=RLCL(1,2)
      MINRC=RLCL(2,2)
      MINCT=RLCL(3,2)
CC  THE SOLUTION IS FOUND.
CC  RESCALE THE PARAMETERS
 5220 MINCT=MADY*MINCT+YMED-MINRC*MADY*XMED/MADX
      MINRC=MINRC*MADY/MADX
      MINMT2=MINMT2*MADY*MADY
      AL=LENGTH-2
      MINMT2=SQRT(MINMT2/(AL*0.0714957))
      DO 5300 I=1,LENGTH
      X(I)=X(I)*MADX+XMED
      Y(I)=Y(I)*MADY+YMED
 5300 CONTINUE
      S2RES=0.0
      DO 5310 I=1,LENGTH
      CALL RESIDU(I)
      S2RES=S2RES+RESID
 5310 CONTINUE
      S2RES=SQRT(S2RES/AL)
      RETURN
      END
CC  *READXY* : READ THE OBSERVATIONS X(I),Y(I)
      SUBROUTINE READXY(LENGTH)
      COMMON SOL1D(500),ITEL,X(1000),Y(1000),W(1000),
     V RESID,RESRT,RCLS,CTLS,VARX,VARY
      INTEGER FMT(50)
      READ(1,8200) FMT
 8200 FORMAT(50A1)
      DO 10 I=1,LENGTH
      READ(1,FMT) X(I),Y(I)
```

```
 10      CONTINUE
         RETURN
          END
CC
CC   *SCHALF* : CALCULATES THE LTS IN THE ONE-DIM. CASE.
CC
         SUBROUTINE SCHALF(LENGTH)
         COMMON SOL1D(500),ITEL,X(1000),Y(1000),W(1000),
       V RESID,RESRT,RCLS,CTLS,VARX,VARY
         COMMON /SLTS/ MINMT2,MINRC,MINCT,BSTSSQ
         REAL MINMT2,MINRC,MINCT
         PREC=1.0E-13
         PRTSUM=0.0
         BSTSSQ=0.0
         AL=LENGTH
         K=INT(AL/2.0)
         NK=K+1
         RNK=NK
         DO 100 I=1,NK
         PRTSUM=PRTSUM+W(I)
 100     CONTINUE
         PRTSUM=PRTSUM/RNK
         DO 110 I=1,NK
         BSTSSQ=BSTSSQ+(W(I)-PRTSUM)**2
 110     CONTINUE
         ITEL=1
         SOL1D(ITEL)=PRTSUM
         LNK=LENGTH-K
         DO 120 I=2,LNK
         PRTSUM=0.0
         PRTSSQ=0.0
         IK=I+K
         DO 130 J=I,IK
         PRTSUM=PRTSUM+W(J)
 130     CONTINUE
         PRTSUM=PRTSUM/RNK
         DO 140 J=I,IK
         PRTSSQ=PRTSSQ+(W(J)-PRTSUM)**2
 140     CONTINUE
         IF(ABS(PRTSSQ-BSTSSQ).GT.PREC) GOTO 2000
         ITEL=ITEL+1
         SOL1D(ITEL)=PRTSUM
         GOTO 120
 2000    IF( PRTSSQ.GT.BSTSSQ) GOTO 120
         ITEL=1
         BSTSSQ=PRTSSQ
         SOL1D(ITEL)=PRTSUM
 120     CONTINUE
         RETURN
         END
CC
CC   *YMAXI*   : CALCULATES THE CONSTANT B IN Y=AX+B
CC               FOR A CERTAIN VALUE OF A.
         FUNCTION YMAXI(LENGTH,RC)
         COMMON SOL1D(500),ITEL,X(1000),Y(1000),W(1000),
       V RESID,RESRT,RCLS,CTLS,VARX,VARY
         COMMON /SLTS/ MINMT2,MINRC,MINCT,BSTSSQ
         REAL MINMT2,MINRC,MINCT
CC
         DO 150 I=1,LENGTH
         W(I)=Y(I)-RC*X(I)
 150     CONTINUE
         CALL ORDEN(W,LENGTH)
         CALL SCHALF(LENGTH)
         YMAXI=BSTSSQ
         RETURN
          END
```

```
CC
CC   *LSTSQR* :  CALCULATES THE LEAST SQUARES  LINE.
CC
        SUBROUTINE LSTSQR(LENGTH)
        COMMON SOL1D(500),ITEL,X(1000),Y(1000),W(1000),
      V RESID,RESRT,RCLS,CTLS,VARX,VARY
        GEMX = 0.0
        GEMY = 0.0
        VARX = 0.0
        VARY = 0.0
        COVXY = 0.0
        AL=LENGTH
        DO 200 I=1,LENGTH
        GEMX=GEMX+X(I)
        GEMY=GEMY+Y(I)
 200    CONTINUE
        GEMX=GEMX/AL
        GEMY=GEMY/AL
        DO 210 I=1,LENGTH
        VARX=VARX+(X(I)-GEMX)**2
        VARY=VARY+(Y(I)-GEMY)**2
        COVXY=COVXY+(X(I)-GEMX)*(Y(I)-GEMY)
 210    CONTINUE
        VARX=VARX/AL
        VARY=VARY/AL
        COVXY=COVXY/AL
        RCLS=COVXY/VARX
        CTLS=GEMY-RCLS*GEMX
        RETURN
         END
CC
CC   *RESIDU* :  CALCULATES THE RESIDUAL Y(I)-A*X(I)-B.
CC
        SUBROUTINE RESIDU(I)
        COMMON SOL1D(500),ITEL,X(1000),Y(1000),W(1000),
      V RESID,RESRT,RCLS,CTLS,VARX,VARY
        RESRT=Y(I)-RCLS*X(I)-CTLS
        RESID=RESRT**2
        RETURN
         END
CC
CC   *SEARCH* :  SEARCHES THE KTH VALUE IN A VECTOR
CC               OF LENGTH 'LENGTH'.
        FUNCTION SEARCH(AA,LENGTH,K)
        DIMENSION AA(LENGTH),A(1000)
        DO 3005 I=1,LENGTH
 3005   A(I)=AA(I)
        L=1
        LR=LENGTH
 3000   IF(L.GE.LR) GOTO 3020
 3010   AX=A(K)
        I=L
        J=LR
 3030   IF(I.GT.J) GOTO 3040
 3060   IF(A(I).GE.AX) GOTO 3080
 3070   I=I+1
        GOTO 3060
 3080   IF(A(J).LE.AX) GOTO 3100
 3090   J=J-1
        GOTO 3080
 3100   IF(I.GT.J) GOTO 3120
 3110   WA=A(I)
        A(I)=A(J)
        A(J)=WA
        I=I+1
        J=J-1
```

```
 3120   GOTO 3030
 3040   IF(J.LT.K)L=I
        IF (K.LT.I) LR=J
        GOTO 3000
 3020   SEARCH=A(K)
        RETURN
         END
CC
CC   *AMDIAN* :  CALCULATES THE MEDIAN OF A VECTOR
CC               OF LENGTH 'LENGTH'.
        FUNCTION AMDIAN(AA,LENGTH)
        DIMENSION AA(LENGTH),A(1000)
        DO 3205 I=1,LENGTH
 3205   A(I)=AA(I)
        AL=LENGTH
        IL=INT(AL/2.0)
        IF(MOD(LENGTH,2).NE.0) GOTO 3230
 3200   AMDIAN= (SEARCH(A,LENGTH,IL)+
      1 SEARCH(A,LENGTH,IL+1))/2.0
        GOTO 3220
 3230   AMDIAN= SEARCH(A,LENGTH,IL+1)
 3220   RETURN
         END
CC
CC   *AMAD*   :  CALCULATES THE MEDIAN ABSOLUTE DEVIATION
CC               OF A VECTOR OF LENGTH 'LENGTH'.
        FUNCTION AMAD(AA,LENGTH,AMED)
        DIMENSION AA(LENGTH),A(1000)
        DO 4005 I=1,LENGTH
 4005   A(I)=AA(I)
        DO 4000 I=1,LENGTH
        A(I)=ABS(A(I)-AMED)
 4000   CONTINUE
        AMAD=AMDIAN(A,LENGTH)
        RETURN
         END
CC
CC   *ALGO*      CALCULATES THE OBJECTIVE FUNCTION
CC               VALUE FOR A CERTAIN SLOPE.
        SUBROUTINE ALGO(TETA,LENGTH,IP)
        COMMON SOL1D(500),ITEL,X(1000),Y(1000),W(1000),
      V RESID,RESRT,RCLS,CTLS,VARX,VARY
        COMMON /SLTS/ MINMT2,MINRC,MINCT,BSTSSQ
        REAL MINMT2,MINRC,MINCT
        A=SIN(TETA)/COS(TETA)
        RLTS=YMAXI(LENGTH,A)
        MINCT=0.0
         DO 4030 III=1,ITEL
        MINCT=MINCT+SOL1D(III)
 4030   CONTINUE
        TEL=ITEL
        MINCT=MINCT/TEL
        MINMT2=RLTS
        MINRC=A
        RETURN
         END
CC
CC   *ORDEN*  :  SORTS A VECTOR OF LENGTH 'LENGTH'.
CC
        SUBROUTINE ORDEN(A,LENGTH)
        DIMENSION A(LENGTH)
        DIMENSION AA(1000)
        DO 40 K=2,LENGTH
        VAL=A(K)
        NK=K-1
        DO 30 L=1,NK
```

```
        IF (VAL.GE.A(L)) GOTO 30
50      L1=L+1
        J=1
        DO 20 M=L1,K
        AA(J)=A(M-1)
       J=J+1
20      CONTINUE
        J=1
        DO 25 M=L1,K
        A(M)=AA(J)
        J=J+1
25      CONTINUE
        A(L)=VAL
        GOTO 40
30      CONTINUE
40      CONTINUE
        RETURN
        END
```

The input data file looks as follows (this is the example of Table 1):

```
    7
(F3.0,F7.0)
 75   18031
 76   18273
 77   16660
 78   15688
 79   23805
 80   15241
 81   13295
```

The first line contains the number of points n , and the second line gives the format in which the data are going to be read. The following lines contain the data itself, one line for each point, first x_i and then y_i . For this example, the output is:

```
L E A S T   T R I M M E D   S Q U A R E S   L I N E
****************************************************

MEDIAN X   =        78.000000
MEDIAN Y   =     16660.000000
MAD X      =         2.000000
MAD Y      =      1419.000000


LTS    Y =      -797.940016 X +      77958.336260
SCALE ESTIMATE    =          284.981923


LS     Y =      -468.821429 X +      53852.785714
SCALE ESTIMATE    =         3490.785314
```

For those interested in the details of the calculation, we shall now describe what the program does. First, it standardizes the data for convenience, ensuring that median$(x_i) = 0$ and median$(|x_i|) = 1$, and the same for y_i . (In the end, the solution is transformed accordingly.) Then, the algorithm consists of replacing (2) by

$$\underset{\hat{a}}{\text{minimize}}\left(\underset{\hat{b}}{\min} \Sigma_{i=1}^{h} [((y_\cdot - \hat{a}x_\cdot) - \hat{b})^2]_{i:n}\right) \qquad (4)$$

and solving these parts separately. The second part of the minimization is quite easy, because for any given $\hat{a}$ it becomes essentially a one-dimensional problem by putting

$$z_i := (y_\cdot - \hat{a}x_\cdot)_{i:n} \quad .$$

This part can be solved exactly because there are only n-h+1 "halves" of the "sample" $z_1 \leq z_2 \leq \ldots \leq z_n$ which need to be considered, being:

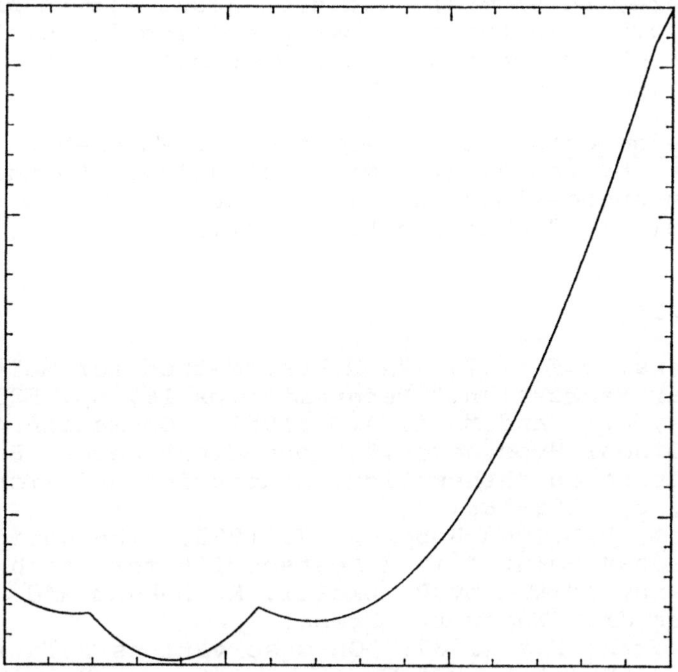

Figure 4. S as a function of $\hat{a}$.

$$H_1 = \{z_1,\ldots,z_h\} \ , \ H_2 = \{z_2,\ldots,z_{h+1}\} \ ,\ldots$$

$$\ldots, \ H_{n-h+1} = \{z_{n-h+1},\ldots,z_n\} \quad .$$

For each of these H_i , one defines:

$$m(i) = \frac{1}{h} \Sigma_{z_j \in H_i} z_j$$

$$s(i) = \Sigma_{z_j \in H_i} (z_j - m(i))^2 \quad .$$

The H_i for which s(i) is minimal is selected, and we put now $\hat{b} := m(i)$. (In case more than one s(i) is minimal, $\hat{b}$ is taken to be the average of the corresponding numbers m(i) .) The minimum s(i) is denoted by $S(\hat{a})$, and now we only have to find $\hat{a}$ for which $S(\hat{a})$ is minimal. This is just a one-dimensional minimization, where the function S is continuous but not everywhere differentiable. For our example of Table 1, a part of the function $S(\hat{a})$ is plotted in Figure 4. In order to find the minimum of S, the program first scans all angles θ in (-1.55,1.55) with a stepsize of 0.02 (here, $tg(\theta) = \hat{a}$), and then scans with a precision of 0.001 in the two most promising areas. In this way, the objective function S does not have to be calculated too often.

Acknowledgements The research of P. Rousseeuw was supported by the Belgian National Science Foundation. Valuable suggestions on this subject were made by F. Hampel, W. Stahel and F. Plastria.

REFERENCES

1. Andrews, D.F. 1974, "A Robust Method for Multiple Linear Regression," Technometrics 16, pp. 523-531.
2. Brown, G.W. and Mood, A.M. 1951, "On Median Tests for Linear Hypotheses," Proceedings Second Berkeley Symposium on Mathematical Statistics and Probability, pp. 159-166.
3. Donoho, D.L. and Huber, P.J. 1983, "The Notion of Breakdown Point," in A Festschrift for Erich L. Lehmann, edited by P. Bickel, K. Doksum and J. Hodges Jr., Wadsworth Press.
4. Edgeworth, F.Y. 1887, "On Observations Relating to Several Quantities," Hermathena 6, pp. 279-285.
5. Hampel, F.R. 1971, "A General Qualitative Definition of Robustness," Ann. Math. Stat. 42, 1887-1896.

6. Hampel, F.R. 1975, "Beyond Location Parameters: Robust Concepts and Methods," Bulletin of the International Statistical Institute 46, pp. 375-382.
7. Hampel, F.R. 1978, "Optimally Bounding the Gross-Error-Sensitivity and the Influence of Position in Factor Space," 1978 Proceedings of the ASA Statistical Computing Section, pp. 59-64.
8. Hill, R.W. 1977, "Robust Regression When There are Outliers in the Carriers," unpublished Ph.D. dissertation, Harvard University.
9. Huber, P.J. 1973, "Robust Regression: Asymptotics, Conjectures and Monte Carlo," Annals of Statistics 1, pp. 799-821.
10. Jaeckel, L.A. 1972, "Estimating Regression Coefficients by Minimizing the Dispersion of the Residuals," Annals of Mathematical Statistics 5, pp. 1449-1458.
11. Krasker, W.S. 1980, "Estimation in Linear Regression Models with Disparate Data Points," Econometrica 48, 1333-1346.
12. Krasker, W.S. and Welsch, R.E. 1982, "Efficient Bounded-Influence Regression Estimation," Journal of the American Statistical Association 77, pp. 595-604.
13. Mallows, C.L. 1975, "On Some Topics in Robustness," unpublished memorandum, Bell Telephone Laboratories, Murray Hill, New Jersey.
14. Marazzi, A. 1980, "Robust Linear Regression Programs in ROBETH," Research Report No. 23, Fachgruppe für Statistik, ETH Zürich.
15. Maronna, R.A., Bustos, O. and Yohai, V. 1979, "Bias- and Efficiency-Robustness of General M-Estimators for Regression with Random Carriers," in Smoothing Techniques for Curve Estimation, edited by T. Gasser and M. Rosenblatt, Springer Verlag, New York.
16. Rousseeuw, P.J. 1982, "Least Median of Squares Regression," Research Report No. 178, Centre for Statistics and Operations Research, Vrije Universiteit Brussel.
17. Rousseeuw, P.J. 1983, "Regression Techniques with High Breakdown Point," The Institute of Mathematical Statistics Bulletin 12, pp. 155.
18. Rousseeuw, P.J., Daniels, B., Leroy, A. 1983, "Applying Robust Regression to Insurance," to appear in Insurance: Mathematics and Economics.
19. Sen, P.K. 1968, "Estimates of the Regression Coefficient Based on Kendall's Tau," Journal of the American Statistical Association 63, pp. 1379-1389.

20.Siegel, A.F. 1982, "Robust Regression Using Repeated Medians," Biometrika 69, 242-244.
21.Theil, H. 1950, "A Rank-Invariant Method of Linear and Polynomial Regression Analysis, I, II and III." Nederlandsche Akademie van Wetenschappen Proceedings Serie A 53, pp. 386-392, 521-525, and 1397-1412.
22."De Verzekering in België", Verslag van de Beroepsvereniging der Verzekeringsondernemingen 1981-1982.

Vrije Universiteit Brussel
Dept. of Math., F 715
Pleinlaan 2
B-1050 Brussels, Belgium

Vrije Universiteit Brussel
CSOO, M203
Pleinlaan 2
B-1050 Brussels, Belgium

QUANTITATIVE MODELS OF PENSION COSTS

Arnold F. Shapiro

The Pennsylvania State University
University Park, PA 16802, USA

ABSTRACT

There is considerable model building currently taking place in the pension area. In order to reduce duplication of effort and to minimize redundancy, it is important that researchers be familiar with past contributions to the pension literature. The purpose of this article is to aid researchers in this regard, by providing an overview of some of the notable contributions to the quantitative models of pension cost.

THE EARLIEST LITERATURE

One of the first papers published on the mathematics of pensions was an article by Hardy (1892). The main thrust of his article was to indicate roughly how the cost of old age pensions could be determined and to emphasis the phenomenal amount of money that the old age pensions would ultimately cost. Concerned primarily with the cost of a national old age pension scheme, Hardy assumed an initially mature group and proceeded to show how to determine the uniform contribution that would be required from each employed person in order to fund a national pension scheme.

Manly (1898) was one of the first articles dealing with the private pension system. The main purpose of the article was to provide solutions to basic problems arising from contributory pension plans. The presentation, which was somewhat elementary by today's standards, but nonetheless a significant contribution at the time, entailed the determination of the premiums and reserves for deferred annuities which allowed refunds in the event of death or withdrawal.

F. de Vylder et al. (eds.), Premium Calculation in Insurance, 333–345.

Manly's first major contribution came in 1901, at which time he laid the foundation for what was to become a standard approach to the computation of expected pension costs.[2] The problems were divided into two classes, those which did not include salary scales and those which did. The former set of problems included the determination of the present value of an annuity which was deferred to age sixty-five, which was payable on retirement at that age, and which was combined with various combinations of a return of premiums with and without interest in the event of death, withdrawal or early retirement. A linear reduction in the pension benefit was assumed in the event of early retirement.

The latter set of problems were of both the defined benefit type and the defined contribution type. In the defined benefit type, Manly determined the present value, at the age of entry, of a pension benefit based on both a career average and final average salary and a number of years of service. In the defined contribution type he gave a method of determining the size of the normal retirement benefit which could be obtained with the accumulation of predetermined contributions.[3]

The next major contribution to the pension literature was by George King (1905). In that paper, he extended the pension literature in five respects. He gave an explanation for transforming raw decrement data into decrement tables, He differentiated between past and future contributions, he investigated the implications of a provision for the return of contributions at other than the valuation rate of interest, he introduced the assumption that "...pensions, taken one with another, are payable continuously throughout the year,"[4] and he devised an approximation for the final average salary.

In the years following the articles of Manly and King most of the advances in the computation of pension cost dealt with the derivation of relevant data and the implementation of that data. Hence, Bacon (1908) developed the concept of salary scales based on time series data, rather than cross-sectional data; Grant (1909) investigated the derivation of mortality data for pensioners; and Thomas (1914) introduced the notion of segregating the various classes of participants by category of employment.

These early studies promoted an understanding of the mathematics involved in the development of expected pension cost models. However, since the scope of the benefits which were then available was somewhat limited, so where the models. Later contributions often dealt with models which were significantly more general.

ACTUARIAL COST METHODS

The contributions to pension mathematics prior to 1915 were due almost entirely to the efforts of European actuaries. Walker (1915) set the stage for what was to become a major area of contribution of the American actuary -- the development of actuarial cost methods -- when he investigated the "initial liability" for benefits based on past service at the commencement of a fund. Interestingly, while Walker, following Manly and Ackland (1912), used the term "initial liability," Hutcheson, in discussing Walker's paper (p. 389), used the term "[initial] accrued liability," a current nomenclature.

It is informative to note the early methods which were proposed for dealing with the initial accrued liability. It appears that the first method used was based upon the assumption that the plan will exist indefinitely into the future since only the interest on the accrued liability was paid.[5] Later actuaries who investigated the mathematics for dealing with the accrued liability took various other approaches: Woodward (1925) suggested that the initial accrued liability be funded by the time the original employees retire; Robbins (1929) suggested that it be funded during the life of the original employees; and Corbett (1936) suggested limiting the initial accrued liability to include only those persons aged forty or over and amortizing it over a twenty-five year period, under the assumption that the error introduced by such a procedure would be negligible. Apparently, it was Woodward (1925, p. 447) who coined the term "normal cost," defining it to be the amount which if set aside annually would be adequate to fund the pension benefits for new employees entering the plan.

It is obvious that a pension plan may be funded by any number of methods, the method chosen in any particular case being dependent upon such things as the capital requirements of a firm and income tax considerations. However, a hallmark in the trend of funding methods in the United States was the publication, in 1945, of the Bulletin on Section 23(p) (1) (A) and (B) of the Internal Revenue Code, which described a number of methods of calculating and amortizing the liabilities of a pension fund. Most of the contributions to pension mathematics following that publication dealt with quantifying the opinions expressed in that Bulletin.

Two of the more important articles of this period were the articles of Seal (1952) and Trowbridge (1952). The main purpose of Seal's article was to discuss and quantify the various funding methods described in the Bulletin as "acceptable" for tax-exempt self-insured plans. Trowbridge, using a somewhat more general approach, first discussed the broad characteristics of a funding method and then proceeded to show the quantitative nature of a variety of these methods. In addition, he discussed and exemplified the accrual of funds under each of the methods which he described.

Most of the methods in common use in the United States also are used in the United Kingdom, although, apparently, the aggregate cost method is the most common. A comparison is given by Colbran (1982). There are variances, however. The qualitative nature of some of these are described by Lyon (1960).

The foregoing articles tended to stress the uniqueness of the various methods of pension funding. Other articles stressed the formulation of general funding models. There have been at least four significant articles on general funding methods.

Trowbridge (1963) wrote an article in which he developed a family of funding methods based on the rate at which the unfunded present value of benefits were being funded. The funding family which he proposed was characterized by an annual contribution, payable in advance, which was composed of (1) the discounted value of the annual interest on the unfunded liability and (2) a payment to reduce the unfunded liability. By varying the size of the contribution to reduce the unfunded liability, the ultimate contribution and fund of any actuarial cost method could be reproduced. In that article he also emphasized the importance of the ratio of the present value of the benefits to the size of the accumulated fund, the "fund ratio", and its progression from year to year.

During the discussion of Trowbridge's paper (1963, p. 171), Nesbitt suggested that the unfunded present value family could be further generalized by distinguishing between the initial accrued liability, if any, and the liability which accrues after that time. Taylor (1967), acting on Nesbitt's suggestion, designed a funding family characterized by contributions, payable in advance, which were composed of a duration dependent contribution to reduce the unfunded liability and a duration dependent contribution to reduce the initial accrued liability. By the appropriate choice of both portions of the contribution the ultimate contribution and fund of any of the actuarial cost methods could be reproduced, including Trowbridge's unfunded present value family.

A third study to interrelate the various actuarial cost methods was that of Cooper and Hickman (1967), in which they developed a general family of actuarial cost methods based on the rate at which retirement income is being purchased, for a member of the pension group, at a given attained age. Here, the family of actuarial cost methods which they developed permitted the purchase of retirement income at either an accelerating or decelerating rate.

The Cooper and Hickman article formed the basis of an exposition on contribution theory by Bowers, Hickman and Nesbitt (1979). Here, a general model for a pension plan involving growth with respect to the population, salaries, and retirement benefits was used to study contribution patterns that may arise under different actuarial cost methods. A byproduct of their analysis was the development of a function to denote the average age when normal cost payment can be viewed as centered. This function turns out to be closely related to the cumulant generating function of probability and statistics, and also provides a bridge that permits pension funding ideas to be stated in terms of compound interest functions.

DYNAMIC SIMULATION OF PENSION COSTS

An important area of investigation is the simulation of pension fund growth since simulated results provide an estimate of the liability accruals under the plan. This is important since it gives the employer a better understanding of the ultimate cost of employee benefits.

One of the first published accounts of the growth of a pension fund was an article by M'Lauchlan (1908). That article illustrated the necessity of accumulating large invested funds during the early years of a funds existence so as to provide for the heavy liability which ultimately will be maturing for payment.

The format advocated by M'Lauchlan for simulating pension fund growth is still current, and it is appropriate, therefore, to mention the procedure that he used. He started with tables of rates of mortality, withdrawals, retirement, and salary increases, an assumed guaranteed annual interest rate, and one thousand entrants at the single entry age twenty. Then, under the assumption that the benefits based upon five percent of salary had been accurately determined, he tabulated the fund which accumulated from contributions and interest and the payments incurred by virtue of withdrawals, death in service, pensions and expenses, over decennial periods, and the value of the accumulated fund at the end of each period.

Manly (1911) extended M'Lauchlan's investigation by showing the effect on pension liability growth of variation in the rates of mortality, withdrawal, retirement and salaries. Manly's primary concern was to show that the ultimate trend of the simulated pension fund depends upon the nature of the assumptions made.

Fund projections have continued to be a major tool in the investigation of pension plan costs. Llewellyn (1938), projected pension payments for a plan through 1979, and compared the present value of that projection with a then-current valuation; Niessen (1950) authored an article titled "Projections -- How to Make Them and How to Use Them;" Trowbridge (1952) traced the fund and contributions, under various actuarial cost methods, through to their "ultimate" state; Griffin (1966) investigated the impact of both decreasing and increasing populations on the adequacy of various actuarial cost methods; and Winklevoss (1974) investigated this impact on the cost of vesting in pension plans.

These types of analysis culminated in the forecast valuation methods of the recent literature. Like the original work of M'Lauchlan, the methods involve the development of expected future population and payroll figures, the development of expected future benefit disbursements, and the development of future cost figures. The distinguishing feature of current methods, however, is the use of open group techniques, in the sense that future new entrants are anticipated. Contributions to this area include that of Fleischer (1975), who discussed the use of the forecast valuation method for the first year actuarial valuation of pension plans, and Schnitzer (1977), who discussed its use in the valuation of an ongoing plan.

The foregoing studies were concerned with empirical simulation of pension costs. An analytical approach to this problem was undertaken by Bowers, Hickman and Nesbitt, singularly and jointly. They authored an introductory paper (Bowers et. al. (1976)) which expounded on the mathematical principles applicable to pension funding under dynamic conditions of population growth, inflation, and automatic adjustment of benefits. The model was deterministic in nature, in the sense that it did not deal with contingency reserves, and, for simplicity, was continuous and based only on retirement benefits.

One outgrowth of this study was a paper by Nesbitt (1982) which dealt with vesting. This paper analyzed the situation where, in addition to the retirement benefit available at the normal retirement age, there was a termination benefit equivalent to the reserve in regard to the terminating participant. A plan with this feature was said to be exactly vested.

A second outgrowth was the paper which dealt with contribution theory which was mentioned previously.

The authors extended this discussion in yet another paper (Bowers et. al. (1982)) wherein they discussed a generalized amortization method for the unfunded accrued liability, the impact of gains and losses, and an application of their model to a variable annuity system.

The exponential growth case generally is considered to be rather special and unrealistic. Nevertheless, it has been used in practical models. Treuil (1981), for example, used an exponential growth model to study alternatives related to the tail end of long-term projections for social insurance and pension systems, under the assumption that a time is being approached when an assumption of stabalized conditions seem as appropriate as any other.

STOCHASTIC SIMULATION OF PENSION COSTS

Stochastic models of pension costs are of relatively recent vintage. Stone (1948) investigated the use of probability generating functions to develop probabilities, at various durations after employees had begun to retire, that the total actual pension payments would differ from the expected total payments. Taylor (1952) investigated the size of the contingency reserve needed to insure, with a given probability, that the funds on hand would be sufficient to pay all promised pensions. These studies dealt exclusively with the retired population.

Variability in pension cost estimates for active plan participants was considered in the studies of Seal (1953), Knopf (1957), and Shapiro (1977 and 1979). Seal investigated the impact of death benefits in a trusteed plan, using a normal approximation to the binomial distribution to introduce variance minimization into the design of pension plans. Knopf investigated the feasibility of a fully-trusteed pension plan using a simplified Monte Carlo approach. Shapiro considered the credibility of projected pension costs using a model based on the direct application of a conditional Bernoulli process and later extended the study to include an unconditional distribution.

The foregoing articles were concerned primarily with the cost to the employer; the impact on the employee received little attention. An important thrust towards filling this gap in the pension literature was provided by Balcer and Sahin (1979). Their contribution was the development of a probabilistic model for tracking the accumulated benefits which were vested during the working lifetime of an individual. A characteristic of their model was that it took into account the fraction of jobs offering pension plans and the probability of portability.

To the extent that pensions may be regarded as annuities, there have been a considerable number of other relevant studies. Piper (1933), for example, developed contingency reserves for life annuities based on the mean and variance associated those annuities. Menge (1937), Hickman (1964), and Boyle (1974), elaborated on the Piper concept: Menge and Boyle using discrete functions and Hickman using continuous functions. It is worth noting that Hickman extended the development to include loss functions and a probabilistic consideration of multiple decrement theory.

The number of lives which persist to a given age from an initial group of lives is generated by a Bernoulli process. Nevertheless, the considerable labor required to generate appropriate distributions under this process resulted in the development of various approximation methods. Hence, Piper assumed a large group of lives and used a normal distribution, as did Seal; Taylor suggested fitting a Pearson Type III distribution to the total present value of life annuity costs; Boermeester (1956) applied a Monte Carlo approach to the problem, as did Knopf; Fretwell and Hickman (1964) investigated upper bounds for the cost using the inequalities of Tchebychef and Uspensky; and Bowers (1967) investigated the use of the Cornish-Fisher expansion to develop probabilities of sufficient reserves, based on correction factors applied to a standard normal table.

COMMENT

The foregoing discussion provides only a cursory overview of the literature of quantitative pension cost models. Nonetheless, it is hoped that the discussion was sufficiently detailed to be useful to the pension cost researcher. To the extent that this article is helpful in this regard, it has served its purpose.

NOTES

[1]It is assumed that the reader is familiar with basic provisions of a pension plan. Readers lacking this knowledge are referred to Allen et. al (1981) and McGill (1979), which deal with specific characteristics of United States plans and governing legislation; Seres and Selley (1980) and Pilch and Wood (1979), which deal with plans in Great Britain; and Archibald (1981), which deals with the continental European situation.

[2]Other actuaries of the day were endeavoring to formulate an appropriate approach for determining the cost of pensions. Prior to the publication of Manly's article, for example, McGowan (1902) discussed the subject. That article was considerably less detailed than Manly's, and contained many flaws from a practical point of view, not the least of which was an unsatisfactory approach to salary scales.

[3]It should not necessarily be assumed that the data used by Manly (1901) were representative of those used in practice at the time, but it is interesting to note that for the purpose of his paper the accumulation interest rate was assumed to be four per-cent per annum, the average age at entry was twenty, the normal retirement age was sixty-five, the defined benefit was two per-cent of total salary, and the defined contribution was five per-cent of salary.

[4]This assumption greatly facilitates the mathematical analysis of pensions since it allows the use of continuous functions.

[5]It was mentioned, for example, in connection with the funding of the past service liability of a pension plan governed by the 1866 Metropolitan Superannuation Act of London (Manly and Ackland (1912), and discussion by Tinner, p. 378).

BIBLIOGRAPY

Allen, E. T., Jr., Melone, J. J., and Rosenbloom, J. S. 1981, Pension Planning. 4th ed. Homewood, Ill.: Richard D. Irwin, Inc.

Archibald, G. C. 1981, "Retirement Benefits in Continental Europe," TFA 37, pp. 339-398.

Bacon, J. 1908, "A Pension Fund Problem; With Some Remarks on the Deduction of Salary-scales," JIA 42, pp. 36-60.

Balcer, Y. and Sahin, I. 1979, "Probabilistic Models for Pension Benefits," JRI 46, pp. 99-123.

Bizley, M. T. L. 1949, "The Comparison of Salary Scales," JIASS 9, pp. 105-120.

Boermeester, J. M. 1956, "Frequency Distribution of Mortality Costs." TSA 8, pp. 1-9.

Bowers, N. L. 1967, "An Approximation to the Distribution of Annuity Costs," TSA 19, pp. 295-309.

Bowers, N. L., Hickman, J. C., and Nesbitt, C. J. 1976, "Introduction to the Dynamics of Pension Funding." TSA 28, pp. 177-204.

Bowers, N. L., Hickman, J. C., and Nesbitt, C. J. 1979, "The Dynamics of Pension Funding: Contribution Theory," TSA 31, pp. 93-119.

Bowers, N. L., Hickman, J. C., and Nesbitt, C. J. 1982, "Notes on the Dynamics of Pension Funding," I:M&E 1, No. 4, pp. 261-270.

Boyle, P. P. 1974, "Note on the Variance of a Widow's Pension," JIA 101, pp. 103-107.

Colbran, R. B. 1982, "Valuation of Final Salary Pension Schemes," JIA 109, pp. 359-385, and discussions, pp. 386-416.

Cooper, S. L., and Hickman, J. C. 1967, "A Family of Accrued Benefit Actuarial Cost Methods," TSA 19, pp. 53-59.

Corbett, H. R. 1936, "Pension Funds: Alternative Funding Methods of Valuation," RAIA 25, pp. 27-32.

Fleischer, D. R. 1975, "The Forecast Valuation Method for Pension Plans." TSA 27 , 93.

Fretwell, R. L., and Hickman, J. C. 1964, "Approximate Probability Statements About Life Annuity Costs," TSA 16, pp. 55-60.

Grant, M. D. 1909, "Staff Pension Funds: An Investigation of the Mortality of Canadian Civil Service Pensioners," TASA 11, pp. 163-171.

Griffin, F. L., Jr. 1966, "Concepts of Adequacy in Pension Plans," TSA 18, pp. 46-63.

Hardy, R. P. 1892, "On the Formula for Determining the Value of Benefits, According to the Principle of Collective Assurance," JIA 30, pp. 79-87.

Hickman, J. C. 1964, "A Statistical Approach to Premiums and Reserves in Multiple Decrement Theory," TSA 16, pp. 1-16.

King, G. 1905, "On Staff Pension Funds," JIA 39, pp. 129-192.

Knopf, M. 1957, "A Practical Demonstration of the Risk Run by a Very Small Company with a Trusteed Pension Plan," PCAPP 6, pp. 230-43.

Llewellyn, G. F. 1938, "On a Method of Forecasting the Future Pension Outgo of a Staff Superannuation Fund," JIA 69, pp. 172-178.

Lyon, C. S. S. 1960, "Controlled Funding Methods for Group Pension Schemes," JIASS 16, pp. 130-146.

Manly, H. W. 1898, "On the Solution of Some Problems Which Frequently Arise Out of the Rules of Pension Funds and Friendly Societies," Transactions of the Second International Actuarial Congress, pp. 860-867.

Manly, H. W. 1901, "On the Valuation of Staff Pension Funds," JIA 36, pp. 209-276.

Manly, H. W. 1902, "On the Valuation of Staff Pension Funds," JIA 37, pp. 193-244.

Manly, H. W. 1911, "On Staff Pension Funds: The Process of the Accumulation of the Funds; The Identity of a Valuation with the Future Progress of a Fund; The Manner of Dealing with Funds which are Insolvent; and Sundry Observations," JIA 45, pp. 149-211.

Manly, H. W., and Ackland, T. G. 1912, "On the Superannuation an Pension Funds of certain Metropolitan Borough Councils, their Establishment, Administration, and Actuarial Investigation," JIA 46, pp. 327-370.

Marples, W. F. 1945, "An Analysis of a Pension Fund," JIA 73, pp. 66-86.

McGill, Dan M. 1979, Fundamentals of Private Pensions. 4th ed. Homewood, Ill.: Richard D. Irwin, Inc.

McGowan, J. 1902, "An Enquiry into the Methods and General Principles to be Adopted in Calculating the Rates of Contribution, Percentage Deduction from Salary, in the Case of Superannuation Funds and Pension Funds Generally," JIA 37, pp. 15-37.

Menge, W. O. 1932, "Forces of Decrement in a Multiple-Decrement Table," RAIA 21, pp. 41-46.

Menge, W. O. 1937, "A Statistical Treatment of Actuarial Functions," RAIA 26, pp. 65-88.

M'Lauchlan, J. J. 1908, "The Fundamental Principles of Pension Funds", TFA 4, pp. 195-227.

Nesbitt, C. J. 1982, "Exploration of Pension Funding in Case of Exact Vesting," I:M&E 1, No. 4, pp. 255-260.

Niessen, A. M. 1950, "Projections -- How to Make Them and How to Use Them," TSA 2, pp. 235-253.

Piper, K. B. 1933, "Contingency Reserves for Life Annuities," TASA 34, pp. 240-249.

Pilch, M. and Wood, V. 1979, Pension Schemes -- A Guide to Principles and Practice, Gower Press, Farnborough.

Robbins, R. B. 1929, "The Accrued Liability of Retirement Plans," RAIA 18, 41.

Schnitzer, R. J. 1977, "Characteristics and Operation of Projection Valuation Methods for Pension Plan Funding," TSA 29, pp. 269-298.

Seal, H. L. 1952, "Acceptable Funding Methods for Self-Insured Pension Plans," PCAPP 2, pp. 17-44.

Seal, H. L. 1953, "The Mathematical Risk of Lump-Sum Death Benefits in a Trusteed Pension Plan," TSA 5, pp. 135-142.

Seres, J. S. D. and Selley, J. W. 1980, Pensions: a Practical Guide, Oyez Publishing Ltd.

Shapiro, A. F. 1977, "The Relevance of Expected Persistency Rates when Projecting Pension Costs," JRI 44, pp. 623-638.

Shapiro, A. F. 1979, "A Bayesian Approach to Persistency Rates in the Projection of Retirement Costs," TSA 31, pp. 337-365.

Stone, D. G. 1948, "Actuarial Note: Mortality Fluctuations in Small Self-Insured Pension Plans," TASA 49, pp. 82-91.

Taylor, J. R. 1967, "The Generalized Family of Aggregate Actuarial Cost Methods for Pension Funding," TSA 19, pp. 1-12.

Taylor, R. H. 1952, "The Probability Distribution of Life Annuity Reserves and Its Application to a Pension System," PCAPP 2, pp. 100-150.

Thomas, E. C. 1914, "On the Valuation of Benefits Dependent upon Promotion to a Higher Status," JIA 48, pp. 217-249.

Treuil, P. 1981, "Fund Development of an Earnings-Related Social Insurance Plan Under Stabalized Conditions," TSA 33, pp. 231-246.

Trowbridge, C. L. 1952, "Fundamentals of Pension Funding," TSA 4, pp. 17-43.

Trowbridge, C. L. 1963, "The Unfunded Present Value Family of Pension Funding Methods," TSA 15, pp. 151-169.

Walker, D. A. 1915, "A Staff Pension Fund," TASA 16, pp. 109-144.

Winklevoss, H. E. 1974, "Cost Sensitivity Analysis of Mandatory Funding and Vesting Standards in Pension Plans," JRI 41, pp. 57-73.

Woodward, J. H. 1925, "The Valuation of Liabilities Under Industrial Pension Plans," TASA 26, pp. 443-479.

CREDIBILITY: ESTIMATION OF STRUCTURAL PARAMETERS

Ben Zehnwirth

School of Economic and Financial Studies,
Macquarie University, Australia.

Invited Lecture given at the Nato Advanced Study Institute on Insurance Premiums, Katholique Universiteit Leuven, July 18-31, 1983.

1. INTRODUCTION AND SUMMARY

Estimation of structural parameters has recently received considerable attention in the actuarial literature. Notable papers include De Vylder (1978, 1981), Dubey and Gisler (1981) and Norberg (1981). The main emphasis appears to be on 'linear' minimum variance unbiased estimators. The technique involves a marginalization approach based on the method of moments - up to second order moments of the marginal distributions of the observable random variables are equated with their sample moments.

In this article we develop some of the well known estimators treated in the abovementioned papers. However, by contrast the present development is based on a unified approach that relies heavily on a connection between linear Bayes theory and Normal based Bayes theory implicit in Hartigan (1969) and a connection between Normal based Bayes theory and Normal based frequentist theory (Goldberger and Theil (1961)).

In the next section we elaborate on these two important connections. They are applied in section 3 to Hachemeister's celebrated regression model and in section 4 to a simple Poisson random coefficient model.

2. GAUSS-MARKOV THEORY VERSUS LINEAR BAYES THEORY

The aim of the present section is to demonstrate that mixed generalized least squares (GLS) estimation has a fruitful (linear)

F. de Vylder et al. (eds.), Premium Calculation in Insurance, 347–359.

Bayes interpretation. The presentation up to expression (2.11) essentially follows Duncan and Horn (1972). In the econometric literature the combining technique culminating in (2.4) and (2.5) is known as mixed estimation. See, for example, Theil (1971, p.347).

Consider a linear model that describes a data base. Adopting the customary notation the model is given by

$$\underset{\sim}{Y} = X\underset{\sim}{\beta} + \underset{\sim}{\varepsilon} \tag{2.1}$$

where $\underset{\sim}{\varepsilon}$ is a zero mean error term with covariance matrix Σ_o. $\underset{\sim}{Y}$ is a vector of observations and X is a known design matrix (of full rank). Moreover, suppose we also have past data to which a linear model with the same parameter $\underset{\sim}{\beta}$ is applicable. Based on the past data we have a GLSE $\underset{\sim}{\beta}_o$ say, with covariance matrix Γ. Write,

$$\underset{\sim}{\beta}_o = \underset{\sim}{\beta} + \underset{\sim}{\delta} \tag{2.2}$$

where $\underset{\sim}{\delta}$ is a zero mean error term with covariance matrix Γ.

In order to utilize both sets of data, combine formatulations (2.1) and (2.2) to yield.

$$\begin{pmatrix} \underset{\sim}{Y} \\ \\ \underset{\sim}{\beta}_o \end{pmatrix} = \begin{pmatrix} X \\ \\ I \end{pmatrix} \underset{\sim}{\beta} + \begin{pmatrix} \underset{\sim}{\varepsilon} \\ \\ \underset{\sim}{\delta} \end{pmatrix} . \tag{2.3}$$

The covariance matrix of $(\underset{\sim}{\varepsilon}', \underset{\sim}{\delta}')'$ is block diagonal with block diagonal entries Σ_o and Γ. For this mixed formulation, the GLSE is given by

$$\underset{\sim}{\tilde{\beta}} = (X'\Sigma^{-1}X + \Gamma^{-1})^{-1}(\Gamma^{-1}\underset{\sim}{\beta}_o + X'\Sigma_o^{-1}\underset{\sim}{Y}) \tag{2.4}$$

with covariance matrix,

$$C[\underset{\sim}{\tilde{\beta}}] = (X'\Sigma_o^{-1}X + \Gamma^{-1})^{-1} . \tag{2.5}$$

In order to simplify expressions (2.4) and (2.5) we use the following two matrix identities that also prove useful in section 3.

(11): $(A^{-1} + B'C^{-1}B)^{-1} = A - AB'D^{-1}BA$

where $D = C + BAB'$ and B is assumed to have full rank.

(12): For any two matrices A and B such that A and B have identical row and column dimension

$$B(I + AB)^{-1} = (I + BA)^{-1} B.$$

Applying the above identities to formula (2.4) leads to

$$\tilde{\beta} = (I - Z)\, \beta_o + Z\hat{\beta} \qquad (2.6)$$

where $\hat{\beta}$, the GLSE for formulation (2.1), is given by

$$\hat{\beta} = (X'\Sigma_o^{-1}X)^{-1}\, X'\Sigma_o^{-1}\, Y \qquad (2.7)$$

and the weighting matrix Z is given by

$$Z = (X'\Sigma_o^{-1}X + \Gamma^{-1})^{-1}\, X'\Sigma_o^{-1}X \qquad (2.8)$$

Alternatively,

$$\tilde{\beta} = \beta_o + K(Y - X\,\beta_o) \qquad (2.9)$$

where K, the so called Kalman gain, is given by

$$K = \Gamma X' \,(X\Gamma\, X' + \Sigma_o)^{-1} \qquad (1.10)$$

Similarly,

$$C\,[\,\tilde{\beta}\,] = \Gamma - KX\Gamma\,. \qquad (2.11)$$

To both credibility and control theorists (2.9) and (2.11) are familiar - they have a linear Bayes interpretation. In order to derive this interpretation consider the linear model (2.1) with the additional a priori information

$$E\,[\,\beta\,] = \beta_o, \quad C\,[\,\beta\,] = \Gamma\quad . \qquad (2.12)$$

Then the inhomogeneous linear Bayes rule (ILBR) for β obtained by minimizing the Bayes risk

$$E\,[\,(a + AY - \beta)'\;\; G(a + AY - \beta)\,] \qquad (2.13)$$

over all vectors a and matrices A of appropriate dimensions, is the GLSE $\tilde{\beta}$ given by expression (2.6). The expectation is taken with respect to both Y and β . We point out that $\tilde{\beta}$ is independent of G an arbitrary positive definite matrix.

We are now in a position to generalize the Bayes model described by expressions (2.1) and 2.12).

Let θ be an unobservable random quantity (possibly a vector) taking values in a set Θ . Assume $\underset{\sim}{Y}$ is an observable random vector for which

$$E[\underset{\sim}{Y}|\theta] = X\underset{\sim}{\beta}(\theta), \qquad C[\underset{\sim}{Y}|\theta] = \Sigma(\theta),$$

$$E[\underset{\sim}{\beta}(\theta)] = \underset{\sim}{\beta}_o, \qquad E[\Sigma(\theta)] = \Sigma_o$$

and

$$C[\beta(\theta)] = \Gamma.$$

X is a known design matrix (of full rank) and $\underset{\sim}{\beta}(.)$ and $\Sigma(.)$ are (random) functions whose domain is Θ and ranges are $\mathbb{R}^s$ and $\mathbb{R}^{n^2}$ respectively. For the sake of notational convenience, write $\underset{\sim}{\beta} = \underset{\sim}{\beta}(\theta)$.

THEOREM: For the general Bayes model described above, the ILBR for $\underset{\sim}{\beta}$ that minimizes expression (2.13) is $\underset{\sim}{\tilde{\beta}}$, the GLSE for the constant $\underset{\sim}{\beta}$, Σ-linear model (2.3). Moreover, expression (2.5) gives the covariance of $\underset{\sim}{\tilde{\tilde{\beta}}} - \underset{\sim}{\beta}$ for both the classical and Bayesian formulations.

It is interesting to point out that in the classical formulation (2.3), β and Σ_o are fixed with $\underset{\sim}{Y}$ and $\underset{\sim}{\tilde{\beta}}_o$ random whereas in the Bayes formulation Y, $\underset{\sim}{\beta}$ and $\Sigma(\theta)$ are random with $\underset{\sim}{\beta}_o$ and Σ fixed.

The ILBR $\underset{\sim}{\tilde{\beta}}$ can be deduced as the solution to the optimization exercise (2.13) by adopting the normal equations from De Vylder (1976) or Goldstein (1975). More generally, if $\underset{\sim}{Y}$ and $\underset{\sim}{\beta}$ are any two random vectors then the ILBR for $\underset{\sim}{\beta}$ based on $\underset{\sim}{Y}$ is

$$\underset{\sim}{\beta}^* = E[\underset{\sim}{\beta}] + C[\underset{\sim}{\beta}, \underset{\sim}{Y}] (C[\underset{\sim}{Y}])^{-1} (\underset{\sim}{Y} - E[\underset{\sim}{Y}]). \tag{2.14}$$

We remark that $\underset{\sim}{\beta}^*$ represents the conditional mean $E[\underset{\sim}{\beta}|Y]$ if $\underset{\sim}{Y}$ and $\underset{\sim}{\beta}$ are jointly Gaussian. Indeed, if the sampling distribution of $\underset{\sim}{Y}$ (given $\underset{\sim}{\beta}$) is multiparameter exponential and $\underset{\sim}{\beta}$ takes the conjugate prior distribution then $\underset{\sim}{\beta}^*$ is the conditional mean $E[\underset{\sim}{\beta}|\underset{\sim}{Y}]$. Accordingly, in that case, the ILBR $\underset{\sim}{\tilde{\beta}}$ is (exact) Bayes rather than just the best inhomogeneous linear approximation. Diaconis and Ylvisaker (1979) prove a partial converse of this result, namely, that if $\underset{\sim}{\tilde{\beta}}$ is Bayes then the prior distribution of $\underset{\sim}{\beta}$ is necessarily the conjugate prior.

Leamer (1978, chap. 3) treats a special case of our more general Bayes model. His case drops out by taking $\theta = (\theta_1, \theta_2)$, $\Theta \subset \mathbb{R}^2$, $\underset{\sim}{\beta}(\theta) = \underset{\sim}{\beta}(\theta_1)$, $\Sigma(\theta) = \Sigma(\theta_2)$ and assuming that the sampling distribution of $\underset{\sim}{Y}$ is Gaussian.

$$\begin{pmatrix} Y_1 \\ \cdot \\ \cdot \\ \cdot \\ Y_q \end{pmatrix} = \begin{pmatrix} X_1 & & \\ & \ddots & \\ & & X_q \end{pmatrix} \begin{pmatrix} \beta_1 \\ \cdot \\ \cdot \\ \cdot \\ \beta_q \end{pmatrix} + \varepsilon \tag{3.1}$$

where ε is a zero mean error vector with $C[\varepsilon]$ block diagonal with jth diagonal entry given by $\sigma_o^2 P_j^{-1}$, and

$$\begin{pmatrix} \beta_o \\ \cdot \\ \cdot \\ \cdot \\ \beta_o \end{pmatrix} = \begin{pmatrix} \beta_1 \\ \cdot \\ \cdot \\ \cdot \\ \beta_q \end{pmatrix} + \delta \tag{3.2}$$

where δ is a zero mean error vector with $C[\delta]$ block diagonal with diagonal entries Γ.

Combining (3.1) and (3.2) yields the ILBR $\tilde{\beta}_j$ given by

$$\tilde{\beta}_j = \beta_o + K_j(Y_j - X_j\beta_o) \tag{3.3}$$

with error covariance matrix

$$C[\tilde{\beta}_j - \beta_j] = \Gamma - K_j X_j \Gamma \tag{3.4}$$

where

$$K_j = \Gamma X_j' (X_j \Gamma X_j' + \sigma_o^2 P_j^{-1})^{-1} \tag{3.5}$$

and

$$Z_j = (\Gamma^{-1} + \sigma_o^{-2} X_j' P_j X_j)^{-1} \sigma_o^{-2} X_j P_j X_j .$$

In order to estimate the structural parameter σ_o^2 it suffices to inspect (3.1) only.

The GLSE of β_j based on (3.1) is

$$\hat{\beta}_j = (X_j' P_j X_j)^{-1} X_j' P_j Y_j \tag{3.6}$$

with

$$C[\hat{\beta}_j] = \sigma_o^2 (X_j' P_j X_j)^{-1} \tag{3.7}$$

Using (2.1), the usual (unbiased) estimator of σ_o^2 is now given by

$$\hat{\sigma}_o^2 = [\sum_{j=1}^{q} (Y_j - X_j \hat{\beta}_j)' P_j (Y_j - X_j \hat{\beta}_j)]/(q(n-s)) \tag{3.8}$$

We now turn to the estimator of the structural parameter β_o. This has received considerable attention in the Econometrics literature. See Judge et. al. (1980), for example.

Since $\{\hat{\sigma}_o^2, \hat{\beta}_j\}$ $j=1 \ldots q$ is sufficient for the model (3.1), write

$$\begin{pmatrix} \hat{\beta}_1 \\ \cdot \\ \cdot \\ \cdot \\ \hat{\beta}_q \end{pmatrix} = \begin{pmatrix} \beta_1 \\ \cdot \\ \cdot \\ \cdot \\ \beta_q \end{pmatrix} + \varepsilon^* \tag{3.9}$$

where ε^* is a zero mean error vector with $C[\varepsilon^*]$ block diagonal with the jth diagonal entry given by $\sigma_o^2 (X_j' P_j X_j)^{-1}$. On combining (3.9) with (3.2) we obtain

$$\begin{pmatrix} \hat{\beta}_1 \\ \cdot \\ \cdot \\ \cdot \\ \hat{\beta}_q \end{pmatrix} = 1 \otimes \beta_o + \delta^* \tag{3.10}$$

where δ^* is a zero mean error vector where $C[\delta^*]$ is block diagonal with jth diagonal block given by $\sigma_o^2 (X_j' P_j X_j)^{-1} + \Gamma$.

The GLSE of β_o based on formulation (3.10) is

$$\hat{\beta}_o = \sum_{j=1}^{q} \bar{Z}_j \hat{\beta}_j \tag{3.11}$$

with

$$C[\hat{\beta}_o] = (\sum_{j=1}^{q} Z_j)^{-1} \tag{3.12}$$

where

$$\bar{Z}_j = (\sum_{j=1}^{q} Z_j)^{-1} Z_j . \tag{3.13}$$

De Vylder (1978) found that the optimal choice of weights W_j in $\Sigma W_j \hat{\beta}_j$ $(\Sigma W_j = I)$ in the sense of minimizing tr $C[\Sigma W_j \hat{\beta}_j]$ is

The classical-Bayes connection holds true if $\beta(\theta_1)$ and $\Sigma(\theta_2)$ are assigned independent conjugate prior distributions. Furthermore, Leamer remarks that a conjugate prior does not make a distinction between sampling and non-sampling prior information - the same can be said of a linear Bayesian who need only specify the first two prior moments. It is of interest to point out again that the present development does not involve parametric distributional assumptions, nor complete specification of a prior - only second order moments, albeit, at the price of admitting only linear estimators.

However, in view of the foregoing remarks 'linear' may be omitted from 'linear Bayes' in many circumstances. Accordingly, the cost to be incurred for admitting only linear estimators is usually not high.

3. HACHEMISTER'S REGRESSION MODEL

Hachemister (1975) treats in great detail a regression model that may be regarded as an empirical Bayes formulation of the general linear Bayes model (2.1) and (2.12) with $\Sigma(\theta)$ having a simple description given below. The following assumptions apply:

(A1): The elements $\theta_1, \theta_2, \ldots, \theta_q$ are iid.

(A2): The pairs $(\theta_1, \underset{\sim}{Y}_1), \ldots, (\theta_q, \underset{\sim}{Y}_q)$ are independent.

(A3): There exist functions $\underset{\sim}{\beta} : \Theta \to \mathbb{R}^s$

and $\sigma^2 : \Theta \to \mathbb{R}^1$ for which $E[\underset{\sim}{Y}_j | \theta_j] = X_j \beta(\theta_j)$ and $C[\underset{\sim}{Y}_j | \theta_j] = \sigma^2(\theta_j) P_j^{-1}$ where X_j is a known design matrix and P_j a known diagonal matrix.

For notational convenience, write $\underset{\sim}{\beta}_j = \beta(\theta_j)$ and let $\underset{\sim}{\beta}_o = E[\underset{\sim}{\beta}_j]$, $\Gamma = C[\underset{\sim}{\beta}_j]$ and $\sigma_o^2 = E[\sigma^2(\theta_j)]$. Also write $P_j =$ diag $(p_{1j}, \ldots, p_{nj})$.

The quantities $\theta_1, \ldots, \theta_q$ may be interpreted as q independent risks from the same population of risks. Corresponding to risk Θ_j is associated a vector of observations $\underset{\sim}{Y}_j = (Y_{1j}, \ldots, Y_{nj})$ where Y_{tj} is the "loss ratio" in year t in respect of risk θ_j.

According to the analysis of the preceding section the constant $\underset{\sim}{\beta}_j$ formulation of the above model is

$W_j = \bar{Z}_j$. Here this is evident from the Gauss-Markov formulation. The estimator $\hat{\beta}_o$ is non-operational since it depends on Γ and σ_o^2. We estimate σ_o^2 using (3.8). An estimator of Γ is given by

$$\hat{\Gamma} = (I - \sum_{j=1}^{q} \bar{Z}_j^2)^{-1} \{ \sum_{j=1}^{q} \bar{Z}_j (\hat{\beta}_j - \hat{\beta}_o)(\hat{\beta}_j - \hat{\beta}_o)'$$

$$- \hat{\sigma}_o^2 \sum_{j=1}^{q} \bar{Z}_j (I - \bar{Z}_j)(X_j' P_j X_j)^{-1} \} \quad (3.14)$$

See Norberg (1981) and De Vylder (1981) for more details regarding the estimation of Γ.

We now remark that in combination with equation (3.8) equations (3.11) to (3.14) are solved iteratively.

3.1 Bühlmann-Straub Model

In the present sub-section we consider the celebrated Bühlmann-Straub (1970) model which may be regarded as a special case of Hackemeister's model.

Hachemeister's model may be reduced to the Bühlmann-Straub (1970) model by taking

$\beta_j = \mu_j$, $X_j = 1$, $\beta_o = \mu_o$ and $\Gamma = \psi$. Note that μ_j, μ_o and ψ are all scalars.

In the present context we have

$$Z_j = p_{.j}(p_{.j} + \sigma_o^2/\psi)^{-1}, \quad \tilde{\mu}_j = \mu_o + Z_j(\bar{Y}_{.j} - \mu_o) \text{ and}$$

$$\hat{\mu}_j = \sum_{i=1}^{n} p_{ij} Y_{ij} / p_{.j}$$

where

$$p_{.j} = \sum_{i=1}^{n} p_{ij}.$$

It follows that,

$$\hat{\sigma}_o = \sum_{j=1}^{q} \sum_{i=1}^{n} p_{ij}(Y_{ij} - \hat{\mu}_j)^2 / (q(n-s))$$

$$\hat{\psi} = (1 - \sum_{j=1}^{q} \bar{Z}_j^2)^{-1} \{ \sum_{j=1}^{q} \bar{Z}_j (\hat{\mu}_j - \hat{\mu}_o)^2 - \sigma_o^2 \sum_{j=1}^{q} \bar{Z}_j (1 - \bar{Z}_j) p_{.j}^{-1} \}$$

and

$$\hat{\psi}_o = \sum_{j=1}^{q} \bar{Z}_j \hat{\mu}_j.$$

These last equations are evaluated iteratively.

4. SIMPLE POISSON EXAMPLE

Suppose λ_i represents the claim rate of an individual risk labelled i.

Assume :

1. $\lambda_1, \lambda_2, \ldots, \lambda_n$ are iid from some density $f_\Lambda(\cdot)$.
2. Conditional on λ_i fixed the observation X_i is Poisson with mean $e_i \lambda_i$. $\{e_i\}$ are known constants representing exposures.

Let $\lambda_o = E[\lambda_i]$ and $\sigma_\lambda^2 = Var[\lambda_i]$. Denote by $\hat{\lambda}_i = X_i e_i^{-1}$, the maximum likelihood estimator of λ_i, were λ_i assumed a constant parameter.

The mixed linear constant coefficient formulation of the random coefficient model may be written.

$$\begin{pmatrix} \hat{\lambda}_1 \\ \cdot \\ \cdot \\ \cdot \\ \hat{\lambda}_n \end{pmatrix} = \begin{pmatrix} \lambda_1 \\ \cdot \\ \cdot \\ \cdot \\ \lambda_n \end{pmatrix} + \underset{\sim}{\varepsilon} \qquad (4.1)$$

where $\underset{\sim}{\varepsilon}$ is zero mean error vector with $C[\underset{\sim}{\varepsilon}] = \lambda_o \, \mathrm{diag}^{-1}(e_1, \ldots, e_n)$, and

$$\begin{pmatrix} \lambda_o \\ \cdot \\ \cdot \\ \cdot \\ \lambda_o \end{pmatrix} = \begin{pmatrix} \lambda_1 \\ \cdot \\ \cdot \\ \cdot \\ \lambda_n \end{pmatrix} + \underset{\sim}{\delta} \qquad (4.2)$$

where $C[\underset{\sim}{\delta}] = \sigma_\lambda^2 I$.

Hence, the ILBR for λ_i is

$$\tilde{\lambda}_i = \lambda_o + Z_i(\hat{\lambda}_i - \lambda_o) \qquad (4.3)$$

where the credibility factor Z_i is given by

$$Z_i = e_i \sigma_\lambda^2 (\sigma_\lambda^2 + \lambda_o)^{-1} \quad . \qquad (4.4)$$

The error-variance (Bayes risk) of $\hat{\lambda}_i$ is

$$C[\tilde{\lambda}_i - \lambda_i] = \{(\lambda_o \, e_i^{-1} + \sigma_\lambda^{-2}\}^{-1}. \qquad (4.5)$$

Combining (4.1) and (4.2) gives

$$(\hat{\lambda}_1,\ldots,\hat{\lambda}_n)' = \underset{\sim}{1}\,\lambda_o + \underset{\sim}{\varepsilon}^* \tag{4.6}$$

where $\underset{\sim}{\varepsilon}^*$ is a zero mean error vector with $C[\underset{\sim}{\varepsilon}^*] = \lambda_o \operatorname{diag}^{-1}(e_1,\ldots,e_n) + \sigma_\lambda^2 I$.

The GLSE for λ_o based on (4.6) is

$$\hat{\lambda}_o = \sum_{i=1}^{n} \overline{Z}_i\, \hat{\lambda}_i \tag{4.7}$$

where,

$$\overline{Z}_i = \left(\sum_{i=1}^{n} Z_i\right)^{-1} Z_i \quad . \tag{4.8}$$

The error-variance of $\hat{\lambda}_o$ is

$$C[\hat{\lambda}_o] = \left(\sum_{i-1}^{n} Z_i\right)^{-1} . \tag{4.9}$$

We can estimate σ_λ^2 by

$$\hat{\sigma}_\lambda^2 = \left(1 - \sum_{i=1}^{n} \overline{Z}_i^2\right)^{-1} \left\{\sum_{i=1}^{n} \overline{Z}_i(\hat{\lambda}_i-\hat{\lambda}_o)^2 - \hat{\lambda}_o \sum_{i=1}^{n} \overline{Z}_i(1-\overline{Z}_i)e_i^{-1}\right\} \tag{4.10}$$

As is customary, 0 is used as an estimate if $\sigma_\lambda^2 \leq 0$. In order determine the estimator $\hat{\lambda}_o$, equations (4.7) to (4.10) are solved iteratively.

We are now in a position to briefly indicate that if $f_\Lambda(.)$ is a gamma density then $\hat{\lambda}_o$ given by equation (4.7) is the maximum likelihood estimator of λ_o.

The gamma density with shape parameter $\alpha > 0$ and scale parameter $\beta > 0$ is

$$f_\Lambda(\lambda) = \frac{\lambda^{\alpha-1}\, e^{-\lambda/\beta}}{\Gamma(\alpha)\quad \beta^\alpha}, \quad \lambda > 0$$

The observations $X_1,\ldots,X_n$ are independent with the marginal probability function of X_i given by

$$\Pr[X_i=x] = \binom{x+\alpha-1}{\alpha-1}(1+e_i\beta)^{-\alpha}\left(\frac{e_i\beta}{1+e_i\beta}\right)^x .$$

The log-likelihood function may be expressed

$$\ell = \ell(\alpha,\beta;x) = -\alpha\Sigma\ln(1+e_i\beta) + x_i \;\Sigma\ln\left(\frac{e_i\beta}{1+e_i\beta}\right)$$

whence

$$\frac{\partial\ell}{\partial\beta} = \alpha\Sigma \frac{e_i}{1+e_i\beta} + \Sigma x_i\left(\frac{1}{\beta} - \frac{e_i}{e_i\beta}\right) .$$

Noting that $\lambda_o = \alpha\beta$ and $\sigma_\lambda^2 = \alpha\beta^2$ and equating the last equation to zero gives equation (4.7).

We give a practical interpretation of the model under discussion: a portfolio may be sub-divided into a number of groups that are assumed homogeneous within themselves. The quantity λ_i represents the claim rate in respect of the ith group. The ILBR $\tilde{\lambda}_i$ is the optimal linear estimator of λ_i with associated 'error ^{2}' $C[\hat{\lambda}_i-\lambda_i]$. The quantity λ_o is the mean claim rate for the whole portfolio. Its linear minimum variance unbiased estimator is $\hat{\lambda}_o$ with associated 'error ^{2}' $C[\hat{\lambda}_o]$.

Illustration

The following data is taken out of Oakenfull et. al. (1982)

Group (i)	Exposures (e_i)	No. of Claims (X_i)
1	55	13
2	55	0
3	55	0
4	50	2
5	50	0
6	40	3
7	40	3
8	40	2
9	30	0
10	25	0

Equations (4.7) to (4.10) were solved iteratively to give

$$\hat{\lambda}_o = 0.04871 \quad \text{and} \quad \text{s.e.}(\hat{\lambda}_o) = 0.02493.$$

Additionally, we have the following results

Group (i)	$\hat{Z}_i$	$\tilde{\lambda}_i$	s.e.$(\tilde{\lambda}_i)$
1	0.850764	0.208357	0.027449
2	0.850764	0.007269	0.027449
3	0.850764	0.007269	0.027449
4	0.838254	0.041409	0.028577
5	0.838254	0.007879	0.028577
6	0.805674	0.069891	0.031323
7	0.805674	0.069891	0.031323
8	0.805674	0.049749	0.031323
9	0.756662	0.011853	0.035051
10	0.721546	0.013563	0.037495

Note that

$$\text{s.e.}(\hat{\lambda}_o) = \left(\sum_{i=1}^{10} \hat{Z}_i \right)^{-\frac{1}{2}}$$

and

$$\text{s.e.}(\tilde{\lambda}_i) = (\hat{\lambda}_o \; \hat{Z}_i \; e_i^{-1})^{\frac{1}{2}} \; .$$

We remark that s.e.$^2(\tilde{\lambda}_i)$ is an estimate of the Bayes risk of $\tilde{\lambda}_i$.

REFERENCES

Buhlmann, H. and Straub, E. (1970). Glaubwurdigkeit fur Schadensatze. Mitteilungen der ve reinigung Schweizerischer Verischerungs Mathematiker, 70, pp.111-133.

De Vylder, F. (1976). Geometrical Credibility. Scand.Actuarial J. pp. 121-149.

De Vylder, F. (1978). Parameter estimation in credibility theory. ASTIN, 10, pp.99-112.

De Vylder, F. (1981). Practical credibility theory with emphasis on optimal parameter estimation. ASTIN, 12, pp.115-132.

Diaconis, P. and Ylvisaker, D. (1979). Conjugate priors for exponential families. Ann. Statist., 7, pp.269-281.

Dubey, A. and Gisler, A. (1981). On parameter estimators in credibility. Mitteilungen der ve reinigung Schweizerischer Verischerungs Mathematiker, 2, pp.187-212.

Duncan, D.B. and Horn, S.D. (1972). Linear dynamic recursive estimation from the viewpoint of regression analysis. J. Amer. Statist. Assoc., 67, pp.815-821.

Goldberger, A.S. and Theil, H. (1961). On pure and mixed statistical estimation in Economics. Intern. Economic Review, 2, pp.65-78.

Goldstein, M. (1975). Approximate Bayes solutions to some nonparametric problems. Ann. Statist. 3, pp.212-517

Hachemeister, C.A. (1975). Credibility for Bayesian models with application to trend. In Credibility: Theory and Applications, Ed. Kahn, P.M. Academic Press, N.Y.

Hartigan, J.A. (1969). Linear Bayesian Methods. J.R. Stat. Soc. B., 35, pp. 379-421.

Jewell, W. (1975). The use of collateral data in credibility theory: a hierarchical model. Giornall dell Instituto Italiano degli Attuari, 38, pp. 1-16.

Judge, G. et. al. (1980). The theory and practice of Econometrics, John Wiley, New York.

Leamer, E.E. (1978). Specification searches: Ad hoc inference with non-experimental data. John Wiley. N.Y.

Norberg, R. (1981). On optimal parameter estimation in credibility. XV ASTIN Colloquium, Loen.

Oakenfull, E., Salbe, I. and Zehnwirth, B. (1984). A note on the estimation of mean intensity of insect infestation. Forthcoming: Biometrics.

Theil, H. (1971). Principles of Econometrics. John Wiley, N.Y.

SHORT COMMUNICATIONS

POPULATION AND SOCIAL SECURITY PROJECTIONS FOR BANGLADESH

John A. Beekman

Ball State University, Muncie, Indiana, 47306, U.S.A.

Abstract. Confidence intervals for year 1986 and year 2001 populations for Bangladesh are developed. Rural-urban shifts in population are analyzed through Markov chains, and reveal a significant projected redistribution to urban areas. A study is presented of the actuarial cost of a potential social security program designed to meet the financial needs of an increasing set of urban dwellers. This program would be patterned after a portion of India's Social Security Program. The program would provide employment-injury protection, death benefits, and retirement benefits, and would be financed through employee and employer contributions. Aggregate projected contributions and benefits for the years 1995, 2005, 2015, and 2025 illustrate the possible program. A discussion of reserve funds to provide for adverse deviations from actuarial and demographic assumptions is included.

I. POPULATION PROJECTIONS

In 1974 the population of Bangladesh was approximately 76 million. The 1981 Census figure was around 90 million, yielding an average annual compound rate r of growth of 2.36%. If this growth rate continues for 5 years, the 1986 projected population would be 101 million, approximately. With the same value for r, the population forecast for 2001, $P_{20} = P_0\ (1.0236)^{20}$ would be about 143.5 million. In the paper (9) Professor Nathan Keyfitz used 1,100 comparisons of population forecasts with the realized numbers to develop a method for confidence intervals for population forecasts. In summary, one provides two numbers L_t and U_t within which the population at time t would appear with two-thirds probability.

F. de Vylder et al. (eds.), Premium Calculation in Insurance, 363–380.

The bounds L_t and U_t are computed from the formulas

$$L_t = P_0(1+r-\sigma)^t,$$

$$U_t = P_0(1+r+\sigma)^t,$$

In Table 3 of (9), values for σ based on 810 United Nations forecasts for jumping-off years 1958, 1963, and 1968 are displayed according to slow, medium, and fast rates of increase. Thirty countries, increasing from 1.8 percent to 2.6 percent per annum, showed a σ value of 0.48 percentage points. If we use that value for Bangladesh,

$$L_5 = 90(1.0188)^5 \doteq 98.8 \text{ million}$$

$$U_5 = 90(1.0284)^5 \doteq 103.5 \text{ million}$$

Thus the odds are 2 to 1 that its population in 1986 will be between 98.8 and 103.5 million. For the 20 year forecasts,

$$L_{20} = 90(1.0188)^{20} \doteq 130.6 \text{ million}$$

$$U_{20} = 90(1.0284)^{20} \doteq 157.6 \text{ million}$$

Thus, a 66.67% confidence interval for the year 2001 population would be 130.6 through 157.6 million.

There are several factors which could bring the value of r down. If the average age of marriage would rise, fewer births would occur. This idea was quantified in (7) by M. H. Kabir and K. S. Ahmed. In 1981, the mean number of children born to Bangladesh women was 6.4. The mean age at marriage for females in Bangladesh was 16 years in 1981. An increase of two years in the average age of marriage would result in a reduction of about 15% in the mean number of children. This lower number would be 5.44. A second factor which could produce fewer births is the continued movement of people from farms to cities. Some attempt to quantify this appears in the author's earlier paper (3). For these and other reasons, the r value of 2.70% in 1974 fell to the 2.36% value in 1981. If we solve the equation $2.70(1-x)^7 = 2.36$, the value of $x = .02$. Let us assume the annual rates are reduced in that manner, in 5 year blocks. Thus, an alternate value for P_{20} is found from the equation

$$P_{20} = P_0(1.0236)^5[1+.0236(.98)^5]^5[1+.0236(.98)^{10}]^5\cdot$$
$$[1+.0236(.98)^{15}]^5, \text{ or } P_{20} = P_0(1.0236)^5(1.0213)^5(1.0193)^5\cdot$$
$$(1.0174)^5 = P_0(1.08412)^5 \doteq 134.8 \text{ million.}$$

If we view the composite product as $(1.0204)^{20}$, then with $r = .0204$, $\sigma = .0048$, $L_{20} = 90(1.0156)^{20} \doteq 122.7$ million, $U_{20} = 90(1.0252)^{20} \doteq 148.1$ million.

Rural-urban shifts in population have other implications. Let us use the Markov chain model of (3). It has two states of nature, i.e., r is the state of being a rural dweller and u is the state of being an urban dweller. The four numbers m_{ij}, $i = u$ or r, $j = u$ or r represent the probabilities of moving from state i to state j in one year. They can be displayed in a matrix of transition probabilities:

$$M = \begin{pmatrix} m_{uu} & m_{ur} \\ m_{ru} & m_{rr} \end{pmatrix}$$

where the elements in each row add to 1. We assume that the initial distribution is P_u persons in urban places and P_r in rural places, arranged in a column vector:

$$P^{(0)} = \begin{pmatrix} P_u \\ P_r \end{pmatrix}.$$

The superscript (0) on P denotes the initial time. The vector notation

$$P^{(t)} = \begin{pmatrix} P_u^{(t)} \\ P_r^{(t)} \end{pmatrix}$$

denotes the urban and rural components of the population t years from the initial time. Immigration and emigration will not be considered in this model. $P^{(t)} = \tilde{M}^t P^{(0)}$ where $\tilde{M}$ is the transpose of M. Thus

$$P^{(1)} = \begin{pmatrix} m_{uu} & m_{ru} \\ m_{ur} & m_{rr} \end{pmatrix} \begin{pmatrix} P_u \\ P_r \end{pmatrix} = \begin{pmatrix} m_{uu}P_u + m_{ru}P_r \\ m_{ur}P_u + m_{rr}P_r \end{pmatrix}.$$

A very thorough study of the migration patterns of Bangladesh occurs in (15) by John Stoeckel, A.K.M. Alauddin Chowdhury, and K.M.A. Aziz. Page 242 of that paper provides the preliminary values of $m_{ru} = .024$, and $m_{ru} - m_{ur} = .011$. That yields $m_{ur} = .013$.

Page 243 then reduces the net gain in the urban area from 1.1 to 0.7 per hundred rural population. Thus we will reduce our rates to yield m_{ru} = .015, m_{ur} =.008, with $m_{ru} - m_{ur}$ = .007. The resulting matrices are

$$M = \begin{pmatrix} .992 & .008 \\ .015 & .985 \end{pmatrix}, \text{ and } \tilde{M} = \begin{pmatrix} .992 & .015 \\ .008 & .985 \end{pmatrix}.$$

Powers of $\tilde{M}$ are as follows:

$$\tilde{M}^2 = \begin{pmatrix} .992 & .015 \\ .008 & .985 \end{pmatrix} \begin{pmatrix} .992 & .015 \\ .008 & .985 \end{pmatrix} = \begin{pmatrix} .984184 & .029655 \\ .015816 & .970345 \end{pmatrix};$$

$$\tilde{M}^5 = \begin{pmatrix} .961798 & .071629 \\ .038202 & .928371 \end{pmatrix}; \quad \tilde{M}^{10} = \begin{pmatrix} .927792 & .135391 \\ .072208 & .864609 \end{pmatrix}; \text{ and}$$

$$\tilde{M}^{20} = \begin{pmatrix} .870574 & .242675 \\ .129426 & .757325 \end{pmatrix}.$$

The model which reflects both total growth and migration is

$$p^{(t)} = \tilde{M}^t(1 + r)^t p^{(0)}.$$

According to Table 9 of (8), the 1981 percentages of rural and urban people were 89.93% and 10.07%. Applying these percentages to 90,000,000 yields $P_u^{(0)}$ = 9,063,000, and $P_r^{(0)}$ = 80,937,000. The forecast for 1986 would be

$$p^{(5)} = \tilde{M}^5(1.0236)^5 \begin{pmatrix} 9,063,000 \\ 80,937,000 \end{pmatrix} = \begin{pmatrix} .961798 & .071629 \\ .038202 & .928371 \end{pmatrix} \cdot$$

$$\begin{pmatrix} 10,184,117 \\ 90,949,117 \end{pmatrix} = \begin{pmatrix} 16,309,658 \\ 84,823,576 \end{pmatrix}.$$

The modified growth rates forecast for 2001 would be

$$P^{(20)} = \tilde{M}^{20}(1.0204)^{20} \begin{pmatrix} 9,063,000 \\ 80,937,000 \end{pmatrix}$$

$$= \begin{pmatrix} .870574 & .242675 \\ .129426 & .757325 \end{pmatrix} \begin{pmatrix} 13,573,160 \\ 121,214,926 \end{pmatrix}$$

$$= \begin{pmatrix} 41,232,272 \\ 93,555,814 \end{pmatrix}.$$

The effect of $\tilde{M}^{20}$ is to change the composition of the total population rather than the numerical size. For simplicity, we have assumed that the lower birth rates produced by migration into towns and cities is reflected in the total growth rate of 1.204 . Perhaps the most frightening aspect of these forecasts is the projected growth in urban dwellers from 9,063,000 in 1981 to 16,309,658 in 1986 to 41,232,272 in 2001. Reference (15) stresses on page 244 "that population redistribution to urban areas is a problem of more immediate concern than the high rate of natural increase . . .; the present data illustrate the importance of implementing programs which would meet the needs of the urban social structure created by the rapid increases in population, such needs as occupational opportunities and health, housing, and educational facilities."

As stated in (7), Bangladesh life expectancy at birth was about 50 years in 1981. This is a considerable gain over the former life expectancies, e.g., in the late 1930's, $\mathring{e}_0 \doteq 25$ years. Bangladesh has rather high infant mortality. A part of Table 32 on page 81 of (6) reports the following:

Age x	Male q(x)	Female q(x)
0	.1474	.1293
1	.1119	.1104

Life expectancies improve after the early years, at least by comparison with other countries. Thus, the same Table 32 provides the following:

	Male $\overset{\circ}{e}_{10}$	Female $\overset{\circ}{e}_{10}$
Life Expectancy at Age 10	58.9	55.7

Page 94 of (6) states that the methods which produced these values embody minimum estimates of mortality. However, the data also related to 1974, so current values probably are at least as high. One can expect life expectancies to grow further, and as people live longer, there will be increased financial and health needs. These needs will be greater in the cities where the traditional care of older people by their children can not be easily accomplished. A national Social Security Program should be considered. The journal Social Security Bulletin contains articles on social security programs throughout the world. In the August, 1980 issue, an article (12) by Joseph G. Simanis appeared. It contained the following sentances. "By 1979, there were 134 countries reporting some form of social security. Traditionally, a work-injury program has been among the first phases of social security established in any given country, and this has continued to be the norm. The countries with only one program each have a work-injury plan. Old-age, invalidity, and survivors programs are the next most prevalent, existing in 123 countries. In third place are the sickness and maternity insurance programs found in 75 countries."

Fundamental to any such program is an actuarial analysis of its cost. One way of looking at the cost of pensions is provided by Problem 40 of Chapter III of (10) which defines

$$\text{"Burden of old-age pensions} = \frac{\text{Population over age 65}}{\text{Population aged 20 to 65}}.\text{"}$$

Let that fraction be denoted by R(65), or $R^z(65)$ to denote the ratio in year z. Pages 41-43 of (14) provide "Mid-year (July 1) population projection of Bangladesh by age group and sex, 1975-2025". These produced the following ratios:

Year	1975	1980	1985	1990	1995	2000
$R^z(65)$	.0606	.0621	.0624	.0640	.0650	.0668

Year	2005	2010	2015	2020	2025
$R^z(65)$	.0703	.0762	.0843	.0948	.1084

Clearly these ratios are growing, and the numbers T^z_{65} of people 65 and older is rapidly increasing. Thus the numbers in years 1975, 1980, on up to 2025 are:

Year	1975	1980	1985	1990
T^z_{65}	2,070,000	2,443,000	2,862,000	3,345,000

Year	1995	2000	2005	2010
T^z_{65}	3,850,000	4,460,000	5,227,000	6,209,000

Year	2015	2020	2025
T^z_{65}	7,446,000	8,971,000	10,889,000

Because of the need to partially provide for the financial needs of an increasing retired set of people, let us now focus on a potential Social Security Program. As discussed in the next section, it seems logical to pattern it after a part of India's Social Security Program. Since that program allows retirement at age 55, we will now redo the above analysis.

Year	1975	1980	1985	1990
$R^z(55)$	.1715	.1724	.1691	.1705
T^z_{55}	5,305,000	6,144,000	7,052,000	8,096,000

Year	1995	2000	2005	2010
$R^z(55)$	.1747	.1819	.1932	.2104
T^z_{55}	9,378,000	10,958,000	12,886,000	15,240,000

Year	2015	2020	2025
$R^z(55)$	.2336	.2675	.2951
T^z_{55}	18,131,000	21,870,000	25,372,000

Both sequences $\{R^z(55)\}$ and $\{T^z_{55}\}$ are growing to such an extent that one wonders about the feasibility of a social security program. That is the subject of Sections II and III.

It is interesting to compare $R^z(65)$ and $R^z(55)$ values for

Bangladesh with those developed for Canada in Calvert (5); also see page 23 of Brown (4). The ratios were continued in analyses of the Canada/Quebec Pension Plan. With our notation, they become:

Year	1976	2001	2031
$R^z(65)$	.1560	.1841	.3337
$R^z(55)$	.3610	.3896	.6584

II. SOCIAL SECURITY PROJECTIONS

Assume that Bangladesh adopts a portion of India's Social Security Program. As described on pages 831, 837, 857, and 858 of (11), that system covers employees of firms with 10 or more workers. Government employees have a different system, and high-paid employees are not included. Employees and employers contribute at the rates of 9.50% and 12.00% of wages. Upon retirement after age 55, a lump-sum return of combined employer-employee contributions (less administrative expenses) plus interest is granted. As is true of India, the family system will continue to play an important role in Bangladesh. Aged and disabled persons will be provided for by the younger members of the family. Under such a system, lump-sum payments upon retirement are desired more than annuities. In India these single payments can be converted into life annuities, but our cost analyses will use lump-sum payments. Lump-sum death benefits of a uniform amount are also provided in India, and we will include such benefits in a Bangladesh program. Employment-injury protection is provided in India, and this will be included in the possible Bangladesh program. In India a portion of the employer contribution (2.50%) is used to pay the administrative expenses of the plan and we will use the same method here. The program for India also includes an unemployment insurance feature, sickness and maternity benefits, and medical care benefits. For the present, those features will not be considered.

Formulas for aggregate contributions, and benefits will now be developed. These are motivated by the 27 principal elements and factors used in long-range cost estimates for the U.S. Old-Age, Survivors, and Disability Insurance System, as described on pages 305-307 of (11).

Assume that in 1990 people between ages 20 and 54 contribute to the program, along with their firms. Assume that in 1991 death benefits, and employment-injury protection commence, whereas first retirement benefits start in 1995. Thus the cohort born in 1940 would contribute for 5 years, be eligible for death and injury payments for five years, and retirement in the fifth year. The 1941 birth cohort would contribute from 1990 through 1995, be eligible for death and injury payments during those years and that part of 1996 prior to retirement in 1996. The 1942 cohort would contribute from 1990 through 1996, be eligible for death and injury payments during those years and that part of 1997 prior to retirement in 1997.

If S stands for the average annual salary for the ${}_5P_{20}^{z}$ people in age group 20-24 in year z, then the average salary for the group ${}_5P_{25}^{z}$ could be written $S(1 + h)^5$ where h reflects increments in pay. Likewise, $S(1 + h)^{10}$, $S(1 + h)^{15}$, $S(1 + h)^{20}$, $S(1 + h)^{25}$, and $S(1 + h)^{30}$ represent average salaries for the sets ${}_5P_x^z$, $x = 30, 35, 40, 45$, and 50. Let k be the combined contribution, as a percentage of wages, of the employee and the employer to the program. Let $m(x,z)$, $f(x,z)$, $20 \leq x \leq 55$, $z \geq 1990$, be functions representing percentages of the 5 year block male and female populations who participate in the Social Security Program. During the year z the aggregate contributions would be

$$\sum_{\alpha=4}^{10} \left\{ \frac{kS}{100} (1 + h)^{5\alpha-20} \left[m(5\alpha,z) \left({}_5^{m}P_{5\alpha}^{z} \right) + f(5\alpha,z) \left({}_5^{f}P_{5\alpha}^{z} \right) \right] \right\}$$

The lump-sum retirement benefits will increase with passing time. Let i be the assumed interest rate for accumulation of contributions. Assume that values for ${}^{m}P_{55}^{z}$ and ${}^{f}P_{55}^{z}$ can be obtained, and that $m(55,z)$ and $f(55,z)$ apply to these single age figures. The lump-sum retirement benefits in year 1995 will total the following:

$$\left[m(55,1995) \left({}^{m}P_{55}^{1995} \right) + f(55,1995) \left({}^{f}P_{55}^{1995} \right) \right] \frac{kS}{100} \left\{ (1 + i)^5 + (1 + h)(1 + i)^4 + (1 + h)^2(1 + i)^3 + (1 + h)^3(1 + i)^2 + (1 + h)^4(1 + i) \right\} .$$

These factors would be appropriate for contributions on July 1, and retirement on July 1. For a general year z the bracketed expression becomes

$$\left\{ (1 + i)^{z-1990} + (1 + h)(1 + i)^{z-1991} + (1 + h)^2(1 + i)^{z-1992} + \ldots + (1 + h)^{z-1991}(1 + i) \right\}$$

with the understanding that all of the exponents $(z - 1990)$, $(z - 1991)$, $(z - 1992)$, . . . are ≥ 0. These benefits will increase with passing time, partially because the $(1 + i)^k$ factors will be greater. Also the $m(x,z)$ and $f(x,z)$ functions will increase with z. Because of the preponderance of rural dwellers, and a requirement of working for a 10 employee firm, $m(x,1990)$ might be as low as .01 for all x values, and $f(x,1990)$ might be .0001 for all x values.

In addition to the retirement benefits, there would be death benefits. We will assume that they are of a uniform amount A. The population projections satisfy the recurrence relations

$$ {}_5^{m}P_{x}^{z+5} = {}_5^{m}P_{x-5}^{z} - \sum_{i=0}^{4} {}_5^{m}D_{x-5}^{z+i} $$

$$ {}_5^{f}P_{x}^{z+5} = {}_5^{f}P_{x-5}^{z} - \sum_{i=0}^{4} {}_5^{f}D_{x-5}^{z+i} $$

where the D symbols reflect the numbers of deaths in 5 year intervals in various calendar years. The total death benefits paid in year z would be

$$ A\left\{\sum_{\alpha=4}^{10}\left[m(5\alpha,\ z)\left({}_5^{m}D_{5\alpha}^{z}\right) + f(5\alpha,\ z)\left({}_5^{f}D_{5\alpha}^{z}\right)\right]\right\}. $$

Because of the rapidly growing population, it would be inaccurate to assume

$$ {}_5^{\delta}D_{x-5}^{z+i} = .2\left\{{}_5^{\delta}P_{x-5}^{z} - {}_5^{\delta}P_{x}^{z+5}\right\},\quad i = 0, 1, 2, 3, 4 $$

for $\delta = m, f$. For simplicity, we will assume that

$$ {}_5^{\delta}D_{x-5}^{z+i} = (1 + r)^{i}\left({}_5^{\delta}D_{x-5}^{z}\right),\quad i = 0, 1, 2, 3, 4. $$

Thus we have

$$ {}_5^{\delta}D_{x-5}^{z}\left\{1 + (1 + r) + (1 + r)^2 + (1 + r)^3 + (1 + r)^4\right\} = $$

$$ {}_5^{\delta}P_{x-5}^{z} - {}_5^{\delta}P_{x}^{z+5},\ \text{or}\ {}_5^{\delta}D_{x-5}^{z} = \frac{r}{(1 + r)^5 - 1}\left\{{}_5^{\delta}P_{x-5}^{z} - {}_5^{\delta}P_{x}^{z+5}\right\} $$

The employment-injury protection will be assumed to be of constant amount B. The aggregate benefits paid in year z would be

$$ B\left\{\sum_{\alpha=4}^{10}\left[m(5\alpha,\ z)\left({}_5^{m}I_{5\alpha}^{z}\right) + f(5\alpha,\ z)\left({}_5^{f}I_{5\alpha}^{z}\right)\right]\right\} $$

where ${}_5^{\delta}I_{5\alpha}^{z}$ represents those employees injured between ages 5α and $5\alpha + 5$ during year z, for $\delta = m, f$.

Initially, the aggregate contributions would exceed the benefits. At least, it would be wise for the contributions to be greater so that some reserves could be created. Those reserves could then earn interest and in future years part of the benefits could be financed from that interest. As is widely known, the U.S. Senate Special Committee on Aging has considered possible solutions to those problems. On June 16, 1981, the American Academy of Actuaries made recommendations to the Special Committee, (1). One recommendation was a minimum reserve level of 25 percent of annual outlays to protect the program and its recipients.

As reported on page 2 of (2), 389 actuaries who responded to a questionnaire from the U.S. Government Accounting Office, felt that "the Old Age-Survivors Insurance and Disability Income Trust Fund should have a balance of more than six months of expenditures." Most respondents preferred a reserve level of twelve months of expenditures.

A contingency fund recommendation for the Canada/Quebec Pension Plan is discussed on pages 24 and 26 of Brown, (4). It contains comments on C/QPP funding included in the report of the Royal Commission on the Status of Pensions in Ontario. The commission recommended that "the Canada Pension Plan should be funded on a pay-as-you-go basis, with a contingency fund maintained at the level required to satisfy twice the year's benefit and administrative cost pay-out three years in advance."

As the program matures, the proportions of workers covered will all grow, the benefits provided will become more comprehensive, and the administrators of the Social Security Program will need some reserves to cushion the perturbations in cash flow.

III. SAMPLE CALCULATIONS

Let us now compute the aggregate contributions during the year 1995. We will let $k = 21.5$, $m(x,1995) = .01$ for all x, $f(x, 1995) = .0001$ for all x, and $h = .03$. Using page 42 of (14), we obtain

$$.215S\,\{58{,}470 + 546 + (1.03)^5[51{,}680 + 483] + (1.03)^{10}\cdot [49{,}400 + 449.8] + (1.03)^{15}[38{,}830 + 369] + (1.03)^{20}\cdot [31{,}930 + 302.8] + (1.03)^{25}[25{,}980 + 247.7] + (1.03)^{30}\cdot [20{,}670 + 199.2]\} \doteq 88{,}438S\ .$$

To compute the lump-sum retirement benefits in 1995, let us assume that i = .05 and h = .03. The formula for the aggregate sum would be:

$$.19S \left[.01 \left({}^{m}P^{1995}_{55}\right) + .0001 \left({}^{f}P^{1995}_{55}\right)\right] \{(1.05)^5 + (1.03)\cdot (1.05)^4 + (1.03)^2(1.05)^3 + (1.03)^3(1.05)^2 + (1.03)^4(1.05)\}.$$

The quantity within braces equals 6.143. For simplicity and conservatism, we will approximate the data at age 55 as .25 of the projected total for ages 55-59. The above formula produces a value of 4708S.

The death benefits in 1995 would be

$$A \left\{ \sum_{\alpha=4}^{10} \left[.01 \; {}^{m}_{5}D^{1995}_{5\alpha} + .0001 \; {}^{f}_{5}D^{1995}_{5\alpha} \right] \right\}.$$

As explained earlier,

$${}^{\delta}_{5}D^{z}_{x-5} \doteq \frac{r}{(1+r)^5 - 1} \left\{ {}^{\delta}_{5}P^{z}_{x-5} - {}^{\delta}_{5}P^{z+5}_{x} \right\}, \quad \delta = m, f.$$

For r = .0236, $\dfrac{r}{(1+r)^5 - 1} = .1908.$

The number of employment-injury benefits will depend on the age, sex, and occupation composition of the workers. Tables 2-1 and 2-7 of (16) provide one comparison of mortality and disability rates. The disability to mortality ratios varied from 36 percent to 62 percent at ages 20 (5) 55. For simplicity and conservatism, we will use B = .65A.

Salaries grow from S to $S(1.03)^{30} = 2.43S$. An appropriate death benefit might be 7.5S.

The combined death and employment-injury benefits for 1995 would thus be:

$$1.65(7.5S) \left\{ \sum_{\alpha=4}^{10} \left[.01 \; {}^{m}_{5}D^{1995}_{5\alpha} + .0001 \; {}^{f}_{5}D^{1995}_{5\alpha} \right] \right\} =$$

$$12.375S\{2251.44 + 19.94\} = 28{,}108S.$$

The total of retirement, death, and employee-injury benefits would be 32,816S. The administrative costs would be 10,283S. The total costs of the program would thus be 43,099S, as contrasted with contributions of 88,438S.

We will now redo the calculations for the years 2005, 2015, and 2025. The fractions covered for all x's will be assumed to be:

Year	2005	2015	2025
m(x, z)	.100	.250	.400
f(x, z)	.001	.005	.020

The aggregate contributions for each of those years will be computed. Those retiring in the years 2005, 2015, and 2025 will have contributed for at most 15, 25, and 35 years respectively. For simplicity, we will assume that each group contributed a uniform number of years, and we shall assume 8 years for 2005 retirees, 13 years for 2015 retirees, and 18 years for 2025 retirees. The combined death and employment-injury benefits for the years 2005, 2015, and 2025 will be computed. For each year the cost of the three benefits plus administrative expenses will be compared with the contributions during the year.

The contributions were computed as for 1995, and used pages 42 and 43 of (14). The results were:

Year 2005	1,121,340S
Year 2015	3,374,993S
Year 2025	6,210,661S.

In computing the retirement benefits, various sums of factors were used. For 2005, the sum was

$$\{(1.05)^8 + 1.03(1.05)^7 + (1.03)^2(1.05)^6 + (1.03)^3(1.05)^5 + (1.03)^4(1.05)^4 + (1.03)^5(1.05)^3 + (1.03)^6(1.05)^2 + (1.03)^7 \cdot (1.05)\} = 11.062\ 805.$$

For 2015, the sum was

$$11.062\ 805\ (1.05)^5 + (1.03)^8(1.05)^5 + (1.03)^9(1.05)^4 + (1.03)^{10} \cdot (1.05)^3 + (1.03)^{11}(1.05)^2 + (1.03)^{12}(1.05) = 21.901\ 628.$$

For 2025, the sum was

$$21.901\,628\,(1.05)^5 + (1.03)^{13}(1.05)^5 + (1.03)^{14}(1.05)^4 + (1.03)^{15}(1.05)^3 + (1.03)^{16}(1.05)^2 + (1.03)^{17}(1.05) = 36.974\,562,$$

With these factors, the retirement benefits were:

Year 2005	116,901S
Year 2015	811,828S
Year 2025	2,901,875S

The combined death and employment-injury benefits for 2005 would be:

$$1.65(7.5S)\left\{\sum_{\alpha=4}^{10}\left[.1\,{}^{m}_{5}D^{2005}_{5\alpha} + .001\,{}^{f}_{5}D^{2005}_{5\alpha}\right]\right\}$$

where

$${}^{\delta}_{5}D^{z}_{5\alpha} \doteq .1908\left[{}^{\delta}_{5}P^{z}_{5\alpha} - {}^{\delta}_{5}P^{z+5}_{5(\alpha+1)}\right], \quad \delta = m, f.$$

For the other years, the .1 and .001 factors become:

Year 2015	.25	.005
Year 2025	.40	.020.

Pages 42 and 43 of (14) yield the following values:

Year	Male D	Female D
2005	260,251	210,643
2015	261,014	225,335
2020	277,805	215,222

By also considering the values at 1995, the following approximate values were determined:

2025	280,000	230,000

The death and employment-injury benefits are the following:

Year	Male Benefits	Female Benefits	Total Benefits
2005	322,061S	2,607S	324,668S
2015	807,512S	13,943S	821,455S
2025	1,386,000S	56,925S	1,442,925S

Of the 21.5 percent total employer-employee contributions, 2.5 percent would be used for the administration of the program.

A summary for the years 1995 (10) 2025 is contained below:

Year	Total Benefits	Administrative Expenses	Total Expenses	Contributions
1995	32,816S	10,283S	43,099S	88,438S
2005	441,569S	130,388S	571,957S	1,121,340S
2015	1,633,283S	392,441S	2,025,724S	3,374,993S
2025	4,344,800S	722,170S	5,066,970S	6,210,661S

IV. RESERVE FUNDS

As explained in Section II, the Bangladesh Social Security Program should include reserves. These are needed to provide for adverse deviations from actuarial and demographic assumptions in

(1) amount of contributions

(2) interest earnings

(3) fertility and mortality

(4) employment-injuries

(5) pension payments.

These reserves would be created by (A) surpluses of contributions over benefits and administrative expenses, and (B) interest earned on existing reserves. For sake of example, the amounts available for reserves in the various years would be:

Year	1995	2005	2015	2025
Reserve Contributions	45,339S	549,383S	1,349,269S	1,143,691S

So that the reader can see both reasons for holding reserves, we will take the year 2005 amount and increase it with interest at 5 percent for one year. That yields 576,852S. But now let us assume that the contributions are 5 percent smaller in 2005 because of lower fertility in 1985, and that pension benefits are 20 percent larger because of mortality improvements between 1985 and 2005. That would draw the original 549,383S down to 405,002S.

A major assumption in the actuarial and demographic projections is the number of people eligible for retirement. Any country has trouble with errors in this regard, and such errors are greater when the country's census and vital registration systems are new. Chapter 3 of (13) is concerned with "Errors in Census Statistics and Vital Statistics and Their Adjustments," in the U.S.. Some of the findings in (6) relate to problems of age reporting which a Bangladesh Social Security Program would face. Page 26, (6), reports on "an increasing tendency to exaggerate age as age increases." Page 9, loc. cit., discusses the incompleteness of birth and death registration on a national basis. An exception to that incompleteness occurs in Matlab thana, a rural area of southern Bangladesh. There the Demographic Surveillance System registers completely the births, deaths, marriages, and migrational movements of over 200,000 people. As further described on page 49 of the same reference "In 1976, registration of vital events was started for a population of 40,000 in another study area, the Teknaf thana of Chittagong district."

V. FINANCING A SOCIAL SECURITY PROGRAM

There are many problems and solutions in financing a social security system. These are discussed in great detail in Chapters 4 and 10 of (11). One key idea is to project a low, medium, and high cost estimate for any future year. As described on page 300 of loc. cit., "These alternative estimates are based on assumptions that vary only in the economic factors of wage and price changes, unemployment and labor force participation rates, and interest rates." The investment of the assets of the trust funds for the U.S. Social Security program has always been in obligations of the federal government (pages 290-293, loc. cit.). The rationale for such a procedure seems most reasonable, and has influenced the choice of a relatively moderate interest rate for accumulation of contributions in this paper. There is good reason for presenting estimates of future benefits as percentages of taxable payroll. Table 10.10 on page 577 of (11) presents estimated costs of U.S. Old-Age, Survivors, and Disability Insurance, Hospital Insurance, and Administrative Expenses, as percentages of taxable payroll, for calendar years 1980 (5) 2005, 2010 (10) 2040, and 2055. Several of these ideas, and indeed others too, have not been pursued in this paper; but these ideas could be treated in a later paper.

Acknowledgements. The author gratefully acknowledges the support of a Ball State University Summer Research Grant in the summer of 1982. Dr. Md. Humayun Kabir and the Institute of Statistical Research and Training, University of Dacca, Bangladesh were very helpful to the author in the preparation of this paper.

REFERENCES

(1) American Academy of Actuaries, "Statement to U.S. Senate on Social Security Legislation," Washington, D.C., June 16, 1981.

(2) American Academy of Actuaries Newsletter, Washington, D.C., July, 1982.

(3) Beekman, John A., "Several Demographic Projection Techniques," Rural Demography 8 (1981), pp. 1-11.

(4) Brown, Robert L., "Actuarial Aspects of the Changing Canadian Demographic Profile," Transactions, Society of Actuaries 34 (1982), pp. 13-30.

(5) Calvert, G. N., Pensions and Survival, Financial Post Books, 1977.

(6) Estimation of Recent Trends in Fertility and Mortality in Bangladesh, Report No. 5 of Committee on Population and Demography, National Academy Press, Washington, D.C., 1981.

(7) Kabir, Md. Humayun, and Kazi S. Ahmed, "Late Marriage Fewer Births," The Bangladesh Observer, January 15, 1982, Dacca.

(8) Kabir, Md., and Ashraful Aziz Chowdhury, "Population Growth and Food Production in Bangladesh," ISTR, University of Dacca, 1982.

(9) Keyfitz, Nathan, "The Limits of Population Forecasting," Population and Development Review 7 (1981), pp. 579-593.

(10) Keyfitz, Nathan, and John A. Beekman, Demography Through Problems, to be published by Springer-Verlag, New York and Heidelberg.

(11) Myers, Robert J., Social Security, 2nd Ed., R. D. Irwin, Inc., Homewood, Illinois, 1981.

(12) Simanis, Joseph G., "Worldwide Trends in Social Security, 1979," Social Security Bulletin 43, August, 1980, pp. 6-9.

(13) Spiegelman, Mortimer, Introduction to Demography, Rev. Ed., Harvard University Press, Cambridge, Mass., 1968.

(14) Statistical Year Book (1980), Government of People's Republic of Bangladesh, Dacca.

(15) Stoeckel, John, A.K.M. Alauddin Chowdhury, and K.M.A. Aziz, "Out-Migration from a Rural Area of Bangladesh," Rural Sociology, 37 (1972), pp. 236-245.

(16) Winklevoss, Howard E., Pension Mathematics With Numerical Illustrations, R. D. Irwin, Inc., Homewood, Illinois, 1977.

STABILITY OF PREMIUM PRINCIPLES UNDER MAXIMUM ENTROPY PERTURBATIONS

B. CHAN

Department of Statistical and Actuarial Sciences
The University of Western Ontario
London, Ontario, Canada N6A 5B9

ABSTRACT. Applications of premium principles imply implicitly certainty about the distributions of the risks. In this note, we study the variations of premiums when the distributions of the risks are changed within certain constrains.

1. INTRODUCTION

In this note, we consider premium principles π acting on risks S where

$$S = X_1 + X_2 + \cdots + X_N \quad . \tag{1.1}$$

Here, X_i are identically distributed as X; X_i and N are independent. For example, the *variance premium principle* is defined by

$$\pi(S) = E(S) + \beta \; \mathrm{Var}(S) \quad . \tag{1.2}$$

The application of this principle reflects explictly the insurer's attitude on risks and *implicitly certainty about the expectation and the variance* of S. Another example, the *Esscher premium principle* [2], is defined by

$$\pi(S) = \phi_S(t) \tag{1.3}$$

where

$$\phi_S(t) = M_S'(t) \;/\; M_S(t) \quad ,$$

and

$$M_S(t) = E(e^{tS})$$

is the moment generating function for S. The application of this principle implies *implicitly certainty about the distribution* of S. In practice, however, the insurer's knowledge of S is seldom

F. de Vylder et al. (eds.), Premium Calculation in Insurance, 381–386.

certain nor complete. In this note, we study the stability of the value $\pi(S)$ when knowledge of S is partial.

2. VARIANCE PREMIUM PRINCIPLE

The most commonly used model on S is the compound Poisson model: N is assumed to be Poisson with mean λ, and consequently $Var(N) = \lambda$. Applying (1.2), we obtain

$$\pi(S) = \lambda\, E(X) + \beta\lambda\, E(X^2)\ . \tag{2.1}$$

Here, we have used

$$E(S) = E(N)\, E(X)\ , \tag{2.2}$$

$$Var(S) = Var(N)[E(X)]^2 + E(N)\, Var(X)\ . \tag{2.3}$$

However, if we have doubts about the distributions of N and are only certain that $E(N) = \lambda$, then the value of $\pi(S)$ can be highly unstable. In fact, it is easy to construct examples of N with $E(N) = \lambda$ and $Var(N) = \infty$. (One such example is $Prob\{N=k\} = \alpha\, k^{-3}$, $k=1,2,\cdots$, with α chosen to make $\sum_{k=1}^{\infty} \alpha\, k^{-2} = \lambda$.) By (2.2) and (2.3), $\pi(S)$ becomes infinite when $Var(N) = \infty$. Thus, an *insurable* ($\pi(S)$ is finite) risk S under the Poisson assumption on N can become *uninsurable* ($\pi(S)$ is infinite) when the assumption on N is weakened to $E(N) = \lambda$.

Instead of considering such worst possible N, suppose we restrict ourselves to negative binomial N of parameters p,α and with $E(N) = \alpha(1-p)/p = \lambda$. Thus

$$Var(N) = \alpha(1-p)/p^2 = \lambda + (\lambda^2/\alpha)\ . \tag{2.4}$$

The negative binomial distribution is indeed a popular candidate for the claim number distribution. The geometric distribution is a special case ($\alpha=1$); the Poisson distribution can be thought of as the limit when $\alpha \to \infty$. We shall discuss an interesting information theoretic aspect of the geometric distribution in the next section. By (1.2), (2.2) and (2.3), when N is negative binomial,

$$\begin{aligned}\pi(S) &= \lambda\, E(X) + \beta\lambda\, E(X^2) + (\beta\lambda^2/\alpha)[E(X)]^2 \\ &= \lambda\, E(X) + \beta\lambda\, E(X^2) + (\beta/\alpha)[E(S)]^2\ .\end{aligned} \tag{2.5}$$

A comparison of (2.1) and (2.5) shows that the uncertainty of N being Poisson would cost a surcharge of $(\beta/\alpha)[E(S)]^2$ if we assume that N can be no worse than a negative binomial distribution of the same mean. The Poisson to negative binomial perturnation would cost more, but the risk remains insurable. In section 4, we shall find that this is not the case for the Esscher premium principle.

3. MAXIMUM ENTROPY DISTRIBUTIONS

Entropy measures the lack of information. See [1] for a brief introduction and [4] for more details on information theoretic approach to insurance models and to statistics. The following two propositions point out that the geometric and the exponential distributions are distributions with maximum entropy (highest uncertainty) among distributions of the same fixed mean. They give an information theoretic reason for the choice of geometric and exponential distributions in sections 2 and 4. The presentations there are done for the generality of a negative binomial N; the geometric ($\alpha=1$) case can be read out effortlessly.

The entropy of a distribution defined on the non-negative integers is given by (3.1). The entropy of a distribution defined on the non-negative reals is given by (3.7).

Proposition 1.
Among N on $0,1,\cdots$ with $E(N) \leq \lambda$,

$$\sum_{k=0}^{\infty} p_k \, \ell n(1/p_k) \tag{3.1}$$

is maximized at

$$p_k = \lambda^k / (1+\lambda)^{k+1} .$$

Proof:
Let H be the entropy, the sum (3.1). Preliminary motivation:

$$\begin{aligned} H - \alpha\, 1 - \beta\, \lambda &= \sum_{k=0}^{\infty} p_k \left(\ell n(1/p_k) - \alpha - \beta k \right) \\ &= \sum_{k=0}^{\infty} p_k \, \ell n((1/p_k)\, e^{-\alpha-\beta k}) \\ &\leq \sum_{k=0}^{\infty} p_k \left[(1/p_k) e^{-\alpha-\beta k} - 1 \right] \\ &= \sum_{k=0}^{\infty} e^{-\alpha-\beta k} - 1 \end{aligned} \tag{3.2}$$

Since $\ell n\, x \leq x-1$ with equality if and only if $x = 1$, the inequality in (3.2) becomes an equality if an only if

$$p_k = e^{-\alpha-\beta k} \tag{3.3}$$

for all k. When p_k take the form of (3.3), the constrains

$$\sum_{k=0}^{\infty} p_k = 1$$

and

$$\sum_{k=0}^{\infty} k\, p_k = \lambda$$

lead to two equations on α and β, for which

$$e^{\alpha} = 1 + \lambda \tag{3.4}$$

$$e^{-\beta} = \lambda \,/\, (1+\lambda) \tag{3.5}$$

are the solutions. We may now proceed to the proof.

Choose α and β by (3.4) and (3.5). With this choice of α and β, the right side of (3.2) has a value of $1 - 1 = 0$. By (3.2), $H - \alpha - \beta\lambda$ has an upper bound of 0; the maximum of H is attained when the inequality in (3.2) becomes an equality. This is achieved by choosing p_k as prescribed by (3.3). Thus,

$H - \alpha - \beta\lambda = 0$, and

$$H = \ell n[\,(1+\lambda)^{1+\lambda} \,/\, \lambda^{\lambda}] \tag{3.6}$$

We have just proved the case for $E(N) = \lambda$. The case for $E(N) \leq \lambda$ can be easily verified by observing that the right side of (3.6) is an increasing function of λ for $\lambda \geq 0$. Q.E.D.

Proposition 2.
Among X on $\mathbb{R}^+$ with $E(X) \leq 1/\beta$

$$\int_0^{\infty} f(x)\, \ell n(1/f(x))\, dx \tag{3.7}$$

is maximized at

$$f(x) = \beta\, e^{-\beta x} .$$

The proof is similar to that of the previous proposition and will be omitted.

4. ESSCHER PREMIUM PRINCIPLE

When the risk S is described by (1.1),

$$M_S(t) = M_N(\,\ell n\, M_X(t)\,),$$

and

$$\phi_S(t) = \phi_N(\,\ell n\, M_X(t)\,)\,\phi_X(t) . \tag{4.1}$$

Thus, when N is Poisson with mean λ,

$$\phi_S(t) = \lambda\, M_X'(t) . \tag{4.2}$$

When N is negative binomial with parameters p, α and mean $\alpha(1-p)/p = \lambda$,

$$\phi_S(t) = \frac{\lambda\, M_X'(t)}{1 - (\lambda/\alpha)(M_X(t)-1)} \tag{4.3}$$

We will use (4.2) and (4.3) to study the Esscher premium $\pi(S)=\phi_S(t)$ under various assumptions on X.

4.1. Constant X, $X \equiv 1/\beta$

By (4.2), when N is Poisson with mean λ,

$$\phi_S(t) = (\lambda/\beta)\, e^{t/\beta} \tag{4.4}$$

$$\text{for all } t\ . \tag{4.5}$$

By (4.3) when N is negative binomial with $E(N) = \lambda$ and $\mathrm{Var}(N) = \lambda + (\lambda^2/\alpha)$,

$$\phi_S(t) = \frac{(\lambda/\beta)\, e^{t/\beta}}{1 - (\lambda/\alpha)(\, e^{t/\beta} - 1\,)}\ , \tag{4.6}$$

and $\phi_S(t)$ is defined only for

$$(\lambda/\beta)(\, e^{t/\beta} - 1\,) < 1\ . \tag{4.7}$$

4.2. Exponential X, $E(X) = 1/\beta$

By (4.2), when N is Poisson with mean λ,

$$\phi_S(t) = \lambda\beta/(\beta-t)^2\ , \tag{4.8}$$

and $\phi_S(t)$ is defined only for

$$t < \beta\ . \tag{4.9}$$

By (4.3), when N is negative binomial with $E(N) = \lambda$ and $\mathrm{Var}(N) = \lambda + (\lambda^2/\alpha)$,

$$\phi_S(t) = \frac{\lambda\beta\ /\ (\beta-t)^2}{1 - [\ \lambda t\ /\ \alpha(\beta-t)\ ]}\ , \tag{4.10}$$

and $\phi_S(t)$ is defined only for

$$(\ 1 + [\lambda/\alpha]\ )\ t < \beta\ . \tag{4.11}$$

As contrast with the variance premium principle, the Poisson to negative binomial perturbation would not only cost more, but for some t, i.e., some Esscher premium principles, $\phi_S(t)$ could be finite (insurable) in the Poisson case but infinite (uninsurable) in the negative binomial case. Compare (4.5) and (4.7), (4.9) and (4.11). This phenomenon is also true with the exponential premium principle and other mixtures of Esscher principles [3].

5. CONCLUSION

In this note, we raise the question of stability of $\pi(S)$ when the assumption of a Poisson N is let loose. We propose using the negative binomial as a reasonable perturbation for Poisson, and we gave information theoretic arguement to support the special case (α=1) of the geometric distribution. We found that under the variance premium principle, the Poisson to negative binomial perturbation would lead to a surcharge; but under the (mixture of) Esscher premium principle, it could change from insurable to uninsurable.

REFERENCES

1. Berliner, B. and Lev, B. On the Use of the Maximum Entropy Concept in Insurance, *Proceedings of the 21th International Congress of Actuaries*, T1, 47-61 (1980).

2. Bühlmann, H. An Economic Premium Principle, *Astin Bulletin* 11, 52-60 (1980).

3. Gerber, H. and Goovaerts, M. On the Representation of Additive Principles of Premium Calculation, *Scandinavian Actuarial Journal*, 221-227 (1981).

4. Kullback, S. *Information Theory and Statistics*, Wiley, New York, 1959.

PRACTICAL RATING OF VARIABLE ACCIDENT EXCESS-LOSS PREMIUMS

Hugo Clemeur

1. GENERAL INTRODUCTION

In numerous circumstances, Accident Excess-Loss Treaties are concluded against a provisional premium which, at the end of the re-insurance year, is adapted not only according to the final premium income of the ceding company for the year, but also according to the loss experience of the year. Sometimes, in addition to the year's loss experience, the losses experienced in previous years are also taken into account.

The idea of a premium made variable according to loss experience, is not proper to the Excess-loss reinsurance. A few examples exist where even in Stop-loss treaties a fluctuating premium is negociated. Much more frequent are the proportional treaties (quota share and surplus) where either a commission scale, a profit commission or even both are applied. The commission scale varies according to the loss ratio's experience , while profit commissions are calculated as a portion of the reinsurer's profit, usually after having made allowance for administration costs and possibly previous losses.

In excess-loss treaties, the general form of the variable premium rate can mostly be expressed as:

$$VP_t^{XL}(a,b) = \mathrm{MIN}\left[\mathrm{MAX}\left(a, \frac{\overline{B}C}{1-\alpha}\right), b\right] \qquad (1)$$

F. de Vylder et al. (eds.), Premium Calculation in Insurance, 387–397.

where $\overline{BC}$ = burning cost of the year t or average burning cost over a number n of past years, including t.

α = loading for safety and administration costs

a, b = lower and upper limits of the variable premium rate.

In most of the cases, $n=1$, meaning that the variable premium has no "memory", and takes only into account last year's experience.

In practice, a provisional excess-loss premium is paid at the beginning of the reinsurance period, usually the minimum premium-rate multiplied by a reasonable guess of the cedant's next year's premium-income in the branch concerned. Basically, one year later, the provisional premium is then adapted both to the insurer's real premium income, and the losses experienced. The excess-loss premium does not become final before total settlement of all losses has occured, which may take quite a while, especially for long tail business. For short tail business, a loss portfolio withdrawal may have been agreed upon, which allows for a fast final calculation of the excess-loss-rate.
For long tail, such provision would be regarded as rather unusual. In some cases, the burning cost is averaged over a number of years before being applied to the excess-loss-premium-rate formula. Two possibilities in this respect are to be examined separately:

1. A single rate is applied for successive groups of n years. The rating is calculated according to the claims experience during these n years and everything occurs as if one had chosen another time-unit of n years, instead of one year. During the first years of each period, a provisional premium (or rate) is determined each year, until the end of the year, after which a readjustment is made retro-actively for each year of the period. The average burning cost taken into account is usually

$$\overline{BC} = \frac{1}{\sum_{i=1}^{n} P_i} \sum_{i=1}^{n} BC_i \, P_i$$

$\overline{BC}$ = average burning cost

P_i = cedant's premium income in the ith year

BC_i = burning cost of the ith year

$\widetilde{BC}$ is then multiplied successively by P_i $(i=1,2\cdots n)$ which determines the corrections to be made to the provisional premiums of previous years, subject to limitations by lower and upper limits a or b if applicable.
From the theoretical point of view, there is no essential difference. This type of formula may prove usefull for example for the cover of events that tend to show some cyclical behaviour.
As disadvantage, it should be noted that cancellation is only possible every n^{th} year.

2. The Excess-loss rate is each year determined according to some "moving average" formula, allowing for the past experience over say n years. This may be compared to some form of experience rating, which is a subject well studied in the direct insurance field. However, experience rating formulas may not be too popular particularly in excess-loss reinsurance, especially when settlements tend to strech over long periods of time, implying frequent and increasingly cumbersome calculations and readjustments.
Furthermore, real experience rating type formulas in reinsurance make the reinsurer play the role of a banker, financing losses rather than compensating them between clusters of similar risks.
Finally, difficulties may arise to define adequate terms for fixing premium rates in the first years, and also for providing a fair deal for both insurer and reinsurer if the treaty is cancelled after a few years.
These thoughts, along with the fact that the average duration of a treaty tends to decrease in the present rapidly changing and difficult economic circumstances may lead to the conclusion that experience rating formulas in Excess-loss reinsurance are not optimal in many respects.

2. PRACTICAL DETERMINATION OF UPPER AND LOWER LIMITS IN VARIABLE EXCESS-LOSS RATES

We now return to the case where $n=1$, which is the most traditional.
Suppose that by some way, we dispose of the distribution function of the burning cost of a given year, that we consider as a stochastic x, say $dF(x)$, $0 \leq x < \infty$
In order to determine the bounderies a and b of the variable Excess-loss rate described in (1), we could start from the requirement that

$$E\left[VP^{XL}(a,b)\right] \geq \frac{E(x)}{1-\alpha} \qquad (2)$$

It is possible to argue about the fact whether the loading α is to be the same for a variable Excess-loss rate as for a fixed Excess-loss rate, but we will return to this question later. From here on, we disregard until further notice all loadings so that (2) becomes

$$E\left[VP^{XL}(a,b)\right] \geq E(x) \qquad \begin{array}{l} a < b \\ 0 < x \leq Max(x) \end{array}$$

Logically, for each a it must be possible to find a lower boundary for b determined by:

$$E\left[VP^{XL}(a,b)\right] = E(x)$$

Hence we derive:

$$\int_0^a a\,dF(x) + \int_a^b x\,dF(x) + \int_b^\infty b\,dF(x) = E(x) \qquad (3)$$

or

$$\int_0^a (a-x)\,dF(x) = \int_b^\infty (x-b)\,dF(x)$$

After differentiation, we obtain that

$$\frac{\partial b}{\partial a} = -\frac{\int_0^a dF(x)}{\int_b^\infty dF(x)} = \frac{F(a)}{1-F(b)}$$

It follows that $\frac{\partial b}{\partial a}$ for all $0 \leq a < b \leq Max(x)$

Similarly, we obtain that

$$\frac{\partial^2 b}{\partial a^2} = \frac{\left(\frac{\partial b}{\partial a}\right)^2 F'(b) - F'(a)}{1-F'(b)}$$

Considering (3), clearly $b\left[a=E(x)\right] = E(x)$ provided that $F'(x) \neq 0$ in $a < x < b$, which we will always assume from here on.

$$\left[\frac{\partial b}{\partial a}\right]_{a=E(x)} = -\frac{F\left[E(x)\right]}{1-F\left[E(x)\right]} < 0$$

$$\left[\frac{\partial^2 b}{\partial \bar{a}^2}\right]_{a=E(x)} = F'[E(x)] \cdot \frac{\left\{\frac{F[E(x)]}{1-F[E(x)]}\right\}^2 - 1}{1-F[E(x)]}$$

Given the type of functions $F'(x)$, with relatively important skewness, long tails generally stretching out towards the right end of the abcis, we may assume, without restricting ourselves too much that mode$(x) < E(x)$, and therefore median$(x) < E(x)$ (equalities would imply a symmetrical function $F(x)$, which is utopic). In consequence, $F[E(x)] > \frac{1}{2}$, therefore

$$\left[\frac{\partial b}{\partial a}\right]_{a=E(x)} < -1$$

and

$$\left[\frac{\partial^2 b}{\partial \bar{a}^2}\right]_{a=E(x)} > 0$$

The conclusion is that b is a continously decreasing function of a, convex at least in the neighbourghood to the right of $E(x)$

It is also clear that $\lim\limits_{a \to 0} b = \mathrm{Max}(x)$

For b, the following set of information is given now, provided that we can evaluate $F[E(x)]$ and $F'[E(x)]$ namely:

$$b[a = E(x)] = E(x)$$

$$b[a = 0] = \mathrm{Max}(x)$$

$$\left[\frac{\partial b}{\partial a}\right]_{a=E(x)} = -\frac{F[E(x)]}{1-F[E(x)]}$$

$$\left[\frac{\partial^2 b}{\partial \bar{a}^2}\right]_{a=E(x)} = F'[E(x)] \cdot \frac{\left\{\frac{F[E(x)]}{1-F[E(x)]}\right\}^2 - 1}{1-F[E(x)]}$$

Instead of b, consider now $h = b - E(x)$ as a function of $k = E(x) - a$, then

$$h(k=0) = 0$$

$$h[k = E(x)] = Max(x) - E(x)$$

$$\left[\frac{\partial h}{\partial k}\right]_{k=0} = -\left[\frac{\partial b}{\partial a}\right]_{a=E(x)}$$

$$\left[\frac{\partial^2 h}{\partial \bar{k}^2}\right]_{k=0} = \left[\frac{\partial^2 b}{\partial \bar{a}^2}\right]_{a=E(x)}$$

A polynomial of the 3th degree would quite well fit the conditions with a possible exception for the case were $Max(x) \to \infty$
Therefore, a simple alternative is presented.
We approach $h(k)$ by: $h(k) = A\, tg\, Bk + Ck^2$

This function not only shows a convenient shape, but also offers a few appreciable computational advantages, and it handels quite easely the case where $Max(x) \to \infty$

For A, B and C, we obtain easily that:
B is the solution of $Max(x) - E(x) = \frac{1}{B}\left[\frac{\partial h}{\partial k}\right]_{k=0} tg\left[B.E(x)\right] + \frac{1}{2}\left[\frac{\partial^2 h}{\partial \bar{k}^2}\right]_{h=0} \overline{E(x)}^2$

$$A = \frac{1}{B}\left[\frac{\partial h}{\partial k}\right]_{k=0}$$

$$C = \frac{1}{2}\left[\frac{\partial^2 h}{\partial \bar{k}^2}\right]_{k=0}$$

If $Max(x) \to \infty$, then evidently $B \to \frac{\pi}{2E(x)}$

Illustration of the method

Suppose that after adjustment of the raw data, following series of BC is observed for a given priority: (in %) 0,1,1,1,3,4,4,6,10,20. We assume $E(x) = \bar{x} = 5$

$F[E(x)]$ is estimated as

$$\hat{F}[E(x)] = \frac{\#\text{ of observations} < \bar{x}}{\#\text{ of observations} > \bar{x}}$$

Note: observations $= \bar{x}$ are considered for 1/2 in dividend and divider.
Hence:

$$\left[\frac{\partial h}{\partial \bar{k}}\right]_{k=0} = \frac{7}{3} = 2,33\ldots$$

For $F'[E(x)]$ we consider the histogram
and estimate $\hat{F}'[E(x)] = 0,1$

hence $\left[\frac{\partial^2 h}{\partial \bar{k}^2}\right]_{k=0} = 0,1 \times 14,8 = 1,48$

$Max(x)$, considering the portfolio and type of business is estimated at 50
Hence, h is estimated as $h = A \, tg \, Bk + Ck^2$

where B = solution of $45 = \frac{2,33\ldots}{B} tg \, 5B + 0,74 \times 25$

which gives easily: $B = 0,24643\ldots$
Therefore: $A = 9,47\ldots$
and $C = 0,74\ldots$

Which gives following tabel:

K	Lower limit a	upper limit b
0	5	5
0,5	4,5	6,35
1	4	8,10
2	3	13,00
3	2	20,30
4	1	50,00

If the given limits for the VP^{XL} where applied, the reinsurer would have experienced following profits with the given series of $\bar{BC}$'s

K	profit of reinsurer
0,5	+ 0,2
1	+ 0,2
2	+ 2,0

This table shows that the approximations are quite reasonable and conservative as profits of the reinsurer prove to be slightly positive. Given the fact that insurers tend not to show to the prospective reinsurer statistics with too many losses, this feature is not necessarily a disadvantage.

A difficult point arises when hardly any information is given about $F[E(x)]$ or $F'[E(x)]$ or even $E(x)$

For such cases, we would like to give the following considerations:

1) If no information is available about $F[E(x)]$, one is forced to make assumptions about the slope and skewness of $F(x)$

For unimodal curves of moderate assymetry however, there is an interesting empirical relationship between mode, mean and median: mean-mode = 3 (mean-median) Considering that in XL situations, the mode often is small compared to the mean $E(x)$,we could also restate this as:

$$E(x) \approx 3\left[E(x) - \text{median}\right]$$

On the other hand

$$F\left[E(x)\right] = \int_0^{E(x)} dF(x) = \int_0^{\text{Median}} dF(x) + \int_{\text{Median}}^{E(x)} dF(x)$$

$$= \frac{1}{2} + \left[E(x) - \text{median}\right] . F'(x') \qquad \text{median} \leq x' < E(x)$$

Therefore, we can approximate this bj:

$$F\left[E(x)\right] = \frac{1}{2} + \frac{E(x)}{3} F'\left[E(x)\right]$$

In the given example, given $F'\left[E(x)\right] = 0,1$ and $E(x) = 5$,
$F\left[E(x)\right] = \frac{1}{2} + 0,17 = 0,67$ (the "true" value found in the example being 0,7)

2) One must admit that it is quite unusual to dispose of $F'[E(x)]$ together with $E(x)$. Under the usual circumstances even, one hardly disposes of a vague clue about $E(x)$: therefore imagination (and experience) must complete the given information in order to be able to give some kind of a variable XL rating within the limits of reason. Under those circumstances, we could assume that x has a gamma type of distribution

$$dF(x) = \frac{a^{\gamma}}{\Gamma(\gamma)} x^{\gamma-1} e^{-ax} dx$$

where $E(x) = \frac{\gamma}{a}$

hence $F'\left[E(x)\right] = \frac{\gamma^{\gamma}}{E(x)} \cdot e^{-\gamma}$

It is well known that the curve of the gamma-distribution is unimodal, except for values of γ less than or equal to unity, where it is J-shaped.

For $\gamma = 1$ $F'\left[E(x)\right] = \frac{0,3679..}{E(x)}$

For $\gamma = 2$ $F'\left[E(x)\right] = \frac{0.5413..}{E(x)}$

And for $\gamma = 3$ $F'\left[E(x)\right] = \frac{0.6721}{E(x)}$

Hence assuming that the distribution of x is J-shaped (with mode in 0) or has a mode only hardly over 0, a reasonable guess for $F'\left[E(x)\right]$ would probably be $\frac{0,4}{E(x)}$

Therefore, for $E(x) = 5$, $\hat{F}'\left[E(x)\right] = 0,08$ (according to the example, the value found was 0,1)

3) If no information is available about $E(x)$, let alone $F'\left[E(x)\right]$ or $F\left[E(x)\right]$, the most reasonable thing to do for the reinsurer would probably be to stay away from the business.
Sometimes however, a variable XL premium rate is considered by some reinsurers as an alternative in the cases where precisely they find it hard to assess a fixed value to the XL premium rate. We will come back to this point further on.

3. FINAL REMARKS

a) Profit and safety margins in variable XL premium rates.

It is easy to see that with variable XL premiums, the variance of the results for the reinsurer is considerably less than with fixed XL premium. One could argue that in such a case, the safety loading, to be applied to

the variable XL premium (and also to its limits) should be reduced in due proportion. It cannot be denied indeed that, as the risk assumed by the reinsurer diminishes, a reduced safety loading would be fair. On the other hand, one must not forget the substantial increase in administrative costs for the reinsurers, for handling variable XL premiums due to the necessity of repeated adjustments and recalculations (especially in the long-tail business) of these premiums which, after a few years, may cause certain amounts of confusion and concern in the accounting departments both of cedant and reinsurer.
These higher costs should be properly anticipated, and eventually reserved for. As a result, the loading applied to the BC doesn't necessarily decrease as compared to the fixed XL premium case.

b) Variable XL premium rates as an alternative in cases where little or nothing is known about $E(x)$

In cases where no information is available about $E(x)$, a reinsurer may find it tempting to present a variable XL premium to the ceding company. By this method, he hopes to reduce the additional risk induced by the lack of knowledge about the average burning cost to be expected. This is an interesting approach, which we will try to analyse here.

As we have seen before, for a given x and $dF(x)$, a set of boundaries for variable XL premiums can be found, where the upper limit is defined by the lower limit or vice versa.
On the other hand, it is easy to show that a given set of limits a and b can "match" a variable BC with an infinitive set of distribution functions.

In other words the functional equation

$$\int_0^a a\, dF(x) + \int_a^b x\, dF(x) + \int_b^\infty b\, dF(x) = E(x)$$

has an infinite set of solutions $F(x)$, and therefore x can have an infinite set of mean values $E(x)$. A great lot of these solutions will however have nothing to do with the distribution functions usually met in actuarial sciences. On the other hand, the range within which $E(x)$ can vary is limited, as necessarily $a \leq E(x) \leq b$ if $F'(x) \neq 0$. Furthermore, considering the skewness of $F'(x)$

it can be deducted that $E(x)$ is to be situated in the first half of the interval (a,b) as $a \leq E(x) \leq \frac{a+b}{2}$ as shown in chapter 2

As a conclusion, suppose that $E(x)$ is known to be situated somewhere between m_1 and m_2. A variable XL premium taking into account the incertitude about $E(x)$ could be safely calculated in the following way:
Given $a \leq m_1$, calculate b under the assumption that $E(x)=m_2$
This calculation however may prove too conservative, and therefore not realistic in an open market.
If $E(x)$ itself has a distribution function F_E one could also calculate b for all $m_1 \leq E(x) \leq m_2$ and given $a < m_1$,

and average $$\bar{b} = \int_{m_1}^{m_2} b\left[E(x)\right] dF_E$$

Note that $$\bar{b} \geq b\left[\int_{m_1}^{m_2} E(x)\, dF_E\right]$$

If previous calculation is not possible or too consuming, a rule of thumb could consist of simply choosing $a = m_1$ and

$$b = a + 2(m_2 - m_1) = 2m_2 - a$$

Finally in all cases an appropriate safety loading must be applied to both limits and burning costs experienced.

MOTOR PREMIUM RATING

Stewart Coutts
Excess Insurance Group

Summary: This paper outlines a very simple structure for the practical analysis of motor premiums from a UK point of view. It indicates that it is possible to model different types of claims cost. These results are combined together with some simple economic assumptions, to arrive at premiums. Then the paper develops a "points system" which is similar to a number of premium systems operated by UK Insurance Companies. This "points system" is used to compare different sets of assumptions. Finally, an analysis of surplus is described with an example.

1. INTRODUCTION

The purpose of this paper is to outline some of the statistical and technical aspects of a premium basis for motor insurance. At present, only broad outlines are available in the UK Institute's literature and in particular, they do not highlight the pragmatic aspects of the problem. There have been some papers written in the ASTIN bulletin, e.g. Pitkanen (1974), however, they are generally theoretical based. The first full analysis found in the literature was written by Kahane and Levy (1975). However, the emphasis was statistical modelling rather than discussing the practical problems involved in premium rating. Recently, two groups have published studies concerning the problems of rating. In The Netherlands, a group of actuaries were asked to revise the motor rating structure and this is written up in Netherland Group Report (1982). At the UK General

F. de Vylder et al. (eds.), Premium Calculation in Insurance, 399–448.

Insurance Statistical Group meeting in Stratford-upon-Avon (1982), there were four case studies from different countries on premium rating. Hence, there is little written assistance to help either the actuary new to motor insurance premium rating or the actuarial student trying to appreciate practical problems of premium rating.

The theme of this paper will be to emphasise a detailed within-portfolio analysis, taking into account simultaneously, inter alia, some of the major motor underwriting factors and separate statistical analyses of claims frequency and costs, together with expenses, to arrive at a breakeven premium. By performing this detailed analysis, it is believed that the person responsible for the premium decision will be able to restrict his attention to the sensitive areas where judgement has to be exercised.

This paper will describe a procedure which is applicable in a competitive motor premium market situation, such as in the UK. However, it is contended that in countries where fixed premium rating set by a tariff is in operation, an in-depth analysis is also necessary. This will enable management to judge where in their motor portfolio the business is profitable. In addition, the analysis will bring out warnings that where all company pooled data is used, it can be harmful, in particular to smaller companies.

The structure for this paper is as follows:-

Section 2	Discusses the reason for a detailed data breakdown.
Section 3	Gives the general background to premium rating by asking, why project?
Section 4	Introduces the Group which is involved in the decision process.
Section 5	Gives the formula to be used.
Section 6	Defines the database.
Section 7	Outlines the general principles for claim frequency and cost statistical analyses.
Section 8	Discusses bodily injury analysis.
Section 9	Outlines the inclusion of expense allocation.

Section 10	Discusses economic factors.
Section 11	Performs the actual premium calculation.
Section 12	Discusses the presentation of the premium rates.
Section 13	Discusses a points table analysis.
Section 14	Compares different sets of assumptions used in the premium bases.
Section 15	Marketing aspects of premium rating.
Section 16	Analysis of surplus.

Finally, to keep this paper to manageable length, it has been necessary to restrict some of the discussion of statistical aspects to a minimum and only give references to papers which will give details of these analyses.

2. DATA BREAKDOWN

The first question to consider is whether the premium rates are to be reviewed in an overall fashion, e.g. adding 10% over the whole portfolio, or whether to apply selective increases to different sections of the portfolio. It is argued that with the use of computers, a selective breakdown of underwriting factors should be undertaken, which by aggregating the results can automatically give the overall level of premium adjustment needed. If, however, the system of analysis is only geared to the overall review, it is very much harder to obtain information about the selective parts of the portfolio.

If the data are sub-divided by underwriting rating factors, we are left with figures which are small in exposure and claim numbers, therefore, simple averages will be suspect. Hence, it is suggested that simple statistical modelling be preferred. This will enable:-

1. Extension of actuarial judgement to small databases which implies that the portfolios of small Insurance Companies can and should be analysed. This statement is strongly worded, but it is believed that it can be accomplished since the theory of statistics in the past 10 years has made 'space age' progress in the analysis of small databases, in far more critical and sensitive areas than motor insurance, namely, medical

statistics. An urgent plea is made to the profession to step up research on small databases, in order that small companies can employ actuaries with the relevant expertise to offset uninformed comments such as 'the data is too scanty to support any meaningful analyses'.

2. When the data is analysed in sufficient detail, then the effects of portfolio changes are reduced and judgements on these effects can be made with confidence.

3. The establishment of statistical structures, however simple, provides 'bench marks' that can be used as a basis to monitor actual results as they emerge. This will be discussed in the final section of the paper.

4. As will be indicated later, analysis reveals that the process of premium rating involves many different assumptions. Changes in some of these assumptions can affect the premium rates significantly, e.g. different bodily injury assumptions, or in varying mathematical models. The effects can only be assessed if the data is sufficiently detailed and the structure sufficiently defined.

In spite of these points, it has been argued by non-actuaries that any actuarial input which might alter rates within the motor portfolio is practically irrelevant, when compared with overall marketing considerations. Hence, the statistical process is considered a mere theoretical exercise, the cost of which, given the personnel involved, is hard to justify. In addition, actuaries within the general insurance field have tended to depreciate any detailed analyses, as recently as the UK General Insurance Statistical Group conference at the one-day seminar on Premiums at Stratford-upon-Avon. One argument was that an experienced actuary did not need to perform any detailed premium analysis, since he should be aware of the premium situation and adjust the rates accordingly. It was also argued that past analyses should be a sufficient basis for changes in premium, if selected data were collected in order to monitor the process.

It is suggested that all these remarks are half-truths. True, actuarial rates may differ considerably from market rates. However, it is the management's decision and they should be aware that the rates charged may in fact generate potential losses, the size of which should be quantified. If decisions are made without having all facts available, then this must be considered poor management. A topical example is that the underwriting rating factor 'car age' is generally ignored or given insufficient weight in market structure; in particular, newer cars are being undercharged. On the opposite side, a detailed analysis may reveal unsuspected marketing opportunities within the present structure. Herein lies the area of greatest potential for statistical analysis. In addition, in countries where there is fixed premium rating, set by a tariff, this may eventually break down, e.g. in The Netherlands. Hence, it is advisable to be able to perform a detailed analysis to cover this eventuality.

As for the argument that the experienced actuary has little need of analysis, it is accepted that judgements could often be made without any in-depth investigation, how is this achieved? The existing body of knowledge has been accumulated largely by trial and error. In fact, however, the profession ought to strive to establish detailed procedures. It is contended that the present examination reading does not give sufficient information for this purpose.

The final argument, appealing to past analyses, tacitly assumes that premium structures are stable over time. It is agreed that frequent changes to the system may be undesirable, but regular analyses are necessary to verify assumptions made in the original calculations. In particular, if the database is small, this will necessitate regular checks to judge whether the statistical inferences were reasonable. For the UK in the late 1970's, many companies reviewed their premium levels quarterly because of the rising rate of inflation. Furthermore, the underlying insurance risk pattern may also vary over time.

Finally, cost and time are put forward as reasons for not performing regular analyses. However, with the advent of microprocessor technology, it is believed that the cost has been cut to a minimum and time reduced to an irrelevant factor.

In summary, a breakdown of the data to take into account rating factors, claim frequency and cost is fundamental to establishing premium bases. To analyse data in this way, modern statistical techniques must be understood and employed regularly since the systems are not necessarily stable. To monitor the results, some form of analysis of surplus has to be carried out.

3. WHY PROJECT?

Before discussing what background information is required in practice, to make judgements about premium levels, it would be useful to visualise the time span involved in the premium analysis. This is best explained by means of a simple example.

Consider a company that reviews its premium rates on 1st October 1980 and let us assume that these rates are expected to be in force for one year. New recommendations would have to be made in practice 3 months in advance. This time lag would be needed to alter computer output and prepare documentation for the rate book to be passed to the broker.

The average policy will be effected halfway through the period over which the premium series is expected to be in force, i.e. on 1st April 1981. This policy will be on risk for one year and, should it have a claim, the claim date will on average be 6 months after date on risk, i.e. 1st October 1981.

The material damage costs will be expected to be settled within 3 months from the date of accident. However, the third party bodily injury costs will take on average 2 years to settle. Prospectively, the average future time span is 3¼ years from the date which the premium decision has to be made, i.e. by the 1st July 1980. Hence, data concerning claims and expenses have to be projected to the 1st October 1981 and beyond.

Expense data will be based on information which is reasonably up to date at the 1st July 1980 and is then projected to 1981. The claim data will be built up from claim numbers, material damage costs and third party bodily injury costs. All but the latter costs will be based on recent data, say 1980. However, the most reliable third party bodily injury data is likely to be 5 years old, i.e. claims occurring in 1975. The reason why the average settlement figure of 2 years is not appropriate is because, in practice, the larger and proportionately more important claims take in excess of 5 years from date of accident to settlement. Hence, the total time span to be projected is on average 8¼ years. This large time span necessitates a number of subjective decisions, the main ones being:-

(i) Are third party claims occurring in 1975 relevant in respect of liability claims expected to occur in 1981 and to be settled on average in 1983 or later? The main problem lies in the settlement figures, as court judgements change with changing social conditions. This has been commonly referred to as judgement drift. If the Company has an individual liability claim estimation process (referred to as a manual basis) then it is possible, though not necessarily reliable, to use later liability information based on recent manual estimates as a substitute for settled data.

(ii) Whatever method is employed as the base for projecting claim costs and expenses, a view of past and future inflation has to be taken. In particular, for liability claims, a view of the rate of inflation to bring 1975 values up to 1980 is needed and thereafter a future rate to project these costs into 1983 or later.

(iii) Finally, a view of future levels of claim frequency has to be decided. Factors including future weather conditions, petrol prices and speed restrictions have to be considered. It will be shown later that these aspects may have a relatively large effect on the profitability of results.

4. THE GROUP

In a Company, it is desirable to make premium decisions within a Group. The decision-maker is the person who ultimately says what the premium rates are going to be; he would usually be a General Manager or the Motor Underwriter.

The rest of the Group would act as advisers to the decision-maker, supplying information on various aspects of the business. The size of the Group might range from one person for each main function, to, in a small company, as few as two, the decision-maker and the underwriter.

The following are the main aspects of the business which have to be considered:-

1. General Underwriting principles, which vary from company to company and reflects, i) prior views on occupations which the marketing should attempt to attract, e.g. teachers, civil servants; ii) wording in the policy conditions to take into account, the

introduction of a new rating factor such as protected No Claim Bonus.

2. The overall market position of the Company compared to its competitors, with regard to growth and pricing. Within the Company, production summaries will be available showing lapse and new business figures. Marketing will also be pushing for sales increases over selected parts of the portfolio, by trying to keep any rate increases to a minimum

3. Analyses from the claim negotiator, who will report the latest manual estimates on present third party liability claims.

4. Statistical analyses will produce information for various members of the Group; concerning in particular, past claims frequency, claims cost and production results.

5. General economic factors which will be used by the Group to project into the future, past claim costs and expenses. This will involve, inter alia, a view as to the effect of the Government's current economic policy on inflation rates, salaries and prices.

5. FORMULAE

It might have seemed ideal to relate the time period basis for calculation of premium rates to a cohort of policies all effected during a particular calendar year. However, this would require an oversophisticated database, which is not available. Hence, a pragmatic approach has been taken, namely, the averaging process outlined in Section 3, which uses basically a calendar rate year.

It would be useful at this point to give the premium formula to be applied by combination of risk factors where appropriate.

The risk premium (RP), i.e. premium excluding expenses.

$$RP = fc$$

where f = claim frequency = $\dfrac{\text{number of claims}}{\text{exposure}}$

c = projected claims cost

$$= AD(1+e)^{n_1} + TPPD(1+f)^{n_2} + TPBI(1+g)^{n_3} + M(1+h)^{n_4} \; *$$

* An alternative suggested by Gunner Benktander is to replace $(1+e)^{n_1}$ by

$$\int_0^\infty e^{\delta s} \, dG(s) \doteq e^{\delta t_1} \quad g(s) \text{ is unspecified}$$

Where the inflation rate is e^{δ}, i.e. $(e^{\delta} - 1)$ and t_1 is the average time to settlement.

where AD = average accident damage cost per claim if settled immediately.
TPPD = average third party property damage cost per claim if settled immediately.
TPBI = average third party bodily injury cost per claim if settled immediately.
M = average miscellaneous cost per claim if settled immediately.

and

- e,f,g,h, are inflation rates for respective types of claim cost.
- n_1, n_2, n_3, n_4 are average settlement periods for respective types of claim cost.

The calculation of the breakeven premium (or office premium) P, depends on how expenses are introduced. There are two main variants:-

(a)
$$P = \frac{RP}{1-S}$$

where S is the $\frac{\text{total expenses}}{\text{total premium income}}$ i.e. a fixed percentage of premium

or

(b)
$$P = \frac{RP + \left(cc.f + \frac{nb + L}{t} + r + ed\right)}{1-w} (1+j)^{\frac{1}{2}}$$

where cc = claims cost expenses and f is claim frequency
nb = new business expense
L = lapse expense
t = time period till lapse
r = renewal expense
ed = endorsement expense
w = commission plus another expense related to premium
j = rate of inflation applied to expense

and the costs per unit are inflated to the relevant date of premium.

This method entails variable plus fixed expenses.

A short discussion about these formulae appears in Section 9, but briefly, the total expected expense will be the same for each method and only the division of expense alters.

Two omissions will be noticed:-

(a) No contingency loading (or solvency loading)
It is not certain whether companies explicitly take this factor into account. Although it has been omitted, algebraically its inclusion would be very easy. The problems are of estimating a value and its effect on the final rate.

(b) Investment Income
It is acknowledged that the market does not explicitly take this factor into account. Implicitly, income is taken into consideration when the whole motor account is scrutinised, in that the ultimate trading results can be compared with the underwriting results. This subject deserves a paper on its own and, hence, is left with the simple comment that research should be carried out to investigate the effect the factor has on premiums.

6. DATA (Sub-titled: Now the Story Begins)

It is now generally accepted that past claim information must be matched with the relevant underwriting factors at the time of the claim. In addition, for motor insurance, the number of cars exposed to risk will also be related to the claim period analysed.

There has been some work performed on the relevance of rating factors, in particular from the Belgian school, e.g. Hallin and Ingenbleek (1981), but it is still in its early stages of development.

It will be assumed that all the existing market underwriting factors will continue in use, since it is unlikely that any Underwriter would consider altering them, without other companies following suit. In order to illustrate the principles under discussion, the rest of the paper will be concerned with a worked example. The following rating factors will be used:-

(1) Type of Cover - Comprehensive or Non-Comprehensive
(2) Policyholder's Age - 17-20, 21-24, 25-29, 30-34, 35+
(3) Car Age - 0-3, 4-7, 8+
(4) Vehicle Group - A, B, C, D

Two comments have to be made:-

(a) Some significant rating factors have been omitted from this list (e.g. district, no claim bonus and use), but the analysis can easily be extended to take them into account.

(b) Following discussion with the Underwriter, it may be possible to aggregate some of the detailed data into relevant groupings, in order to make the analysis more manageable. In our example, as an illustration, Policyholder Age 35+ was grouped instead of sub-dividing into 35-50, 50-60 and 60+.

7. THE GENERAL PRINCIPLES OF CLAIMS ANALYSIS

The principal objective of the analysis is to project past claims data for the relevant period. If the example in Section 3 is taken, then on 1st July 1980 a premium for the 1st October 1980 has to be decided. The premium rates will be in force for one year and the average date of claims arising will be 1st October 1981. Hence, all claim information has to be projected to 1st October 1981 and onwards. There are two levels of decision to make, namely,

(a) the overall levels of frequency and claims cost,
and thereafter
(b) the within-portfolio levels, i.e. the relationship between rating factors.

The first stage is dealt with by the Group with very little statistical analysis, but the ultimate decision requires a great deal of judgement. The second stage is basically where the statistical modelling takes over and actuarial judgement comes into play.

The following sections give a brief resume of the process for different parts of the analysis. Details of statistical modelling were contained in a talk given to Post Qualification Course for Actuaries on General Linear Models (GLM) by Prof. A.P. Dawid and Messrs. Baxter, Coutts and Ross in 1980. Further details were given in a paper entitled "General Linear Models in Insurance" at the 1980 Congress of Actuaries in Switzerland. More recently in the ASTIN 1982 a paper by Albrecht discussed all the literature on this type of analysis. Even more accurate models may be achieved by ascertaining the claims amounts distribution by Ziai's methods (1979) and inputting into the powerful GLM algorithm.

7.1 CLAIM FREQUENCY

(a) Overall Levels

Data for past years would be available by, say, sub-groups Comprehensive and Non-Comprehensive, in the following format in Table 1:-

Table 1

Frequency per 1,000 vehicles Year of Accident	Comprehensive Quarter of Accident 1	2	3	4	Year
1977	142	125	115	152	138
1978	152	137	138	154	148
1979	184	134	141	164	156
1980	159	134*			

* includes an IBNR estimate

Discussion on projecting the 1980 results into 1981 would be centred around such items as:-

(i) weather conditions,
(ii) petrol prices,
(iii) road repairs,
(iv) general economic conditions, since these might affect the frequency with which policyholders have their cars serviced.

In our example, it was decided to use the 1979 overall level (i.e. 156) as the projection for 1981.

(b) Within-Portfolio Analysis

All the rating factors mentioned above are considered, with separate analyses for Comprehensive and Non-Comprehensive. The method of smoothing applied was basically the Johnson and Hey (1971) model, slightly adjusted by using logit of proportions as the dependent variable (see Baxter et al (1980) for details) to obtain the smoothed claim frequency for all combinations of rating factors. The smoothed values were adjusted to give a projected overall frequency equal to the 1979 values, such as those shown for Comprehensive in Table 1. By breaking down the data according to all the rating factors, the effect of portfolio changes is minimised.

7.2 PROPERTY DAMAGE AND MISCELLANEOUS COSTS

(a) Overall

For Accidental Damage (or Fire or Theft for Non-Comprehensive), Third Party Property Damage and Miscellaneous Costs, the object is to obtain the average cost of a claim in October 1981 if settled immediately. This value is then projected forward after taking into account the average settlement date. In practice, inflation is assumed to be the main projecting factor for overall levels. Hence, for each respective claims cost, actual values are projected by agreed specific inflation rates up to the projected settlement date. (Inflation rates will be discussed in Section 10).

From detailed analyses, the average settlement periods used were 3 months, 6 months and 3 months respectively.

(b) Within-Portfolio Analysis

To date, statistical modelling using all relevant rating factors has been seriously attempted only in accidental damage claims (see Baxter et al (1980)). A great deal more work is required to get reasonable models to represent the data. Models similar to the Johnson-Hey (1971) do not give reasonable results, since they do not take into account the skew distribution inherent in the accidental damage claims (see Ziai (1979)).

For Third Party Property Damage, separate averages were used based on car grouping. For Miscellaneous, two overall values were used, Comprehensive and Non-Comprehensive. The reason for not modelling these types of claims was simply lack of time.

8. BODILY INJURY

This is by far the hardest part of the statistical analysis, since, on average, only about 5% of all claims per year involve bodily injury costs, yielding for the example portfolio about 1,000 claims. This section deserves a paper to itself, so it is only possible to highlight some of the main problems and present the pragmatic solutions used in this example. However, before discussing details, it is necessary to look at the arguments against pooling data from different companies to arrive at an input to premium calculations.

Since the number of claims for each company is, in practice, small, it has been suggested that all companies should pool their data. That would not, however, solve the problem, for a reason best illustrated by way of an example. Table 2 shows some typical average bodily injury costs per claim for an individual company. They have been adjusted for earnings inflation plus 'judgement drift' (as previously defined) to bring them all up to the value of 1982, the projected date.

Table 2

Year of Accident	Average Bodily Injury Costs per Claim Inflated up to 1982
	£
1972	119
1973	80
1974	115
1975	58
1976	104
1977	90
1978	79
1979	67

If inflation were the only factor operating on these averages, relatively constant values would be expected. However, it is clear that there is wide variation. This is due to the extreme skewness of the underlying distribution of bodily injury claims; there is no reason to suppose that the distribution shifts from year to year.

Corresponding pooled data for all companies is available from British Insurance Association motor statistics but is not available for publication. However, if it were, then the sample size would be far greater and the variability smaller than for any single company. Suppose an average claim of £90 (in 1982 values) resulted from pooled data for years 72 to 79. If the Company based rates on this, large profits would be shown if the 1975 experience were repeated, or large losses if the 1972 figures occurred again. The real problem is that the Company experiences a small sample of the total market experience and therefore its premium rates should make an allowance for its own variability.

(a) <u>Overall</u>

The Group has to decide the basis of projecting bodily injury claims. Assuming that they have decided to use their company's data, they have to select the base period. The possibilities are:-

- (i) Using the latest results (1978 and 1979), which, however, entail predominantly manual estimates.
- (ii) Grouping earlier years, which contain claims fully settled.

In our example, years 72-76 and 72-77 were grouped and then inflated (see Section 10) to the 1982 projections. The average settlement period was established to be 2 years.

(b) <u>Within-Portfolio Analysis</u>

There are two basic decision areas:-

- (i) The selection of rating factors is very difficult. In our example, single factors based on Policyholder Age were considered appropriate. This is a good example where GLM would be very effective, as it could take the shape of the distribution into account. There is another area of research where the profession should lead. Some work by Chang and Fairley (1980) shows that modelling is useful.
- (ii) As the data are broken down into even smaller groups, the effect of large claims becomes very important, since a large claim in a small cell will disproportionately affect the results. Hence some method of smoothing has to be applied. In the example, all claims over £10,000 were cut off at that value and the excess was respread over the whole portfolio. This method is open to criticism, but it was felt appropriate at the time.

  Another factor to be considered is whether the large claims are in fact correlated with particular rating factors or randomly spread.

This completes the statistical analysis of costs and it would be useful to summarise the analysis so far.

Table 3 - Summarises Analyses

Type of Analysis	Rating Factors	Whether Modelling Technique Used	Overall Level
Claim Frequency	All	YES	1979
Accidental Damage	All	YES	Inflation
Property Damage	Comp. & Non-Comp. Car Group	NO	Inflation
Miscellaneous	Comp. & Non-Comp.	NO	Inflation
Bodily Injury	Comp., Non-Comp. Policyholder	NO	1972-76 1972-77) + Inflation)

9. EXPENSES

It is not proposed to go into detail about the collection of data to obtain a breakdown of expenses. A paper by Ian Ruston to this Society (1977) has described the process very well. As far as the Group is concerned, it will obtain this information directly from the accounting department within the Company, which projects future expenses. The only factor it would consider is the future rate of inflation related to expenses (see Section 10).

As far as the premium formula is concerned, two approaches are available.

The first is to consider all expenses as fixed in the short term - see (a) in Section 5. The total value of expenses is obtained from the accountants and then expressed as a percentage of premium income. Commission and other premium related expenses are added to this percentage. In practice, the value should be between 25% to 45%, depending on type of company. The rationale behind this is that in the short term (say 2 years) staff levels (which make up 80% of expenses) are virtually fixed. Thus, it is argued, further sophistication is needless.

The second method (b) is to divide up costs into those which are fixed and expressed as a percentage of premium, and those which are identifiable as separate totals such as claim cost expense, new business expense, lapse expense and endorsement expenses.

For the example, the following were used and then inflated to 1981 levels:-

Fixed as a percentage of premium (incl. commission)	17%
Claim Cost	£14 per claim
New Business	£6 per Policy
Lapse	£1.50 per Policy
Renewal	£1.50 per Policy
Endorsement	£2.25 per Policy

Both methods will give the same overall expense allocation, if the portfolio is similar to that of the base period. However, if the portfolio changes, the second method will give more variable results.

10. ECONOMIC ASSUMPTIONS

The Group will have to form some overall views on inflation. As mentioned earlier, general economic factors and government policies will dominate. Usually, several scenarios will be followed. Different rates of inflation will be applied to different parts of the analysis. In this respect, in the UK, the British Insurance Association Economic Advisory Group is very helpful in reporting and projecting inflation separately for each type of claim. Another problem is to establish which index will be an appropriate indicator for each type of inflation risk.

The following is a summary of the rates and assumptions used by the Group in this example.

Table 4 - Inflation Summary

		SCENARIOS	
	INFLATION INDEX	HIGH	LOW
Property Claims	Earnings + Material Goods	11%	7%
Bodily Damage	Earnings + Judgement Drift	17%	13%
Expenses	Internal, based on Salary Increase	10%	10%

Since inflation has a major effect on the results, the decision maker will have to dominate this Group decision.

11. THE CALCULATION OF PREMIUM RATES

The stage has now been reached when the premium calculation can be performed. This was carried out on microprocessor by applying both formulae in Section 5 to the data generated by Sections 7, 8, 9 and 10. An example of the final premium is given in Appendix 1 Column (4) (break even premium) for all sets of rating factors. The underlying assumptions were low inflation, bodily injury 1972-76 and expenses all fixed.

In a later section, the effects of different sets of assumptions on the premium rates are discussed.

12. PRESENTATION OF RESULTS

The traditional way of presenting rates was via a rate book, where every combination of rating factors was defined with its respective rates. In the early 1970's, the Co-operative Insurance Company introduced a points system and several companies are now using similar systems. In the following sections, we look at the problem of deriving a premium for a points table. Firstly, we must define a points table. Then we will examine why a Company would want to determine premiums in this way. We will then consider the mathematical background and develop an algorithm for deriving a points table. Finally, we will seek to interpret the results.

WHAT IS A POINTS SYSTEM?

The explanation of a points system is probably best achieved by way of an example. Consider the following:-

Table 5 - Points Table

	COMP	NON-COMP
Cover	19	0

	17-20	21-24	25-29	30-34	35+
Policyholder's Age	30	12	6	2	0

	0-3	4-7	8+
Car Age	8	4	0

	A	B	C	D
Vehicle Group	0	5	10	20

It can be seen that we have four rating factors and with each is associated a scale. For each point on the scale, there is a value, expressed in points. The differences in the points on each scale, reduced to percentages, are called relativities.

The procedure used is to look up the points score for each rating factor and then aggregate them. We then use a points conversion table to arrive at the premium to be charged. The conversion table is simply a list of points scores with associated monetary values - see Appendix 2. For example, consider Comprehensive, Policyholder Age 35+, Car Age 4-7 and Vehicle Group A -

19 + 0 + 4 + 0 pts. = 23 pts. i.e. £12.35

13. WHY USE A POINTS SYSTEM?

The advantages of a points system are mainly practical. They include, not necessarily in order of importance:-

(a) It is cheaper to produce than the rate book.

(b) It is easy to revise rates.

(c) The calculation is relatively straightforward.

(d) The method is easy to understand.

There are however problems associated with such a system. Examples are:-

(a) It may be regarded by outsiders as over-simplified.

(b) Fine tuning of the rate structure is no longer possible for marketing purposes.

(c) The points values (as opposed to the monetary values) change only rarely, and changes over time in the underlying variables tend to be neglected.

(d) Because the point system is easy to interpret, any adjustments to the points which reflect statistical experience, are thought by outsiders to be errors.

(e) The resulting premium structure is not exact, because the underlying analysis is ultimately a simplification.

13.1 METHOD OF CALCULATION

We wish to convert the break even premium into a points system. Then, the problem is to compare our fitted points premium (estimated from our model) with the break even premium. That is

$$(P-\hat{P})$$

for each combination of rating factors, where

$\hat{P}$ = fitted premium
P = break even premium

Proceed as follows. Let

$$P_{ijkl} = (1.0325)^{\theta} \qquad (1)$$

P_{ijkl} = premium associated with a given rating factor

where
i = level of cover (C)
j = level of Policyholder's age (PH)
k = level of car age (CA)
l = level of vehicle group (VG)

Then define

$$\theta = M + C_i + PH_j + CA_k + VG_l + In + e_{ijkl} \qquad (2)$$

where M is a constant

In = all interactions between terms

e_{ijkl} is an error term

The choice of the figure 1.0325 is quite arbitrary. This is an example of a multiplicative system.

This complicated analysis is used instead of, say, single tables, in order to explain the relationships between the factors. The interested reader is referred to Prof. A.P. Dawid (Post Qualification Seminar for UK Institute of Actuaries in 1980).

Returning to equation (1), take logs to the base 1.0325, whence

$$\log_{1.0325} P_{ijkl} = \theta$$

We then fit this relationship by weighted least squares:-

$$\min \sum_{ijkl} W_{ijkl} (\log P_{ijkl} - \hat{\theta})^2$$

where $\hat{}$ denotes the least squares estimator

W_{ijkl} is some set of weights to be selected

C_i PH_j CA_k and VG_l etc. are the estimated points for the table.

We now require to know whether we can dispense with any of the terms in (2), particularly the interaction terms (about which more will be said later). We must then estimate values for the factors retained.

The method of Johnson and Hey (1971) can be used to arrive at a points system following equation (2). To ascertain whether any of the terms can be discarded use General Linear Models. However, for simple assumptions, the methods of Orthogonal Weighted Least Squares (OWLS) and of Johnson and Hey and the General Linear Model (Baxter et al (1980)) all give similar results.

OWLS relies on the fact that it is possible to factorise the weights in equation (1), that is

$$W_{ijkl} \propto C_i \, (PH)_j \, (CA)_k \, (VG)_l$$

This is demonstrated in Appendix 3.

This approximation has one highly practical advantage. It does not require the inversion of matrices. Hence, it is possible to analyse a large number of factors without computational capacity becoming a constraint, and this method is adopted here.

13.2 AN ALGORITHM

An algorithm follows for the computation of a points table.

(1) The first step is to choose a set of suitable weights. A suitable choice would be standing business, since, as will later be demonstrated, the method is reasonably robust to the choice of weights selected.

(2) A standard analysis of variance is performed, using

$$\log W_{ijkl} = (\log \text{ of standing business})_{ijkl}$$

(3) The factors $\hat{c}_i$ $(\hat{ph})_j$ $(\hat{ca})_k$ $(\hat{vg})_l$ are selected, after checking for interactions from the Analysis of Variance.

(4) The W_{ijkl} is replaced by the product of the estimates of c, ph, ca, vg from the previous step, i.e. we find $\hat{\theta}_{ijkl}$ from the expression

$$\min \sum \hat{c}_i\, (\hat{ph})_j\, (\hat{ca})_k\, (\hat{vg})_l\, (\log P-\theta)^2 \qquad (3)$$

(5) The Weighted Analysis of Variance of log P is then checked to ensure that there are no significant interactions.

(6) Assuming that no significant interactions appear, then the estimates of the main effects serve as the points within the table.

(7) The fitted values for premiums computed through the model are compared with the break even premiums, in order to assess the goodness of fit.

Step 1

An example will now be presented. Appendix 1 shows the basic data. The rating factors can be seen on the left. (Column (1) shows the Standard Business, which is being used for the weights, i.e. W_{ijkl}.

Step 2

Analysis of Variance on log W_{ijkl} gives the following results:-

Table 6 - ANOVA of log W

Factor	Sum of Squares	Degrees of Freedom	Mean Square Ratio
Cover	1.5	1	258
Car Age	5.9	2	487
Vehicle Group	4.6	3	253
Policyholder's Age	33.0	4	1360
Cover x Car Age	12.4	2	1028
Cover x Vehicle Group	0.0	3	1
Cover x PH Age	3.9	4	161
Car Age x Vehicle Group	2.3	6	62
Car Age x PH Age	0.2	8	3
Vehicle Group x PH Age	0.5	12	7
Residual	0.5	74	
TOTAL	64.8	119	

From this, it can be seen that, in addition to the main effects, there is a significant association between cover and car age. This point will be returned to later.

Step 3

The standing business distribution yields weights for the factors:-

Table 7 - Weights for Standing Business

Cover

Comp.	Non-Comp.
4.8	2.8

Car Age

0.3	4.7	8+
1.8	5.3	5.3

Vehicle Group

A	B	C	D
2.3	6.1	5.5	2.4

Policyholder Age

17-20	21-24	25-20	30-34	35+
0.9	1.9	2.8	4.3	32.3

The values represent the distribution of standing business within the portfolio. For example, there are 32 times as many policies in Age 35+ as in Age 17-20.

Step 4

Using these estimates W_{ijkl}, $\hat{\theta}$ can be estimated from equation (3).

Step 5

The Analysis of Variance on log P_{ijkl} yields:-

Table 8 - ANOVA log P

Factor	Sum of Squares	Degrees of Freedom	Mean Squares	Mean Square Ratio
	(1)	(2)	(3)=(1)÷(2)	(4)=(3)/ Residual (3)
Cover	4 545 659	1	4 545 659	7 788
Car Age	3 254 634	2	1 627 317	2 788
Vehicle Group	4 004 061	3	1 334 687	2 287
Policyholder's Age	2 758 813	4	689 703	1 102
Cover x Car Age	231 341	2	115 670	198
Cover x Vehicle Gp.	84 683	3	28 228	48
Cover x PH Age	40 913	4	10 228	18
Car Age x Vehicle Gp.	5 869	6	978	2
Car Age x PH Age	5 576	8	697	1
Vehicle Gp. x PH Age	21 566	12	1 797	3
Residual	43 191	74	584	
TOTAL	14 996 306	119		

It is important to note that this analysis does not allow us to make any statement about the goodness of fit of the model, since we do not have available an independent estimate of the residual variance. All that the analysis shows is that, relatively, the main effects are far more important than the interaction terms, which can therefore be neglected.

From these results, we estimate the main effects as follows:-

Table 9 - Estimates of Points

Weighted Grand Mean
132.9

Cover

Comp.	Non-Comp.
6.5	-10.9

Car Age

0.3	4.7	8+
11.0	4.1	-7.8

Vehicle Group

A	B	C	D
-9.0	-4.3	0.8	17.2

Policyholder Age

17-20	21-24	25-20	30-34	35+
27.8	20.6	6.9	2.4	-2.3

It would be easier to make comparisons if there were no negative values. This is simply accomplished, by transforming the smallest value in each line to zero, making the same addition to the other values in the line, and adjusting the overall mean subsequently.

For example, with car age, we can alter the value for '8+' to zero by adding 7.8, whence 18.8 for '0-3' and 11.9 for '4-7'. The overall mean is later reduced by 7.8, as in the following table.

Table 10 - Altering the Base to Facilitate Comparison

	Unadjusted	Adjusted
Overall	132.9	102.8
Cover		
Comp.	6.5	17.4
Non-Comp.	-10.9	0
Policyholder Age		
17-20	27.8	30.2
21-24	20.6	23.0
25-29	6.9	9.3
30-34	-2.4	0
35+	-2.3	.1
Car Age		
0-3	11.0	18.8
4-7	4.1	11.9
8+	-7.8	0
Vehicle Group		
A	-9.0	0
B	-4.3	4.7
C	0.8	9.8
D	17.2	26.2

Step 7

We now wish to examine the goodness of fit of the new premium structure. There are three tests:-

(1) The fitted premiums can be compared with the break even premiums for each rating factor (i.e. E/A).

(2) The premium income we would expect, taking account of our standing business, from our fitted premiums can be compared with the break even premiums.

(3) The expected premium income from the fitted premiums can also be compared with that of the present premium structure.

Column (6) of Appendix 1 shows the relationship between the fitted premiums and the break even premiums, expressed as a percentage. It can be seen that, generally, the divergence is only four percentage points, though in a few cases, e.g. Comprehensive cover for an eight-year-old vehicle in Group D driven by a teenager, it is very large. Note also that for Comprehensive vehicle age 0-7, the ratio is less than 100, whilst for the corresponding Non-Comprehensive values, it exceeds 100. This suggests that Comprehensive and Non-Comprehensive require separate tables, as discussed below.

Column (7) of Appendix 1 shows the difference between the fitted premiums and break even premiums, weighted for the standing business. Column 7 reveals that the actual difference by cell in required premium income is in some cases considerable. It would seem that Comprehensive is being undercharged and Non-Comprehensive overcharged. We will return to this. However, overall the total premium income is reasonable, as shown by the following:-

(A) Total Standing Business	90,362
(B) Points Premiums (Fitted using main effects)	£7,898,372
(C) Break Even Premiums	£8,268,308
(D) Difference (B)-(C)	-£369,936
(E) Ratio (B)/(C) (%)	96

A similar comparison may be made with our present premium structure, for example:-

(A) Total Standing Business	90,362
(B) Points Premiums (Fitted using main effects)	£9,012,454
(C) Actual Premiums	£8,812,028
(D) Difference (B)-(C)	£200,426
(E) Ratio (C)/(B) (%)	97

This indicates that a premium increase of about 3% is required to break even.

13.3 PROBLEMS

There are a number of problems which such a method might encounter in practice, such as:-

(1) What are the appropriate weights?

(2) What happens if we later change the assumptions, e.g. as to expenses or the impact of inflation on claim costs?

(3) What if we discover interactions between the main effects at the first stage?

To investigate the effect of changing the weights, an experiment was undertaken. The weights used in the previous example - Column (1) of Appendix 1 - form a series, starting 32, 62, 103... and ending ...151, 167, 858. This series was reversed, i.e. the weights used on the second occasion began 858, 167, 151... and ended 103, 62, 32.

The calculations were repeated with the following results:-

Table 11 - Comparison of Different Portfolios

	Original	Revised
Overall	103	103
Cover		
Comp.	17	19
Non-Comp.	0	0
Policyholder Age		
17-20	30	32
21-24	23	23
25-29	9	9
30-34	0	0
35+	0	2
Car Age		
0-3	19	13
4-7	12	10
8+	0	0
Vehicle Group		
A	0	0
B	5	5
C	10	9
D	26	21

Although some differences are apparent, these are perhaps not so great as might have been expected, given the drastic nature of the change. We conclude, therefore, that the method is fairly robust with respect to changes in weights.

14. COMPARISON OF DIFFERENT SETS OF ASSUMPTIONS

Let us now examine the effect of changes in assumptions. One of the great advantages of this method is the relative ease with which calculations may be performed. Thus, changes in assumptions and repetition of the whole calculation does not present great difficulties. We will investigate the following changes:-

(a) Inflation: We will have two scenarios - High, with inflation at 11% for property damage and 17% for liability settlements, and Low, with inflation at 7% for property damage and 13% for liability settlements.

(b) We will compare data for Third Party Bodily Injury (TPBI) claims based on the period 1972-1976 against data based on the period 1972-1977.

(c) Expenses: We will compare the effect of treating all expenses as fixed, against the effect of distinguishing between fixed and variable expenses.

The results are shown in the following table:-

Table 12 - Summary of Results

Assumptions	(1)	(2)	(3)	(4)	(5)	(6)
Inflation	Low	High	Low	Low	Low	Present
TPBI	72-76	72-76	72-77	72-76	72-76	Points
Expenses	Fixed	Fixed	Fixed	Fixed & Var.	Fixed	System
Weights	Present	Present	Present	Present	Reversed	
Overall	103	106	100	109	103	104
Cover						
Comp.	17	16	19	15	19	18
Non-Comp.	0	0	0	0	0	0
Policyholder Age						
17-20	30	30	33	27	32	29
21-24	23	23	23	20	23	23
25-29	9	9	10	8	9	8
30-34	0	0	0	0	0	0
35+	0	0	1	0	2	0
Car Age						
0-3	19	19	19	16	13	7
4-7	12	12	12	10	10	4
8+	0	0	0	0	0	0
Vehicle Group						
A	0	0	0	0	0	0
B	5	5	5	4	5	5
C	10	10	10	8	9	12
D	26	26	26	23	21	23

The effect of varying the inflation assumption can be seen by comparing Columns (1) and (2). As we might expect, the overall level has risen, but the relativities are virtually unchanged.

Changes in the base year assumptions for TPBI are reflected in Columns (1) and (3). The lower overall level suggests that the 72-77 basis yields a lower overall average. This time there has been a change in the Policyholder Age relativities. The factors affected are Cover and Policyholder Age, which again agrees with our intuitive expectations.

Altering the expense assumptions from Fixed (Column (1)) to Fixed plus Variable (Column (4)), gives results as expected, namely, an increase in overall level combined with a narrowing of the relativities.

Column (5) repeats the results obtained when the weights were reversed. It can be seen that the overall level does not change but the relativities do.

We now turn to the question of what is to be done if associations between main effects are observed at the first stage. It will be recalled that in our worked example, an association was noticed between cover and age of car. One possible solution is to have separate rating structures for the two different types of cover.The results are summarised below:-

Table 13 - Separate Tables for Cover

	Comprehensive	Non-Comprehensive
Overall	118	105
Policyholder Age		
17-20	28	36
21-24	23	25
25-29	10	11
30-34	2	0
35+	0	3
Car Age		
0-3	23	11
4-7	13	9
8+	0	0
Vehicle Group		
A	0	0
B	5	4
C	11	7
D	28	21

These results suggest that there may be valid theoretical grounds for having separate premium structures. However, we must never forget that these decisions are taken on practical as well as theoretical grounds. It is highly probable that despite these results, we will opt for a single premium structure, because of marketing considerations.

A further advantage flows from the ease of computation. We can take another company's rates and perform an analysis upon them. In this way, their rating structure can be compared with our own.

15. MARKETING ASPECTS

The Group will now be in a position to discuss the premium recommendation.

Factors they will have to consider will include inter alia:-

(i) What other companies have done since the last meeting.

(ii) Whether the competitive position will allow all or some of the recommendations to be implemented. For example, in Table 12, compare the actuarial premium recommendations Column (1) with the existing structure Column (6). The outstanding differences occur when comparing the points for Car Age. Hence the actuarial recommendation might indicate an increase in rates for new cars. However, this is unlikely to be acceptable because of the market conditions, which give little weight to this rating factor.

(iii) Whether there are to be rate changes, their timing, and the likely reaction of the market

(iv) Any special marketing campaign proposed, e.g. introducing new factors, offering discounts for older drivers, or introducing NCD protected.

As a final thought, the market place for selling motor insurance is changing, owing to the advent of systems such as CEFAX on television. The time will come (when the next generation of televisions are made) when quotations will be available to the public via the television in their own homes. Insurance companies and brokers will have to consider the implications of this new dimension. This just highlights the dynamic world of marketing.

The issues raised by marketing considerations are very important, and in fact dominate any decision process. However, the subject is too complex to discuss formally in this paper.

16. ANALYSIS OF SURPLUS

Premium rates are simply an attempt to forecast claim and expense experience. For proper control an analysis of surplus should be regularly performed to measure where the forecasts are failing and the effects on the profitability of business.

The following analysis of surplus for motor is based on a note published by Andrew Grant in the GIRO bulletin (1981). Additional papers by Brennan (1968) and Taylor (1974) are also relevant.

In 1977, a projection based on data then available was made for 1978; the main categories of data are listed below. The results actually experienced differ from those projected because the assumptions made in the projection do not wholly agree with the actual experience. We need a method of analysing these differences into their component contributions, which can later be added back to yield the differences between actual and projected results. This problem is similar to an analysis of surplus and the information gained will be useful in fine-tuning the assumptions in the model, as the monetary effect of the differences between the assumptions made in the model and the actual experience is discovered. It would also provide a useful format for analysing these differences for management.

Companies use models of which the simplest use only premiums and loss ratios, while the more complex ones involve assumptions about exposure, average premium rate, average claim cost, claim frequency and expenses. In the model considered, assumptions on the following factors are made for each period and risk group:-

- SB Standing Business (number of vehicles at the end of the year)
- AWP Average Written Premium
- ACF Average Claim Frequency
- ACC Average Claim Cost
- VER Variable Expense Rates (this would include the commission rate)
- FE Fixed Expenses

The Earned Premium EP and the Exposure EX can be easily calculated using the $^1/8$ method for quarterly projections or the $^1/24$ method for monthly projections. The Average Earned Premium AEP can be calculated by dividing the EP by the EX. (At present, tax and investment income are ignored).

The result from the year's Underwriting Profit UP (i.e. excluding any adjustment to Outstanding Claims for prior years) can be expressed in the form

UP =
Earned Premium - Claims - Expenses (inclusive of commission)

The Underwriting Profit UP projected by the model is

$$UP = EX \times AEP - EX \times ACF \times ACC - SB \times VER \times AWP - FE$$

We will denote by ' the figures derived from the actual results. It is possible to derive the AEP' from the Earned Premium EP' and the Exposure EX'. The claims can easily be expressed in the form EX' x ACF' x ACC' as the

Exposure EX' and the Claim Frequency ACF' are known. The Underwriting Profit can, therefore, be expressed as

UP' =
EX' x AEP' - EX' x ACF' x ACC' - SB' x VER' x AWP' - FE'

The differences between the actual experience and that projected are the result of differences in the exposure, claim frequency, level of expenses, inflation rate, etc. We need a method of assessing the numerical contributions from each of the factors. The formulae for a possible analysis are given below.

Effects of

Exposure
(EX' - EX) x (AEP - ACF x ACC) - (SB' - SB) x VER x AWP

Average Premium
EX' x (AEP' - AEP) - SB' x VER x (AWP' - AWP)

Claims Frequency
- (EX' x (ACF' - ACF) x ACC)

Claims Inflation
- (EX' x ACF' x (ACC* - ACC))

Claims Cost
- (EX' x ACF' x (ACC' - ACC*))

Expenses
- (FE' - FE + SB' x VER' x AWP' - SB' x VER x AWP')

(where ACC* is the forecast of the average claims cost using either the known or the latest estimates of claims cost inflation rates).

The formulae above can easily be checked from the expression for Underwriting Profit.

The effect of Claim Frequency and Claim Cost and Claims Inflation can be expressed in several different ways. Two of these are given below:-

	Original	Alternative
Claim Frequency	-EX'x(ACF'-ACF)xACC	-EX'x(ACF'-ACF)xACC'
Claim Cost	-EX'xACF'x(ACC'-ACC*)	-EX'xACFx(ACC'-ACC*)
Claims Inflation	-EX'xACF'x(ACC*-ACC)	-EX'xACFx(ACC*-ACC)

The Original is preferred and is used in in the analysis because the effect of the claim frequency is independent of the actual claim cost. The actual claim cost wil be partly based on the Outstanding Claims estimated and will change over time until the ultimate settlement. Any adjustment to these estimates in the analysis will affect only the result attributed to Claim Cost.

Various other alternative breakdowns of the analysis exist and the choice largely depends on the projection procedure, the data and the factors to be highlighted. As an example, the effect of average claim cost can be divided into its component parts, namely, Accidental Damage, Third Party Bodily Injury, Third Party Property Damage and Miscellaneous. Again the effect of inflation can be highlighted and the model would be similar to that used for making the allowance for inflation in the original projection. If it is required to highlight both factors - that is inflation and types of claim - and adequate data are available, the year's claim experience using the year of account bases can be expressed as

- EX' x (ACF1' x AD' + ACF2' x TPBI' + ACF3' x TPPD' + ACF4' x M')
- (ACF1 x AD + ACF2 x TPBI + ACF3 x TPPD + ACF4 x M)

The analysis would give the following results:-

<u>Claim Frequency</u>
- EX' x (AD x (ACF1' - ACF1) + TPBI x (ACF2' - ACF2) + TPPD x (ACF3' - ACF3) + M x (ACF4' - ACF4))

<u>Claims Inflation</u>
- EX' x (ACF1' x (AD* - AD) + ACF2' x (TPBI* - TPBI) + ACF3' x (TPPD* - TPPD) + ACF4' x (M* - M))

<u>Accidental Damage</u>
- EX' x ACF1' x (AD' - AD*)

<u>Third Party Bodily Injury</u>
- EX' x ACF2' x (TPBI' - TPBI*)

<u>Third Party Property Damage</u>
- EX' x ACF3' x (TPPD' - TPPD*)

<u>Miscellaneous</u>
- EX' x ACF4' x (M' - M*)

In the analysis above, AD, TPBI, TPPD and M represent the average estimated cost per claim, for the Accidental Damage, Third Party Bodily Injury, Third Party Property Damage and Miscellaneous types of claim and their respective claim frequencies are denoted by ACF1, ACF2, ACF3 and ACF4. The * denotes estimates using the updated estimates of the claims inflation rates.

Further, since ACC' will vary from year to year until utlimate settlement, it may be useful to extend the analysis to measure what effect changes in ACC' have on the overall surplus. The extension is straightforwad but is not carried out.

Analysis of Surplus - Example

The following example is taken from the forecasts of part of the Private Car portfolio of a medium sized UK insurance company. The forecasts are made on a year of accident basis, this example being 1978. We give four sets of forecasts for 1978, corresponding to a projection at the end of each quarter in 1977.

The actual results for claim costs included the latest case estimates and therefore this analysis would be subject to some adjustments as the claim payments run off.

(Figures in '000s)	1st Forecast	2nd Forecast	3rd Forecast	4th Forecast	Results
Standing Business	101.4	88.4	87.4	85.4	78.3
Exposure	97.5	87.8	87.8	85.8	80.8
Written Premium	6244.	5702.	6114.	6339.	5794.
Earned Premium	6061.	5684.	5902.	6035.	5600.
No. of Claims	13.1	12.6	12.4	12.2	11.7
Claims Costs	3993.	3938.	3821.	3712.	3348.
Expenses +	2077.	1987.	2184.	2295.	2342.
Profit (Loss) *	(9)	(241)	(103)	28	(90)

* Profit = Earned Premium - Claims Cost - Expenses
\+ including Variable Expenses (15% of Written Premium)

To calculate the effects for the analysis of surplus we need, in addition, to recalculate the forecasts of claims cost using the latest estimates of claims inflation. These averages are given below:-

	1st Forecast	2nd Forecast	3rd Forecast	4th Forecast	Results
Forecasted Average	304.81	312.54	308.15	304.26	286.15
Adjusted Forecast	289.85	300.13	304.13	302.84	

The effect of each factor is then:

(Figures in '000s)	1st Forecast	2nd Forecast	3rd Forecast	4th Forecast
Exposure	-141	-41	-71	-56
Premiums	431	258	121	-81
Claim Frequency	-257	-33	-89	-64
Claims Inflation	175	145	47	17
Claims Costs	43	164	210	195
Expenses	-333	-341	-206	-129

The initial underestimation of the average premium is partly due to the premium increases during 1977 and 1978 which were not allowed for in the earlier forecasts. It would appear to be relatively straightforward to extend the ideas presented in this paper to separate off the effects of premium increases and therefore examine more closely the actual method of projecting premiums.

The effect of claims cost unexpectedly increases between the first and last forecasts, highlighting either a need to examine the methods used to project claims cost, or possibly an unexpected change in the claims experience.

The relatively large effect of expenses is due to two problems. Firstly, there is a slight difference between forecasts and results in the basis for allocating fixed expenses between classes; also, the inflation rates used to project expenses during 1978 could be out of line with those actually experienced. An obvious extension of these methods would be to separate the effect of inflation from effect of expenses in a similar manner to that employed for claims costs.

CONCLUSION

This paper has argued for a detailed breakdown of motor insurance data. This allows for an indepth analysis of claim experience which in turn requires statistical modelling, which assists in understanding the underlying structure. In addition, it is suggested that premium rates can be structured to take into account explicitly the underlying assumptions involved in rating. This will help actuaries concerned with motor rating to appreciate the problems and advise the Underwriter from a position of knowledge.

ACKNOWLEDGEMENTS:

I would like to give credit to an Insurance Company, for allowing the data to be produced. Thank Mick Brockman for his efforts in getting the computer to perform the various calculations and Russell Devitt and Michael Wietzman for their help in preparing the paper.
In addition, I would like to thank numerous colleagues for their instructive criticism, which have helped generate most of the ideas in this paper, in particular Graham Ross and Graham Masters. Finally, a word of congratulations to Jan Stow for reading my writing and typing the paper accurately and quickly.

REFERENCES

Albrecht, P. (1982) — Parametric Multiple Regression Risk Models : Theory and Statistical Analysis 16th ASTIN Colloquium Liege

Baxter, L.A./Coutts, S.M. and Ross, G.A.F. (1980) — Applications of Linear Models in Motot Insurance 21st ICA Vol.2 'Test Hypotheses by Statistical Investigations'

Brennan, A.R. (1968) — A note on the Theory of Analysis of Surplus TIAANZ p.134

Chang, L. and Fairley, W. (1980) — An estimation Model for Multivariate Insurance Rate Classifications : Journal of Risk and Insurance Vol.47

Dawid, A.P. (1980) — Statistical Models for Smoothing and Predictions. (Talk given at a Post Qualification Seminar on the Application of Linear Models to the Analysis of Motor Insurance Data)

Grant, A. (1981) — Analysis of Surplus in Motor Insurance GIRO Bulletin December

Hallin, M. and Ingenbleek, J.F. (1981) — Etude Statistique des facteurs influencant le risque automobile 15th ASTIN Colloquium Loen

Johnson, P.D. and Hey, G.B. (1971) — Statistical Studies in Motor Insurance JIA Vol.97 p.199

Netherlands Group Report (1982) (Chairman Wit, G.W.) — New Motor Rating Structure in the Netherlands (Publication of the ASTIN-GROEP Nederland)

Pitkanen, P. (1974) — Tariff Theory - ASTIN. Colloquium in Turku

Ruston, I.L. (1977) — Expenses in General Insurance Student Society Paper

Taylor, G.C. (1974) — Symmetry Between Components of Analysis of Surplus TIAANZ p.374

Ziai, Y. (1979) Statistical Models of Claim Amount Distributions in General Insurance
Phd Thesis at The City University, London

APPENDIX 1

PREMIUM ANALYSIS SHEET

PRIVATE CAR PREM. ANALYSIS TPBI 72-76 - MAIN EFFECTS

COVER	VEH. AGE	VEH. GRP.	P/H AGE	STANDING BUSINESS (1)	POINTS (2)	FITTED PREMIUM (3)	BREAK EVEN PREMIUM (4)	DIFF. (5) (3)-(4)	RATIO (6) (3)-(4)	SB* DIFF. (7) (2)*(5)
COMP	0-3	01	17-20	32	169	224.33	239.77	-15.44	94	-494.13
			21-24	62	162	178.34	195.06	-17.52	91	-1086.25
			25-29	103	148	114.86	127.60	-12.74	90	-1312.26
			30-34	152	139	85.34	98.41	-13.07	87	-1986.77
			35+	1446	139	85.83	93.85	-8.01	91	-11588.70
		02	17-20	52	174	260.64	274.10	-13.46	95	-699.80
			21-24	146	167	207.21	226.13	-18.92	92	-2762.21
			25-29	287	153	133.45	148.51	-15.06	90	-4321.96
			30-34	472	144	99.16	115.17	-16.02	86	-7559.46
			35+	5063	144	99.73	108.17	-8.44	92	-42741.08
		03+04	17-20	60	179	306.72	318.06	-11.34	96	-680.34
			21-24	188	172	243.84	263.98	-20.14	92	-3785.56
			25-29	444	158	157.04	176.64	-19.60	89	-8702.82
			30-34	820	149	116.69	139.32	-22.63	84	-18559.92
			35+	8142	149	117.36	130.48	-13.12	90	-106860.34
		05+	17-20	21	195	517.16	478.57	38.59	108	810.32
			21-24	153	188	411.13	409.69	1.44	100	220.23
			25-29	291	174	264.79	287.65	-22.87	92	-6654.55
			30-34	630	165	196.74	234.11	-37.36	84	-23538.35
			35+	4498	165	197.87	218.70	-20.83	90	-93672.76

APPENDIX 1

COVER	VEH. AGE	VEH. GRP.	P/H AGE	STANDING BUSINESS (1)	POINTS (2)	FITTED PREMIUM (3)	BREAK EVEN PREMIUM (4)	DIFF. (5) (3)-(4)	RATIO (6) (3)-(4)	SB* DIFF. (7) (2)*(5)
COMP (cont.)	4-7	01	17-20	35	162	179.65	180.96	-1.31	99	-45.98
			21-24	93	155	142.82	147.30	-4.48	97	-416.96
			25-29	126	141	91.98	92.91	-.93	99	-117.00
			30-34	181	132	68.34	70.29	-1.94	97	-351.34
			35+	2081	132	68.74	67.01	1.72	1.03	3588.71
		02	17-20	94	167	208.73	211.33	-2.60	99	-244.19
			21-24	229	160	165.94	171.14	-5.20	97	-1191.07
			25-29	388	146	106.87	108.55	-1.68	98	-652.15
			30-34	744	137	79.41	82.42	-3.02	96	-2244.03
			35+	7911	137	79.86	78.27	1.59	102	12571.22
		03+04	17-20	65	172	245.63	244.76	.87	100	56.69
			21-24	199	165	195.27	201.16	-5.89	97	-1171.37
			25-29	437	151	125.76	130.96	-5.20	96	-2270.47
			30-34	801	142	93.45	101.41	-7.96	92	-6376.06
			35+	6877	142	93.98	94.98	-1.00	99	-6851.97
		05+	17-20	21	188	414.15	375.48	38.68	110	812.18
			21-24	113	181	329.25	320.20	9.05	103	1022.22
			25-29	215	167	212.05	219.23	-7.18	97	-1544.60
			30-34	427	158	157.56	176.52	-18.96	89	-8095.25
			35+	2919	159	158.46	165.21	-6.74	96	-19005.60

APPENDIX 1

COVER	VEH. AGE	VEH. GRP.	P/H AGE	STANDING BUSINESS (1)	POINTS (2)	FITTED PREMIUM (3)	BREAK EVEN PREMIUM (4)	DIFF. (5) (3)-(4)	RATIO (6) (3)-(4)	SB* DIFF. (7) (2)*(5)
COMP	8+	01	17-20	17	150	122.89	114.91	7.98	107	135.67
(cont.)			21-24	39	143	97.70	95.90	1.79	102	69.92
		A	25-29	76	130	62.92	58.92	4.00	107	303.87
			30-34	170	120	46.75	43.27	3.48	108	591.62
			35+	2826	120	47.02	42.20	4.82	111	13618.48
		B	17-20	43	155	142.79	142.39	.40	100	17.16
			21-24	108	148	113.51	115.23	-1.71	99	-185.21
			25-29	219	134	73.11	71.05	2.06	103	450.16
			30-34	487	125	54.32	51.83	2.49	105	1212.97
			35+	6992	125	54.63	50.12	4.52	109	31573.06
		C	17-20	27	160	168.03	164.31	3.72	102	100.31
			21-24	60	153	133.58	134.76	-1.17	99	-70.45
			25-29	158	139	86.03	86.51	-.47	99	-74.85
			30-34	310	130	63.92	65.29	-1.37	98	-423.41
			35+	3346	130	64.29	62.34	1.95	103	6529.18
		D	17-20	3	177	283.31	144.26	139.05	196	417.15
			21-24	19	169	225.23	207.59	17.64	108	335.17
			25-29	61	156	145.06	151.06	-6.00	96	-366.20
			30-34	126	146	107.78	120.93	-13.15	89	-1657.32
			35+	966	147	108.40	114.98	-6.58	94	-6357.65

APPENDIX 1

COVER	VEH. AGE	VEH. GRP.	P/H AGE	STANDING BUSINESS (1)	POINTS (2)	FITTED PREMIUM (3)	BREAK EVEN PREMIUM (4)	DIFF. (5) (3)-(4)	RATIO (6) (3)-(4)	SB* DIFF. (7) (2)*(5)
NON-COMP	0-3	A	17-20	6	152	128.52	110.39	18.14	116	108.82
			21-24	6	145	102.17	73.65	28.53	139	171.15
			25-29	6	131	65.80	28.63	37.17	230	223.05
			30-34	12	122	48.89	28.46	20.43	172	245.17
			35+	52	122	49.17	44.99	4.10	109	217.55
		B	17-20	15	157	149.33	132.62	16.71	113	250.65
			21-24	29	149	118.71	100.67	18.04	118	523.24
			25-29	30	136	76.46	62.24	14.21	123	426.41
			30-34	23	126	56.81	41.68	15.13	136	347.94
			35+	181	126	57.14	52.22	4.92	109	889.68
		C	17-20	17.	162	175.73	151.57	24.16	116	410.67
			21-24	34	154	139.70	111.7	27.83	125	946.16
			25-29	37	141	89.97	71.52	18.46	126	682.92
			30-34	42	131	66.85	50.27	16.58	133	696.35
			35+	291	132	67.24	57.95	9.29	116	2702.05
		D	17-20	12	178	296.29	214.86	81.43	138	977.12
			21-24	25	171	235.55	170.06	65.49	139	1637.22
			25-29	23	157	151.70	108.86	42.84	139	985.37
			30-34	22	148	112.72	80.36	32.36	140	711.95
			35+	121	148	113.37	91.79	21.58	124	2611.02

APPENDIX 1

COVER	VEH. AGE	VEH. GRP.	P/H AGE	STANDING BUSINESS (1)	POINTS (2)	FITTED PREMIUM (3)	BREAK EVEN PREMIUM (4)	DIFF. (5) (3)-(4)	RATIO (6) (3)-(4)	SB* DIFF. (7) (2)*(5)
NON-COMP (cont.)	4-7	A	17-20	85	145	102.9	119.51	-16.59	86	-1410.04
			21-24	78	138	81.82	83.93	-2.11	97	-164.50
			25-29	77	124	52.70	53.69	-.99	98	-76.24
			30-34	83	115	32.16	38.28	.87	102	72.60
			35+	482	115	39.38	42.25	-2.87	93	-1385.02
		B	17-20	147	150	119.59	135.28	-15.69	88	-2307.08
			21-24	194	142	95.07	94.90	.17	100	33.08
			25-29	211	129	61.23	61.52	-.29	100	-60.88
			30-34	293	119	45.49	43.44	2.05	105	601.12
			35+	1723	120	45.76	47.28	-1.52	97	-2621.38
		C	17-20	140	155	140.73	147.00	-6.28	96	-878.93
			21-24	220	147	111.88	103.96	7.91	108	1740.51
			25-29	253	134	72.05	67.95	4.11	106	1038.70
			30-34	300	124	53.54	48.20	5.34	111	1601.84
			35+	1636	125	53.84	52.10	1.66	103	2715.77
		D	17-20	67	171	237.28	213.99	23.29	111	1560.23
			21-24	122	164	188.63	157.69	30.94	120	3774.78
			25-29	130	150	121.49	106.12	15.36	114	1997.20
			30-34	174	141	90.27	78.23	12.04	115	2094.84
			35+	695	141	90.79	82.41	8.37	110	5820.59

APPENDIX 1

COVER	VEH. AGE	VEH. GRP.	P/H AGE	STANDING BUSINESS (1)	POINTS (2)	FITTED PREMIUM (3)	BREAK EVEN PREMIUM (4)	DIFF. (5) (3)-(4)	RATIO (6) (3)-(4)	SB* DIFF. (7) (2)*(5)
NON-COMP (cont.)	8+	A	17-20	346	133	70.41	93.87	-23.46	75	-8117.92
			21-24	303	126	55.97	64.69	-8.72	87	-2642.16
			25-29	392	112	36.05	41.81	-5.76	86	-2256.72
			30-34	504	103	26.79	29.02	-2.23	92	-1125.03
			35+	3217	103	26.94	31.23	-4.29	86	-13815.38
		B	17-20	528	138	81.80	104.49	-22.69	78	-11979.27
			21-24	609	131	65.03	72.08	-7.05	90	-4291.87
			25-29	774	117	41.88	45.99	-4.11	91	-3178.10
			30-34	989	107	31.12	31.74	-.62	98	-613.66
			35+	6036	108	31.30	35.05	-3.75	89	-22608.46
		C	17-20	209	143	96.27	113.85	-17.58	85	-3674.95
			21-24	284	136	76.53	79.14	-2.61	97	-742.18
			25-29	346	122	49.29	50.75	-1.46	97	-505.74
			30-34	558	113	36.62	35.57	1.06	103	589.09
			35+	2835	113	36.83	38.43	-1.59	96	-4517.95
		D	17-20	56	159	162.31	166.29	-3.97	98	-222.43
			21-24	136	152	129.04	120.32	8.72	107	1105.91
			25-29	151	138	83.11	79.74	3.37	104	508.16
			30-34	167	129	61.75	58.38	3.37	106	562.27
			35+	858	129	62.10	62.24	-.13	100	-113.94

TOTALS		
	(A) STANDING BUSINESS	90362
	(B) POINTS PREMIUMS	7898372.00
	(C) BREAK EVEN PREMIUMS	8268307.50
	(D) DIFFERENCE (B)-(C)	-369935.50
	(E) RATIO (B)/(C)	96

APPENDIX 2

POINTS CONVERSION TABLE

Points	Premium	Points	Premium	Points	Premium
	£		£		£
0	5.90	29	14.95	58	37.80
		30	15.45	59	39.00
1	6.10			60	40.25
2	6.30	31	15.95		
3	6.50	32	16.45	61	41.60
4	6.70	33	17.00	62	42.95
5	6.95	34	17.55	63	44.35
		35	18.10	64	45.75
6	7.15			65	47.25
7	7.40	36	18.70		
8	7.65	37	19.30	66	48.80
9	7.90	38	19.95	67	50.40
10	8.15	39	20.55	68	52.00
		40	21.25	69	53.70
11	8.40			70	55.45
12	8.70	41	21.95		
13	8.95	42	22.65	71	57.25
14	9.25	43	23.40	72	59.10
15	9.55	44	24.15	73	61.05
		45	24.95	74	63.00
16	9.85			75	65.05
17	10.20	46	25.75		
18	10.50	47	26.55	76	67.20
19	10.85	48	27.45	77	69.35
20	11.20	49	28.35	78	71.60
		50	29.25	79	73.95
21	11.55			80	76.35
22	11.95	51	30.20		
23	12.35	52	31.20	81	78.85
24	12.75	53	32.20	82	81.40
25	13.15	54	33.25	83	84.05
		55	34.30	84	86.75
26	13.60			85	89.60
27	14.00	56	35.45		
28	14.45	57	36.60	86	92.50

APPENDIX 3

ORTHOGONAL WEIGHTED LEAST SQUARES ANALYSIS

The general problem is to minimise the following expression:-

$$\sum_{ijk} \frac{1}{V_{ijk}} \left[Z_{ijk} - (\hat{S} + \hat{R}_i + \hat{U}_j + \hat{V}_k + (\hat{In})) \right]^2 \qquad \ldots (A.1)$$

where Z_{ijkl} is the dependent variable for cell ijkl

V_{ijkl} is the weights for cell ijkl
S is the overall mean
R_i is the effect of R at level i
U_j is the effect of U at level j
V_k is the effect of V at level k
(In) is the effect of the interaction terms
$\wedge$ is the least squared estimate

e.g. Johnson and Hey (1971) used $\frac{1}{n}$ as the weight.

It has been suggested that the minimisation could be simplified by assuming that the weight $\frac{1}{V_{ijk}}$ factorises, i.e. if V_{ijk} satisfies

$$\frac{1}{V_{ijk}} \propto r_i\, u_j\, v_k \text{ (for all ijk)} \qquad \ldots (A.2)$$

and various concomitant constraints. The design becomes orthogonal, enabling the following estimators to be derived:

$$\hat{S} = \sum_i \sum_j \sum_k \left(\frac{r_i}{r.} \frac{u_j}{u.} \frac{v_k}{v.} \right) Z_{ijk}$$

$$\hat{R}_i = \sum_j \sum_k \frac{u_j}{u.} \frac{v_k}{v.} (Z_{ijk}) - \hat{S}$$

$$(RU)_{ij} = \sum_k \frac{v_k}{v.} (Z_{ijk}) - \hat{S} - \hat{R}_i - \hat{U}_j$$

and similar equations for $(RV)_{ik}$ and $(UV)_{jk}$

where $r. = \sum_i r_i$, $u. = \sum_j u_j$ and $v. = \sum_k v_k$

and the constraints are

$$\sum_i r_i R_i = \sum_j u_j U_j = \sum_k v_k V_k = 0$$

$$\sum_i r_i (RU)_{ij} = \sum_j u_j (RU)_{ij} = 0 \text{ etc.}$$

One method of justifying the factorisation, is by an analysis of $-\log V_{ijk}$. Under the assumption of perfect factorisation

$$-\log V_{ijk} = s' + r'_i + u'_j + v'_k$$

where s' is an arbitary constant and r'_i, u'_j and v'_k correspond to r_i, u_j and v_k

in equation (A.2). Thus if the interactions in an analysis of variance of $-\log V_{ijk}$ are negligible, there is evidence supporting the factorisation and hence the main effects can be used to calculate the weights.

In practice the weights do not always factorise perfectly, but experience shows that the model is robust with respect to the weights chosen.

THE MEAN SQUARE ERROR OF A RANDOMLY DISCOUNTED SEQUENCE OF UNCERTAIN PAYMENTS

Piet de Jong

University of British Columbia, Vancouver, B.C., Canada

Expressions are developed for the mean square deviation between a discounted sequence of uncertain payments and the estimated present value. These formulas incorporate random variation from four sources: (1) from the fact that expected values of individual payments are estimated, (2) from the fact that future discount factors are estimated, (3) random variation in the events that give rise to future payments, and (4) random variation in future discount rates. The results extend and encompass existing expressions based on the assumed constancy of one or more of the random vectors.

INTRODUCTION

In many situations contracts are drawn up which provide for one or more future payments conditional on the occurrence of uncertain events. Premiums for such contracts are based on individually discounted expected payments calculated under some assumed regime of probabilities and future interest rates. Loadings on premiums partially reflect the risk associated with the contract which is often measured through the mean square deviation of the actual from expected. For this reason and as a basis for calculating appropriate reserves, it is important to be able to measure the mean square deviation.

Previous work in this area include the work of Pollard and Pollard (1968), Boyle (1973), Klugman (1978), and the two papers by Bellhouse and Panjer (1977, 1981). In one sense our treatment is more particular then some of these references, in that for concreteness attention is restricted to a sequence of payments

F. de Vylder et al. (eds.), Premium Calculation in Insurance, 449–459

paid at the end of each of n years provided a specified condition holds. Despite this, the arguments can be readily translated to more complicated situations. In a more important sense our treatment is more general than this previous work in that variances or mean square errors are computed without assuming the constancy of one or more of the underlying random variables. Thus, for example, Klugman (1978) considers the standard deviation of an annuity only allowing for variation associated with estimating survival probabilities. By the same token Pollard and Pollard (1968) consider the moments of some actuarial functions considering mortality experience as the only source of random fluctation. Bellhouse and Panjer (1977, 1981) instead consider the role of interest rate fluctuations and, for example, ignore estimation variation. In contrast to these contributions, this paper simultaneously considers all four sources of deviation of the estimated discounted value from the actual present value.

THE MEAN SQUARE ERROR

Consider a contract where a payment is made at the end of each of n years $i=1,2,\ldots,n$ provided a specified condition holds. The actual present value of such a sequence of payments is the random quantity

$$\sum_{i=1}^{n} y_i z_i = y'z \quad , \tag{1}$$

where ' denotes transposition, $y'=(y_1, y_2,\ldots, y_n)$ is a random vector of payments, and $z'=(z_1, z_2,\ldots, z_n)$ a random vector of unknown future discount factors. The expected value of (1), computed under the hypothesis that y and z are uncorrelated is

$$E[y'z] = E[y']E[z] = p'v \quad , \tag{2}$$

where p is a vector of expected payments and v a vector of expected future discount factors. As an estimate of (2) we use $\hat{p}'\hat{v}$ where $\hat{p}$ and $\hat{v}$ are estimates of the expected payment values and future expected discount factors.

Since $\hat{p}'\hat{v}$ is a forecast of y'z it is important to consider the mean square error

$$E[(y'z - \hat{p}'\hat{v})^2] \quad , \tag{3}$$

where the expectation is taken with respect to the joint distribution of all the random variables y, z, $\hat{p}$, and $\hat{v}$. The mean square error (3) relates to the expected deviation between the realized value and the estimated expected value and hence, in a loose sense, the 'risk' associated with a single contract. Sellers of such a contract may attempt to insulate themselves against this risk by diversifying into a variety of hopefully independent contracts. The appropriate degree of diversification, and possibly reinsurance, relative to a seller's financial status can only be gauged in the light of measures such as the mean square error (3).

To expand (3) assume y, z, $\hat{p}$, and $\hat{v}$ are uncorrelated. At this stage we are not concerned with the reasonableness of this assumption, only with technical consequences. Hence (3) becomes

$$\begin{aligned} &E[\{(y'z-p'v)-(\hat{p}'\hat{v}-p'v)\}^2] \\ &\quad = E[(y'z-p'v)^2] + E[(\hat{p}'\hat{v}-p'v)^2] \\ &\quad = Cov(y'z) + E[\{(\hat{p}'\hat{v}-\bar{p}'\bar{v})-(p'v-\bar{p}'\bar{v})\}^2] \\ &\quad = Cov(y'z) + Cov(\hat{p}'\hat{v}) + (p'v\bar{p}'\bar{v})^2 \quad , \end{aligned} \tag{4}$$

where $\bar{p}=E[\hat{p}]$ and $\bar{v}=E[\hat{v}]$ and Cov denotes covariance. The first term on the right equals

$$\begin{aligned} Cov(y'z) &= E[y'zz'y] - \{E[y'z]\}^2 \\ &= tr\{E[yy']E[zz']\} - \{p'v\}^2 \\ &= tr\{[Cov(y)+pp'][Cov(z)+vv']\} - \{p'v\}^2 \\ &= tr\{Cov(y)Cov(z)\} + p'Cov(z)p + v'Cov(y)v \quad , \end{aligned} \tag{5}$$

where tr denotes the trace operation on a matrix. The second term in (4) is similarly expanded. Substituting these expressions back into (4) yields

$$\begin{aligned} E[(y'z-\hat{p}'\hat{v})^2] = &\; tr\{Cov(y)Cov(z)\} + tr\{Cov(\hat{p})Cov(\hat{v})\} \\ &+ p'Cov(z)p + \bar{p}'Cov(\hat{v})\bar{p} \\ &+ v'Cov(y)v + \bar{v}'Cov(\hat{p})\bar{v} \\ &+ (p'v-\bar{p}'\bar{v})^2 \; . \end{aligned} \tag{6}$$

Equation (6) is the basic mean square error expression. The final term relates to the possible bias associated with estimating p and v. The four middle terms are familiar individual contributions arising from the individual randomness associated with each of the four random vectors. The trace expressions are 'interaction' terms which arise despite the assumed zero correlations, although note there is no 'interaction' between the estimators $\hat{p}$ or $\hat{v}$ and the underlying random vectors y or z.

Since $\hat{p}$ and $\hat{v}$ are estimators while y and z are random variables modelling the experience behavior, subsequent discussion is facilitated by writing

(mean square error) = (experience error) + (estimation error),

where

$$(\text{experience error}) = \text{tr}\{\text{Cov}(y)\text{Cov}(z)\} + p'\text{Cov}(z)p + v'\text{Cov}(y)v,$$

and similarly

$$(\text{estimation error}) = \text{tr}\{\text{Cov}(\hat{p})\text{Cov}(\hat{v})\} + \bar{p}'\text{Cov}(\hat{v})\bar{p} + \bar{v}'\text{Cov}(\hat{p})\bar{v} + (p'v-\bar{p}'\bar{v})^2.$$

Further development of (6) requires assumptions regarding the evolution of interest rates, specification of the estimation steps underlying the construction of $\hat{v}$ and $\hat{p}$, and further particularization of payment conditions. Reasonable and practically useful specifications are explored in the next few sections. Before embarking, we return to consider the uncorrelated condition on y, z, $\hat{p}$ and $\hat{v}$, which is a basic assumption underlying the derivation of (6). In the usual contexts the payments y will be unrelated to the other random variables. Similarly it is reasonable to assume that the estimation variation embodied in $\hat{p}$ is unrelated to both the course of future interest rates embodied in z and the estimation variation associated with $\hat{v}$. This leaves z and $\hat{v}$ as the only remaining combination where there may be some basis for arguing dependence. Yet even this dependence vanishes under the widely applicable model for interest rates outlined in the next section.

FUTURE INTEREST RATES

Consider a random walk type model for future interest rates. The motivation is to gain further insight into (6). Besides the empirical support for such a model, it is a first step beyond strictly deterministic specifications, and easily reduces to this situation. It would be actuarially interesting and important to study more complicated interest rate models, for example random walks with drifts, or perhaps stationary time series models considered by Bellhouse and Panjer (1977, 1981) , but this is outside the scope of the present article.

Let d_j for j=1,2,...,n be the future 1 year forces of interest so that

$$z_j = \exp\{-(d_1 + d_2 + \dots + d_j)\} \quad .$$

Assume that the d_j follow a random walk model

$$d_j = d_{j-1} + e_j \quad , \qquad j=1,2,\dots,n$$

where the e_j are independent Normal random variables with mean zero and variance σ^2. Accordingly

$$z_j = \mu^j \prod_{i=1}^{j} \exp\{-(j-i+1)e_i\} \quad ,$$

where $\mu = \exp(-d_0)$ is the one year discount factor corresponding to the known current one year force of interest. From the usual results regarding the moment generating function of Normal random variables

$$v_j = E[z_j] = \mu^j \exp\{\sigma^2 j(j+1)(2j+1)/12\} \ , \ j=1,2,\ldots,n \qquad (7)$$

and after some manipulation for $j,k=1,2,\ldots,n$ with $j \geq k$

$$E[z_j z_k] = v_j u_{kj} \quad ,$$

where

$$u_{kj} = \mu^k \exp\{\sigma^2 k(k+1)(2j+1)/4\} \quad . \qquad j \geq k$$

Substituting these results into the experience error portion of (6) leads to

$$(\text{experience error}) = \sum_{j=1}^{n} v_j \{u_{jj} E[y_j^2] + 2 \sum_{k<j} u_{kj} E[y_j y_k]\} - (p'v)^2. \qquad (8)$$

ESTIMATION OF FUTURE INTEREST RATES

We now develop the estimation error term under the random walk model for the force of interest, combined with a reasonable approach to estimating the possibly unknown parameter σ^2. This approach is to use previous observed e_j's and form their sample variance. Under the Normality hypothesis the resulting estimate of σ^2 is Chi-squared with m degrees of freedom, where m is the number of observations. The variance of a Chi-squared random variable is twice the degrees of freedom and hence $\text{Cov}(\hat{\sigma}^2) = 2\sigma^4/m$. By substituting $\hat{\sigma}^2$ for σ^2 in (7) yields an estimate $\hat{v}$ of v. The covariance matrix of this estimator is approximately derived by using the first order Taylor approximation to the mapping (7) from σ^2 to v. Accordingly

$$\text{Cov}(\hat{v}) \doteq (2\sigma^4/m) \ ww' \quad , \qquad (9)$$

where w is a vector with jth entry $v_j\{j(j+1)(2j+1)/12\}$. Substituting this result into the estimation error portion of (6) leads to (estimation error) $\doteq$

$$(2\sigma^4/m)\{w'\mathrm{Cov}(p)w+(p'w)^2\}+v'\mathrm{Cov}(p)v+\{(p-\bar{p})'v\}^2, \qquad (10)$$

where the bias inherent in $\hat{v}$ is ignored. Both the bias and the error in the appromixation (9) are negligible if $\hat{\sigma}$ is close (in mean square) to σ i.e., provided m is large.

A direct approach to the approximation (10) is to use Bayesian arguments. Thus the estimate $\hat{\sigma}$ may be selected on the basis of a priori considerations. The value of m, or perhaps more cogently 1-1/m, then reflects the confidence in the estimate. A limiting case occurs as $m\to\infty$, i.e. as we become more certain of the appropriate value of σ. In the limit the first term in (10) vanishes. An even more particular situation occurs if we know $\sigma=0$. Equation (6) then yields

$$v'\mathrm{Cov}(\hat{y})v + v'\mathrm{Cov}(\hat{p})v + \{(\bar{p}-p)'v\}^2 \quad , \qquad (11)$$

which is the exact mean square error under a deterministic interest rate regime.

APPLICATION TO ANNUITIES CERTAIN

Consider the situation where n certain payments of amount 1 are made. In this case the probabilities $\hat{p}$ form a vector 1 of ones, $\mathrm{Cov}(\hat{p})=0$ and $\mathrm{Cov}(y)=0$. Substituting these values into (6) leads to

$$1'\mathrm{Cov}(z)1 + 1'\mathrm{Cov}(\hat{v})1 + \{1'(v-\bar{v})\}^2$$

$$\doteq 1'\mathrm{Cov}(z)1 + (2\sigma^4/m)(1'w)$$

$$= \sum_{j=1}^{n} v_j\{u_{jj} + 2\sum_{k<j} u_{kj}\} - (1'v)^2$$

$$+ \{\sigma^4/(6m)\} \sum_{j=1}^{n} v_j j(j+1)(2j+1) \quad , \qquad (12)$$

where the approximation follows upon assuming the random walk model of the previous sections. The first two terms form the exact experience error while the last term is the approximate estimation error. More general expressions for varying annuities certain may also be easily derived using the above approach.

To illustrate use of the formula, consider Table 1. The table presents the square root of both the experience error and estimation error as a percentage of the annuity value. As indicated by (12) the experience error does not vary with m. Moreover as $m \to \infty$ the estimation error goes to zero. The table indicates that the experience error tends to dominate and that both errors are crucially dependent on the term n of the contract with the square root of the mean square error being approximately equal to the annuity value when n=25, d_0=0.07 and σ=0.02.

Table 1: Square roots of experience and estimation errors as a percentage of annuity certain value*

			n	
d_0	σ	m	5	25
0.03	0.01		<u>3.8</u>	<u>29.3</u>
		2	0.1	6.6
		10	0.0	3.0
	0.02		<u>7.5</u>	<u>92.7</u>
		2	0.4	35.4
		10	0.2	15.8
0.07	0.01		<u>3.6</u>	<u>22.6</u>
		2	0.1	4.7
		10	0.0	2.1
	0.02		<u>7.3</u>	<u>69.2</u>
		2	0.4	25.6
		10	0.2	11.4

* Underlined numbers relate to experience error, others to estimation error.

APPLICATION TO LIFE ANNUITIES

If the random vector y is such that y_i=0 implies y_j=0 for j>i, then the vector of payments can be interpreted y as life contingent annuity i.e. an annuity contingent on a single or group of lives. Consider the case of a unit sequence of payments, although similar arguments may be used where payments vary in a known manner. Under this restriction p becomes a vector of survival probabilities and

$$E[y_j] = p_j \;, \qquad j=1,2,\ldots,n$$

$$E[y_j y_k] = p_j \;, \qquad j \geq k$$

and hence under the assumed random walk model for the force of interest

$$\text{(experience error)} = \sum_{j=1}^{n} v_j p_j \{u_{jj} + 2 \sum_{k<j} u_{kj}\} - (p'v)^2. \quad (13)$$

When $\sigma=0$ then $v_j=\mu^j$, $u_{kj}=\mu^k$ for $k\leq j$, and (13) reduces to

$$\sum_{j=1}^{n} \mu^j p_j \{\mu^j + 2 \sum_{k<j} \mu^k\} - \{\sum_{j=1}^{n} \mu^j p_j\}^2 = \{\frac{2\mu}{1-\mu}\} \sum_{j=1}^{n} \mu^j p_j - \sum_{j=1}^{n} \mu^{2j} p_j - \{\sum_{j=1}^{n} \mu^j p_j\}^2$$

which is the usual expression for the variance of a life contingent annuity (see e.g., Neill (1977, p. 390)).

The estimation error in this situation, and under the random walk model, takes the form (10). Where the experience underlying the estimate $\hat{p}$ is large (as in most classical situations) we may assume $\hat{p}=p$ and $\text{Cov}(\hat{p})=0$, in which case

$$\text{(estimation error)} \doteq \{(2\sigma^4)/m\} (p'w)^2 \quad (14)$$

In other contexts p will be estimated from a relatively small data base and perhaps via the usual steps of first constructing raw age specific mortality rates, secondly smoothing these rates, and thirdly transforming the smoothed rates to survival probabilities. Although the mapping which carries the raw mortality rates into the final survival probability estimates is nonlinear (on account of the third step), the approximate covariance matrix of the resulting $\hat{p}$ may be derived from the variances associated with the raw mortality rates, by approximating the nonlinear mapping by a linear one. The appropriate argument is similar to that used to derive the approximation (9). Variances or mean square errors which incorporate variation due to mortality estimation error (but not any other sources of variation) are considered in Klugman (1978).

Table 2 illustrates the size of experience and estimation errors for a life contingent contract. Estimates $\hat{p}_j$ of survival probabilities are taken fom the results of de Jong and Boyle (1983), and are based on the mortality experience of retired male lifes under an actual pension plan. The estimates $\hat{p}$, relating to ages 58 and older, reflect a mortality experience somewhat higher than usual. Experience errors again dominate the estimation errors at all considered combinations. The percentage experience errors are, as expected, higher than the corresponding annuity certain values on account of the stochastic nature of the mortality experience. As d_0, σ, m and n increase, the difference becomes marginal. Moving on to the estimation error, the first error for given m relates to a $Cov(\hat{p})$ matrix corresponding to de Jong and Boyle's (1983) case where the given mortality experience

Table 2: Square roots of experience and estimation errors as a percentage of life contingent annuity value*

d_0	σ	m	n = 5		n = 25	
0.03	0.01		20.6		50.6	
		2	1.5	0.1	8.1	3.8
		10	1.5	0.0	7.4	1.7
	0.02		21.7		95.6	
		2	1.5	0.4	22.5	20.1
		10	1.5	0.2	12.8	9.1
0.07	0.01		20.2		41.3	
		2	1.5	0.1	6.1	2.6
		10	1.5	0.0	5.6	1.2
	0.02		21.2		69.6	
		2	1.5	0.4	15.6	13.9
		10	1.5	0.2	9.2	6.2

* Underlined numbers relate to experience errors, others to estimation error. The latter are reported for the cases $Cov(\hat{p})\neq 0$ and $Cov(\hat{p})=0$. See text for details.

more or less dictates the mortality estimates. The second experience error for given m relates to $Cov(\hat{p})=0$ and hence these cases are directly comparable to Table 1 estimation errors. Finally note that m is not crucially important especially when n=5.

APPLICATION TO LIFE ASSURANCE

If a single unit payment is made at the end of the year upon the expiration of a life or status of lives then p is a vector of 'expiration' probabilities. In this case

$$E[y_j] = p_j \quad , \qquad j=1,2,\ldots,n$$

$$E[y_j y_k] = \delta_{jk} p_j \quad , \qquad j \geq k$$

where δ_{jk} is the Kronecker delta. Substituting these expressions into (8) yields

$$(\text{experience error}) = \sum_{j=1}^{n} p_j v_j u_{jj} - (p'v)^2 \quad .$$

The estimation error is again (10) or (14) with the remarks regarding the derivation of $\mathrm{Cov}(\hat{p})$ made in the previous section being equally applicable here.

Table 3 illustrates relevant figures for the life assurance case. The layout is similar to Table 2. Both estimation and experience errors are large, but this reflects the fact that the present values of the contracts are small. When n=5, there is hardly any variation across the different d_0, σ and m. For n=25 such variation is apparent. Overall, experience errors again dominate estimation errors.

Table 3: Square roots of experience and estimation errors as a percentage of life contingent assurance value*

d_0	σ	m	n = 5		n = 25	
0.03	0.01		548.7		80.6	
		2	80.0	0.2	26.3	13.1
		10	80.0	0.1	23.4	5.8
	0.02		550.6		231.9	
		2	80.0	0.5	67.7	61.9
		10	80.0	0.2	36.5	27.7
0.07	0.01		550.9		85.2	
		2	80.8	0.1	26.4	10.9
		10	80.8	0.1	24.5	4.9
	0.02		552.5		194.0	
		2	80.8	0.5	59.7	53.5
		10	80.7	0.2	33.7	23.9

* Underlined numbers relate to experience errors, others to estimation error. The latter are reported for the cases $\mathrm{Cov}(\hat{p}) \neq 0$ and $\mathrm{Cov}(\hat{p}) = 0$. See text for details.

REFERENCES

Bellhouse, D.R. and Panjer, H.H. 1981, "Stochastic Modelling of Interest Rates with Applications to Life Contingencies - Part II", The Journal of Risk and Insurance, Vol. XLVIII, pp. 628-637.

Boyle, P.P. 1973, "Note on the Variance of a Widow's Pension", Journal of the Institute of Actuaries, 99, pp. 103-107.

de Jong, P. and Boyle, P.P. 1983, "Monitoring Mortality: A State-space Approach", Journal of Econometrics, (in press).

Klugman, S. 1978, "The Standard Deviation of the Present Value of Life Annuity as estimated from a Graduated Mortality Table", Actuarial Research Clearing House 2, pp. 91-99.

Neill, A., 1977, Life Contingencies, London, Heinemann.

Panjer, H.H. and Bellhouse, D.R. 1977, "Stochastic Modeling of Interest Rates with Applications to Life Contingencies", The Journal of Risk and Insurance, Vol. 5 pp. 177-188.

Pollard, A.H. and Pollard, J.H. 1968, "A Stochastic Approach to Actuarial Functions", Journal of the Institute of Actuaries, 95, pp. 79-113.

OPERATIONAL TIME: A SHORT AND SIMPLE EXISTENCE PROOF

Volker Mammitzsch

University of Marburg

A new and elementary proof is given for the well-known fact that each counting process with independent increments having continuous mean function may be transformed into a Poisson process by means of the so-called operational time.

In risk theory, the claim number process $(N_t;t\geq o)$ plays an important role, see e.g. [2]. Usually, $(N_t;t\geq o)$ is supposed to be a counting process with independent increments, i.e.

(1) there holds a.s. $N_o = 0$ and
$t \mapsto N_t$ is right-continuous monotonously increasing step function with jumps =1;

(2) for all $o \leq t_1 < s_1 \leq t_2 < s_2 \leq \cdots \leq t_n < s_n$ the differences $N_{s_1} - N_{t_1}, \ldots, N_{s_n} - N_{t_n}$ are independent.

Moreover, a stochastic process $(X_t;t\geq o)$ is said to have stationary increments if

(3) the distribution of $X_{t+h} - X_t$, $h > o$, $t \geq o$ is independent of t.

The claim number process will not always have this property. However, by a change in time scale this property may be attained which also implies that the transformed random variables are Poisson distributed according to the well-known

<u>Theorem:</u> Let $(X_t;t\geq o)$ a counting process with independent in-

F. de Vylder et al. (eds.), Premium Calculation in Insurance, 461–465.

crements and let $m(t):=EX_t$ be a continuous function in t with $m(t)>o$ for $t>o$, tending to infinity as $t\to\infty$. Let denote $t(\tau) := \inf\{s; m(s)>\tau\}$ and define the process $(Y_\tau;\tau\geq o)$ by $Y_\tau := X_{t(\tau)}$, $\tau\geq o$. Then

(a) $(Y_\tau;\tau\geq o)$ is a counting process with independent increments

(b) $(Y_\tau,\tau\geq o)$ has stationary increments

(c) $P\{Y_\tau=k\} = e^{-\tau}\frac{\tau^k}{k!}$, $k=0,1,2,\ldots$.

Remarks: 1. In some sense, $t(\tau)$ is the inverse function of $m(t)$ as $m(t(\tau))=\tau$; $\tau=m(t)$ is called "operational time", c.f. [6].

The conditions $m(t)>o$ for $t>o$ and $m(t)\to\infty$ as $t\to\infty$ make sure that Y_τ is defined for all $\tau\in(o,\infty)$.

2. Classical proofs of that theorem make use of the intensity function $\lambda(t)$ (see e.g. [5], [9], [4]). However, in this case one finds $EX_t = \int_o^t \lambda(s)ds$ which means that the function $m(t)$ has to be absolutely continuous which is not always true (c.f. [7]).

3. Another proof given by ÇINLAR [3] uses a characterization of Poisson processes due to WATANABE [10] without giving a proof of WATANABE's theorem. Following an idea of BREMAUD [1], HARTMANN [7] gives a different proof which still is rather long and technical, using complicated martingale theory.

4. Applying some arguments of ITO [8], we will give a short proof which relies on elementary standard tools of analysis and probability only.

Proof: 1. Apparently $m(t)$ and hence $t(\tau)$ is monotonously increasing. Moreover $t(\tau)$ is right-continuous. Thus, (a) is trivial.

2. The following inequalities are obvious

(i) $1-x\leq e^x$ for all $-\infty<x<\infty$.

(ii) For each $\varepsilon>o$ there exists a $\delta=\delta(\varepsilon)>o$ such that
$1-x\geq e^{(1+\varepsilon)x}$ for all $-\delta\leq x\leq o$.

3. We claim that $X := X_{t_1} - X_{t_o}$ is Poisson distributed, $o\leq t_o<t_1<\infty$. Indeed, let $t_o\leq t<t+h\leq t_1$, then

$$P(|X_{t+h}-X_t|\geq 1) = P(X_{t+h}-X_t\geq 1)\leq E(X_{t+h}-X_t) = m(t+h)-m(t)$$

which tends to 0 uninformly in t if $h\to o$.

For each $n\in\mathbb{N}$, let $t_o = t_{n,o}<t_{n,1}<\cdots<t_{n,m_n} = t_1$, $t_{n,i}-t_{n,i-1}\leq\frac{1}{n}$, be a subdivision of $[t_o,t_1]$ and let denote

$$X_{n,h} := \begin{cases} X_{t_{n,h}} - X_{t_{n,h-1}} & \text{if } X_{t_{n,h}} - X_{t_{n,h-1}} \quad 1, \\ 1 & \text{otherwise} \quad ; \end{cases}$$

$$X^{(n)} = \sum_{h=1}^{m_n} X_{n,h} \; .$$

Since almost all functions $t \mapsto X_t$ are stepwise with jumps 1, we have

$$P\{X^{(n)} \to X\} = 1.$$

Let $s \mapsto g_X(s) := Es^X$ the generating function of X, $s \in [0,1]$.
For all $o < s \leq 1$ we have $o < s^X \leq 1$, thus by the dominated convergence theorem with $\lambda_{n,h} := P\{X_{n,h} = 1\}$ we find

$$\begin{aligned} 0 < Es^X &= \lim_{n\to\infty} Es^{X^{(n)}} \\ &= \lim_{n\to\infty} \prod_{h=1}^{m_n} Es^{X_{n,h}} \\ &= \lim_{n\to\infty} \prod_{h=1}^{m_n} ((1-\lambda_{n,h}) + s\cdot\lambda_{n,h}) \\ &= \lim_{n\to\infty} \prod_{h=1}^{m_n} (1-\lambda_{n,h}(1-s)) \\ (2(i)) \quad &\leq \underline{\lim}_{n\to\infty} \prod_{h=1}^{m_n} e^{-\lambda_{n,h}(1-s)} \\ &= \underline{\lim}_{n\to\infty} e^{-\lambda^{(n)}(1-s)} \end{aligned}$$

where $\lambda^{(n)} := \sum_{h=1}^{m_n} \lambda_{n,h}$. It follows that $(\lambda^{(n)}; n \in \mathbb{N})$ is bounded.

By choosing an appropriate subsequence, w.l.o.g. we may assume $\lambda^{(n)} \to \lambda < \infty$ as $n\to\infty$. This yields $g_X(s) \leq e^{-\lambda(1-s)}$ for all $s \in (0,1]$.

On the other hand, for every $\varepsilon > o$ let n be such that $\lambda_{n,h} < \delta(\varepsilon)$. Then

$$g_X(s) = \lim_{n\to\infty} \prod_{h=1}^{m_n} (1-\lambda_{n,h}(1-s))$$

$$(2(ii)) \geqslant \overline{\lim_{n\to\infty}} \prod_{h=1}^{m_n} e^{-(1+\varepsilon)\lambda_{n,h}(1-s)}$$

$$= \overline{\lim_{n\to\infty}}\, e^{-\lambda^{(n)}(1+\varepsilon)(1-s)}$$

$$= e^{-\lambda(1+\varepsilon)(1-s)} .$$

Letting $\varepsilon \to o$, we see that $g_X(s) = e^{-\lambda(1-s)}$ for a suitable $\lambda \in [0,\infty)$ and for all $s \in (0,1]$. Therefore, $X = X_{t_1} - X_{t_o}$ is Poisson.

4. As to 3., $Y_{\tau+h} - Y_\tau = X_{t(\tau+h)} - X_{t(\tau)}$ is Poisson distributed for all $h > o$, $t \geqslant o$ with a corresponding parameter

$$E(Y_{\tau+h} - Y_\tau) = EX_{t(\tau+h)} - EX_{t(\tau)} = m(t(\tau+h)) - m(t(\tau)) = \tau+h-\tau = h$$

which is independent of τ. Thus, (b) is true.

Letting $\tau = o$, we find $Y_o = X_o = 0$, and finally

$Y_h = Y_h - Y_o$ is Poisson distributed with corresponding parameter h.

Hence, (c) is shown, too.

Acknowledgements: The present paper is a simplified and more elaborate version of my paper "Ein einfacher Beweis zur Konstruktion der operationalen Zeit" in Blätter der DGVM XVI, April 1983.

I am very much indebted to Prof. STEINEBACH and Dr. RHIEL (Marburg) for their helpful comments in preparing this paper.

REFERENCES

[1] Bremaud, P. 1975 *An extension of Watanabe's theorem of characterization of Poisson processes over the positive real half line.* J. Appl. Prob. 12, p.p. 396-399

[2] Bühlmann, H. 1970 *Mathematical Methods in Risk Theory.* Springer, Berlin-Heidelberg-New York

[3] Cinlar, E. 1975 *Introduction to Stochastic Processes.* Prentice Hall Inc., Englewood Cliffs, N.J.

[4] Cox, D.R. and Isham, V. 1980 *Point Processes.* Chapman and Hall, London

[5] Cramér, H. 1969 *On streams of random events.* Skand. Aktuarietidskr. Ergänzungsband, p.p. 13-23

[6] Feller, W. 1966 *Introduction to Probability Theory and its Applications I.* Wiley, New York

[7] Hartmann, G.-M. 1981 *Über das Problem der operationalen Zeit in der Risikotheorie.* Diplomarbeit, Marburg

[8] Ito, K. 1969 *Stochastic Processes.* Lecture Notes Series 16, Aarhus Universitet

[9] Jung, J. and Lundberg, O. 1969 *Risk processes connected with compound Poisson processes.* Skand. Aktuarietiskr., Ergänzungsband, p.p. 118-131

[10] Watanabe, S. 1964 *On discontinuous additive functionals and Lévy measures of a Markov process.* Japanese J. Math. 34, p.p. 53-70

SIMULATION IN ACTUARIAL WORK. SOME COMPUTATIONAL PROBLEMS

Ermanno Pitacco

Ist. di Matematica Finanziaria - University of Trieste

Two simulation problems are discussed and two algorithms are suggested. The first algorithm concerns the simulation of the yearly numbers of claims of a car driver and can be used in the evaluation of a merit - rating system in motor insurance. The second algorithm concerns the simulation of the random survival times of assured lives and can be useful in calculations related both to private and social insurance.

1. INTRODUCTION

It is well known that simulation methods can be usefully applied to actuarial work. Therefore it seems to be useless to dwell on the reasons for such an application. On the contrary, the aim of this paper is to propound two particular simulation algorithms, the first one in the frame of non-life insurance and the second in the frame of private life and pension insurance.

In non-life insurance, simulation methods can be applied, for instance, to the evaluation of a merit-rating system. Particularly, predictions about the distribution of drivers among merit classes, also in the presence of exceptions to the transition rules such as "correctives", can be easily obtained through the simulation of the numbers of claims, whilst an analytical approach might be very hard.

In life insurance models, the simulation may concern several variables. Here we are interested in simulation of random survival times of the policy holders. It is possible to distinguish between two approaches, a "collective" one and an "individual" one. In the first approach, the simulation output consists of the number of survivors year by year: so we can answer the question: "how

F. de Vylder et al. (eds.), Premium Calculation in Insurance, 467–474.

many claims in year t?" but not the question "which claims in year t?". On the contrary, the individual approach yields the survival time for each assured, annuitant or pensioner, and so we can answer the question "which claims?". This second approach especially fits for small funds or small portfolios, allowing an empirical evaluation of variances of incomes and payments.

2. A NON - LIFE INSURANCE PROBLEM

Let $\{N(t),\ t>0\}$ be an integer positive valued stochastic process and let $N(t)$ be the number of claims of a car driver between 0 and t (the unit of time being the year).
We assume that

$$(1) \qquad \mathrm{Prob}\{N(t)=n\,|\,\theta\} = e^{-\theta t}\,\frac{(\theta t)^n}{n!}\,; \quad n=0,1,\ldots$$

where θ is a possible value of the positive random variable Θ, the unknown parameter of the Poisson process. Furthermore, we assume that Θ has a gamma distribution, with density function given by:

$$(2) \qquad g(\theta) = \frac{\beta^{\alpha}}{\Gamma(\alpha)}\,\theta^{\alpha-1}\,e^{-\beta\theta} \qquad (\alpha,\beta>0)$$

Therefore, $\{N(t)\}$ is a weighted Poisson process with a gamma structure function.
Let N_i denote the number of claims of our car driver in year i.
Then

$$N_i = N(i) - N(i-1)\,;\quad i=1,2,\ldots \quad (N(0)=0).$$

Given the assumptions (1) and (2), the joint distribution of $N_{i_1},\ldots,N_{i_m}$, for any m and any $(i_1,\ldots,i_m)$, is negative binomial. In fact, it can be easily proved that

$$(3) \qquad \mathrm{Prob}\{N_{i_1}=n_1,\ldots,N_{i_m}=n_m\} =$$

$$= \int_0^{+\infty} \mathrm{Prob}\{N_{i_1}=n_1,\ldots,N_{i_m}=n_m\,|\,\Theta=\theta\}\,g(\theta)\,d\theta =$$

$$= \frac{\beta^{\alpha}}{\Gamma(\alpha)}\;\frac{\Gamma\left(\alpha+\sum_{j=1}^{m} n_j\right)}{\prod_{j=1}^{m}(n_j!)\,(\beta+m)^{\alpha+\sum n_j}}\,.$$

Furthermore, each N_i has a marginal-conditional distribution of negative binomial type, the parameters depending on the observed values of the preceding variables N_h, $h=1,\dots,i-1$. On the contrary, for any fixed value θ the variables N_1, $N_2,\dots$ are indipendent and identically distributed (i.i.d.), each with marginal Poisson distribution and parameter θ.
Let us suppose that we are interested in the simulation of N_1, $N_2,\dots, N_m$, for a given number m of years. A well known simulation algorithm for such a sequence is the following one.
Algorithm A1.
Step 1. Draw at random a value θ of Θ, according to the given gamma distribution.
Step 2. Given θ, simulate sequentially the i.i.d. variables N_1, $N_2,\dots,N_m$, according to the Poisson distribution with parameter θ.
The step 2 can be fulfilled by simulation of the interarrival times T_1, $T_2,\dots$, until $T_r \geq m$, where
T_1 = random time of the first claim
T_i = random time elapsed between the (i-1)st and the ith claim, $i=2,3,\dots$
In fact, it is well known that, given $\Theta=\theta$, T_1, $T_2,\dots$ are i.i.d. exponentially distributed with parameter θ. Let F(t) denote the probability distribution function of any T_i

$$F(t) = \text{Prob}\{T_i \leq t\} = 1 - e^{-\theta t} .$$

Thus, the simulation of T_1, $T_2,\dots$ can be made by the inversion method (see Fishman, 1973). In fact, if we denote by U_i a uniform random number in [0,1], the random variable

$$F^{-1}(U_i) = -\frac{\log(1 - U_i)}{\theta}$$

has an exponential distribution with parameter θ.
This simple algorithm has a weak point. Actually, there are some computational difficulties in accomplishing step 1, since we cannot apply the simple inversion method to a gamma distribution. Thus, the simulation of the random variable Θ requires more complicated methods (see Fishman, 1976).
On the contrary, the suggested algorithm (see also Pitacco, 1979) by-passes the above point thanks to a "direct" simulation of the weighted process. First, remark that, for any m, any simulated realization of the subsequence $(N_1,\dots,N_m)$ of the random sequence $(\Theta,N_1,\dots,N_m)$ can be got with the same probability in two ways. The first one consists in the two-steps procedure (algorithm A1). The second one requires only the simulation of $(N_1,\dots,N_m)$ according to its negative binomial distribution. Of course, the equivalence holds thanks to (3).
Thus, the algorithm A2 is simply the following one:
- simulate sequentially N_1, N_2, ... , N_m according to their negative binomial distributions.

In order to fulfil the simulation in the easiest way, it is useful to remember the following result (see Daboni, 1975). If the unconditional joint distributions of $N_1, N_2, \dots$ are negative binomial, the interarrival times $T_1, T_2, \dots$ have the following probability density functions:

$$\varphi_{T_j}(t \mid T_1=t_1,\dots,T_{j-1}=t_{j-1}) =$$

$$= \frac{(\alpha+j-1)\,(\beta+\sum_{h=1}^{j-1} t_h)^{\alpha+j-1}}{(\beta+\sum_{h=1}^{j-1} t_h + t)^{\alpha+j}} .$$

For simplicity, let

$$\alpha(j) = \alpha + j - 1$$

$$\beta(j) = \beta + \sum_{h=1}^{j-1} t_h .$$

Then, the cumulative distribution function of T_j is

(4) $$\Phi_{T_j}(t \mid t_1,\dots t_{j-1}) = 1 - \frac{\beta(j)^{\alpha(j)}}{(\beta(j)+t)^{\alpha(j)}} .$$

Remark that the expressions of $\alpha(j)$, $\beta(j)$ exhibit the adjustment of the parameters due to simulated observations.
The above result allows the sequential simulation of $T_1, T_2, \dots$ by the simple inversion method. In fact, if U_j is a random variable with uniform distribution in $[0,1]$, the random variable

$$\Phi^{-1}_{T_j}(U_j) = \beta(j)\,(1-U_j)^{-1/\alpha(j)} - \beta(j)$$

has the distribution given by (4).
Some numerical results show the efficiency of the method. We have considered several values of α, β and consequently of the expected value $E(N_i) = \alpha/\beta$. Some values of the ratio between the central processor time required by algorithm A2 and the one required by algorithm A1 are displayed in Table 1. According to the value of α, different procedures for the simulation of Θ in algorithm A1 have been used (about that, see Pitacco,1979). We observe that

the relative efficiency of our algorithm decreases as the number of years of simulation increases (and the ratio can be greater than one for very large numbers of years). Of course this is due to the fact that the time saving gained by the non-simulation of Θ is divided into an increasing number of simulations. However, the algorithm reveals a good relative efficiency with regard to numbers of years interesting the practical actuarial projections (usually, ten years at most).

			number of years			
α	β	$E(N_i)=\alpha/\beta$	m=5	m=10	m=15	m=20
0.94	5.875	0.16	0.443	0.572	0.675	0.746
1.2	7.5	0.16	0.721	0.851	0.922	0.979
1.6	10.	0.16	0.664	0.799	0.860	0.933
2.5	15.625	0.16	0.590	0.726	0.812	0.882
3.5	21.875	0.16	0.542	0.677	0.761	0.831
0.94	5.2	0.181	0.463	0.592	0.710	0.769
1.6	14.55	0.11	0.635	0.739	0.825	0.878

Table 1

3. A LIFE INSURANCE PROBLEM

Let x denote the age of an assured life and T_x the survival time, a random variable with possible values $1,2,3,\dots,\omega-x$, where ω is the "maximum" age. Let

$$q_{x+h} = \text{Prob}\{T_{x+h} = 1\} \quad , \quad h=0,1,2,\dots$$

$$_{h/}q_x = \text{Prob}\{T_x = h+1\} \quad , \quad h=0,1,2,\dots$$

Our goal is the simulation of T_x.
Consider the following events and the respective probabilities

$$E_1 = \{T_x = 1\} \qquad P(E_1) = q_x$$

$$E_2 = \{T_x = 2 \mid \bar{E}_1\} \qquad P(E_2) = q_{x+1}$$

$$E_3 = \{T_x = 3 \mid \bar{E}_1\bar{E}_2\} \qquad P(E_3) = q_{x+2}$$

$$\dots \qquad \dots$$

A very simple and well known algorithm works "year by year" and requires the simulation of events $E_1, E_2,\dots,$ until simulated death. Each simulation is very easy but the procedure requires a random number of simulations (and therefore of random numbers

generations) with expected value equal to the expected survival time.
Here we want to describe an alternative algorithm, suggested by myself and a colleague of my Institute (see Crisma-Pitacco,1982).
Let

$$Q_x(t) = \text{Prob}\{T_x \leq t\} = \begin{cases} 0 & t < 1 \\ \sum_{k \leq t-1} {}_{k/}q_x & 1 \leq t < \omega - x \\ 1 & t \geq \omega - x \end{cases}$$

Let U be a uniform random number in [0,1]. The two events

$$\{T_x = j\} \quad ; \quad \{Q_x(j-1) < U \leq Q_x(j)\}$$

have the same probability ${}_{j-1/}q_x$, for $j=1,2,\ldots,\omega-x$. Therefore, if and only if the outcome u of our random drawing belongs to the interval $]Q_x(j-1),Q_x(j)]$, we take j as simulated survival time, that is

$$j = \varphi_x(u)$$

where the integer valued function $\varphi_x(.)$ "inverts" the function $Q_x(.)$ in the sense that $\varphi_x(U)$ has Q_x as cumulative distribution function.
We observe that the calculation of j given u requires a table look-up among the values

$$Q_x(1),\ Q_x(2),\ Q_x(3),\ \ldots$$

Such a table look-up can be accomplished through binary search or other algorithms. Here we suggest the following procedure.
Let $f_x(u)$ be a real valued function, defined in the interval $0 \leq u \leq 1$, that approaches in some sense the function $\varphi_x(u)$. Then, the procedure is the following one:
1) generation of a random number u;
2) calculation of $t = f_x(u)$;
3) beginning from t, sequential search in the table Q_x of the value $j = \varphi_x(u)$.

Which are the desirable requisites for the function f_x? It is intuitive that
a) it must be "near" φ_x;
b) it must be quickly calculable.

With regard to point a), we may observe that for a given family of functions, $\{f_x\}$, a reasonable goal is the minimization of the expected number of steps in the sequential scanning that begins from t. Thus, we may take in $\{f_x\}$ the function $\tilde{f}_x$ for which the following equality holds:

$$E(|\tilde{f}_x(U) - \varphi_x(U)|) = \min_{\{f_x\}} E(|f_x(U) - \varphi_x(U)|) .$$

From a computational point of view, an easier minimization problem is

$$\min_{\{f_x\}} E(f_x(U) - \varphi_x(U))^2$$

We can take in $\{f_x\}$ the function $\hat{f}_x$ that minimizes the above expression (and so we achieve an approximation of $\tilde{f}_x$).
With regard to point b), we may consider polynomial functions. Then, for a fixed degree n, we have

$$f_x(u) = \sum_{j=0}^{n} a_j u^j$$

Since the random variable U is uniformly distributed in $[0,1]$, we get

$$E(f_x(U) - \varphi_x(U))^2 = \int_0^1 (\sum_{j=0}^{n} a_j u^j - \varphi_x(u))^2 du =$$

$$= G(a_0, a_1, \ldots, a_n).$$

The minimization of G requires:

$$\frac{\partial G}{\partial a_i} = 0 \; ; \quad i=0,1,\ldots,n$$

It can be easily shown (see Crisma-Pitacco, 1982) that the optimal values of a_0, a_1, ..., a_n are the solution of the following system:

$$\sum_{j=0}^{n} \frac{1}{j+i+1} a_j = \frac{1}{i+1} \left[\omega - x - \sum_{h=0}^{\omega-x-1} Q_x^{i+1}(h) \right] \; ; \quad i=0,1,..,n$$

Very good results have been obtained by using polynomials of degree three. Some computational results are displayed in Table 2. Moreover, it is interesting to remark that the average number of steps in the sequential search required by a polynomial of degree 3 is 0.5651 for x=25, and 0.1576 for x=40. These low average numbers attest the goodness of fit attained by a low degree polynomial.
The above simulation method to determine the lifetime of an assured life can be used in various actuarial evaluation problems. For instance, it allows the calculation of the yearly levels of an assurance fund whose incomes are premiums and whose payments are assured amounts and annuities. Of course, premiums and payments are functions of the lifetimes of the policyholders. We observe that the "year by year" simulation method suggests a computational organization based on "yearly" additions and subtractions related

age	year	direct simulation		
	by	binary search	polynomial approx.	
x	year		1	3
25	8.536	2.310	1.238	1.061
40	6.082	2.162	1.059	0.990
55	3.749	2.056	0.938	0.962

Table 2. Central processor time in seconds required to perform 10,000 simulations (CDC 170-720 computer)

to premiums and annuities. On the contrary, the "direct" simulation of random lifetimes suggests a computational organization based on recording the start and the stop of each individual sequence of premiums and of annuity payments. The calculations concerning the levels of the fund can be consequently accomplished at the end of the simulation by simple recurrent formulas (see Crisma - Pitacco, 1982). Such a model allows a further substantial time saving. Thus, we can observe that a different key idea concerning a simulation problem may point out some chances of computational reorganization.

REFERENCES

(1) Crisma, L. and Pitacco, E., 1982, Su due metodi simulativi per la valutazione di grandezze interessanti assicurazioni sulla vita, Rivista di matematica per le scienze economiche e sociali, vol. 5, 1, pp. 41-51

(2) Daboni, L., 1975, "Aspetti strutturali ed inferenziali di particolari processi di arrivo di sinistri", Accademia Nazionale dei Lincei, Roma

(3) Fishman, G.S., 1973, "Concepts and methods in discrete event digital simulation", J. Wiley

(4) Fishman, G.S., 1976, Sampling from the gamma distribution on a computer", Communications of the ACM, vol. 19, 7, pp. 407 - 409

(5) Pitacco,E., 1979, Simulazione di un processo mistura di Poisson. Un'applicazione attuariale, Proceedings of Giornate AIRO79, pp. S 33 - S 48

(6) Wallace, N.D., 1974, Computer generation of gamma random variates with non-integral shape parameters, Communications of the ACM, vol. 17, 12, pp. 691 - 695

BAYESIAN SEQUENTIAL ANALYSIS OF MULTIVARIATE POINT PROCESSES

Raimund Rhiel

University of Marburg

Modern martingale theory is applied to treat the topic of multivariate point processes, e.g. risk processes, multivariate counting processes. The general decision theoretic model of sequential and nonsequential procedures is used to estimate the structural parameters. Likelihood ratios and posterior distributions can always be represented by means of compensators and intensities. It is indicated, how compensators and intensities with respect to the unconditional (collective) probability law can be derived from the compensators and intensities with respect to the conditional (individual) probability laws. Some special loss functions and intensities depending multilinearly on the structural parameters are discussed in detail as examples.

1. THE BAYESIAN SEQUENTIAL MODEL

1.1. General definitions and notations; the nonsequential model

Let Ω be an abstract sample space with elements ω. We consider the stochastic process (S_t) of e.g. total claim amounts up to time t as a time-continuous multivariate point process on Ω with values in some separabel Banach space E endowed with the σ-algebra $\mathfrak{E}$ of Borel subsets (e.g. $E=R^n$) . The sample paths of the process S are assumed right continuous with left-hand limits (which usually is abbreviated by the French expression "càdlàg", i.e. continu à droite et limité à gauche). S is a pure jump process without accumulation of jumps in finite intervals, i.e. we have $S_o=0$ and

$$S_t = \sum_{0<u<t} \Delta S_u \quad \text{where} \quad \Delta S_u = S_u - S_{u-} \tag{1}$$

F. de Vylder et al. (eds.), Premium Calculation in Insurance, 475–487.

and S_{u-} denotes the left-hand limit of S at time u. All our future statistical decisions will be based on the observable process S. Therefore we introduce the σ-algebras $\mathfrak{F}_t = \sigma(S_u | u \leq t)$ on the space Ω. One can show that $\mathfrak{F}_t$ is countably generated and that $\mathfrak{F}_t = \mathfrak{F}_{t+} := \bigcap_{s>t} \mathfrak{F}_s$ for each t, i.e. the family ($\mathfrak{F}_t$) is right continuous. The last property results from the fact that S has only a finite number of jumps in finite intervals (Jacod, 1975).
The parameter space or space of structural parameters Θ is supposed to be a Borel measurable subset of some Polish space in order to ensure the existence of conditional distributions. The space Θ will be endowed with the σ-algebra $\underline{\underline{\Theta}}$ of Borel subsets. This σ-algebra is countably generated. On the space $(\Theta, \underline{\underline{\Theta}})$ a prior distribution or structure function τ is given. τ contains the information about the collective Θ. Let be given a family $(P_\theta)_{\theta\varepsilon\Theta}$ of probability measures (hypotheses) on $(\Omega, \mathfrak{F}_\infty)$, such that the mappings $\theta \longmapsto P_\theta(A)$ are measurable w.r.t. $\underline{\underline{\Theta}}$ for all $A \varepsilon \mathfrak{F}_\infty$.
We define a probability measure Q_τ on $\underline{\underline{\Theta}} \otimes \mathfrak{F}_\infty$ by

$$Q_\tau(B \times A) = \int_B P_\theta(A)\, d\tau(\theta) \qquad , B\varepsilon\underline{\underline{\Theta}} \ , \ A\varepsilon \mathfrak{F}_\infty \tag{2}$$

and a probability measure P_τ on $\mathfrak{F}_\infty$ by

$$P_\tau(A) = Q_\tau(\Theta \times A) \qquad , A\varepsilon \mathfrak{F}_\infty \tag{3}$$

The posterior probability of $\theta \varepsilon B$ for $B\varepsilon\underline{\underline{\Theta}}$ having observed $\omega\varepsilon\Omega$ at time $t \geq 0$ is denoted by $Q_t(\omega, B)$. It is characterized by the equation

$$Q_\tau(B \times A) = \int_A Q_t(\omega, B)\, dP_\tau(\omega) \quad \text{for all } A\varepsilon \mathfrak{F}_t \tag{4}$$

Of course, we have $Q_0(\omega, B) = \tau(B)$ P_τ a.e.
In Rhiel (1981) it is shown that $Q_t(\omega, B)$ can be chosen such that

i. $B \longmapsto Q_t(\omega, B)$ is a probability measure on $\underline{\underline{\Theta}}$ for all (ω, t) ,

ii. $t \longmapsto Q_t(\omega, B)$ is càdlàg P_τ a.e. , (5)

iii. $\omega \longmapsto Q_t(\omega, B)$ is measurable w.r.t. $\mathfrak{F}_t$.

Next, we need an action space $\mathcal{A}$ with σ-algebra $\underline{\underline{\mathcal{A}}}$.
Goodness of decisions taken by the statistician will be measured by a loss function $L : \Theta \times \mathcal{A} \longrightarrow R^+$, which has to be measurable w.r.t. $\underline{\underline{\Theta}} \times \underline{\underline{\mathcal{A}}}$. $L(\theta, a)$ gives the loss incurred by taking action a if θ is the true parameter. More generally, we could have introduced loss functions $L : \Theta \times \mathcal{A} \times \Omega \times R^+ \longrightarrow R^+$, where then $L(\theta, a, \omega, t)$ depends on the observed $\omega\varepsilon\Omega$ at time t (cf. Rhiel 1981). Here we will restrict ourselves to loss functions of the first type for simplicity.

In the sequel we look for Bayes optimal decisions based on observations $\omega\varepsilon\Omega$ up to time t. A function $d_t : \Omega \longrightarrow \mathcal{A}$

measurable w.r.t. $\mathcal{F}_t$ and $\mathcal{A}$ is called a decision function (procedure). The Bayes risk $r_\tau(t,d_t)$ of this procedure is defined by

$$\begin{aligned} r_\tau(t,d_t) &= E_{Q_\tau} L(\theta, d_t(\omega)) \\ &= \int_\Theta (\int_\Omega L(\theta,d_t(\omega))\, dP_\theta(\omega))\, d\tau(\theta) \\ &= \int_\Omega (\int_\Theta L(\theta,d_t(\omega))\, Q_t(\omega,d\theta))\, dP_\tau(\omega) \\ &= E_\tau(\int_\Theta L(\theta,d_t(\omega))\, Q_t(\omega,d\theta)) \end{aligned} \tag{6}$$

This expression will be minimal, iff d_t^o is chosen such that

$$\int_\Theta L(\theta,d_t^o(\omega))\, Q_t(\omega,d\theta) = \inf_{a\varepsilon\mathcal{A}} \int_\Theta L(\theta,a)\, Q_t(\omega,d\theta) \quad P_\tau \text{ a.e.} \tag{7}$$

Then we have

$$r_\tau(t,d_t^o) = E_\tau(\inf_{a\varepsilon\mathcal{A}} \int_\Theta L(\theta,a)\, Q_t(\omega,d\theta)) \tag{8}$$

The last minimization can very easily be carried out in all practically relevant cases, e.g. in the following examples.

1.2. Examples of loss functions

1.Example: Let $\Theta = \Theta_1 \cup \ldots \cup \Theta_k$ be a decomposition of Θ into mutually disjoint measurable sets and $\mathcal{A} = \{a_1,\ldots,a_k\}$. The action a_i represents the decision that θ is contained in Θ_i. Then we should have $L(\theta,a_i) = 0$ iff $\theta\varepsilon\Theta_i$ (correct decision). The simplest example would be $L(\theta,a_i) = 1$ iff $\theta\notin\Theta_i$ (wrong decision), hence

$$L(\theta,a_i) = 1 - 1_{\Theta_i}(\theta) \qquad \text{(0-1 error loss)} .$$

Then all wrong decisions are equally weighted; in practical cases of course they should not:
Let e.g. Θ_1 represent the set of the best car drivers of an insurance company and Θ_k the set of the worst drivers. If a driver who really would belong to Θ_k is grouped into Θ_1, then this wrong decision is worse than the wrong decision to put him into the set Θ_{k-1} of the second worst drivers. Therefore it is better to set

$$L(\theta,a_i) = c_{il} \quad \text{if} \quad \theta\varepsilon\Theta_l \;, \text{ where } c_{ii} = 0 \text{ and } c_{il}$$

increasing in $|i - l|$.
The optimal decision $d_t^o(\omega)$ is to decide for $a_i \varepsilon \mathcal{A}$, if $\omega\varepsilon A_i$ with

$$A_i = \{ \int_\Theta L(\theta,a_i)Q_t(\omega,d\theta) \leqq \int_\Theta L(\theta,a_j)Q_t(\omega,d\theta) \text{ for all } j \} \tag{9}$$

In the case of 0-1 error loss we have

$$A_i = \{ Q_t(\omega,\Theta_i) \geqq Q_t(\omega,\Theta_j) \quad \text{for all } j = 1\ldots k \} \tag{10}$$

Having introduced weights c_{il} we get

$$A_i = \{ \sum_{l=1}^{k} c_{il} Q_t(\omega,\Theta_l) \leq \sum_{l=1}^{k} c_{jl} Q_t(\omega,\Theta_l) \text{ for all } j \} \quad (11)$$

(These sets A_i which should be made mutually disjoint belong to $\mathfrak{F}_t$)

2.Example: In parameter estimation problems one frequently uses squared loss error. Let $\mathcal{A}$ be a measurable subset of a euclidian space and

$$L(\theta,a) = (h(\theta) - a)'W(\theta)(h(\theta) - a) , \quad (12)$$

where $h : \Theta \longrightarrow \mathcal{A}$ is the parameter function to be estimated and $W(\theta)$ are symmetric and positive definite matrices.
In this case the optimal decision is

$$d_t^o(\omega) = (\int_\Theta W(\theta)Q_t(\omega,d\theta))^{-1}(\int_\Theta W(\theta)h(\theta)Q_t(\omega,d\theta)) \quad (13)$$

If $W(\theta) = W$, then

$$d_t^o(\omega) = \int_\Theta h(\theta) Q_t(\omega,d\theta) \quad (14)$$

is independent of the matrix W . $d_t^o(\omega)$ is the mean of h w.r.t. the posterior distribution.

1.3. The sequential model and the stochastic decision process χ

In a sequential setting we allow random times (stopping times) T rather than fixed times t . A stopping time T is a mapping $T : \Omega \longrightarrow R^+$ with $\{T \leq t\} \epsilon \mathfrak{F}_t$ for each $t \geq 0$. The σ-algebra $\mathfrak{F}_T = \{ A \epsilon \Omega \mid A \cap \{T \leq t\} \epsilon \mathfrak{F}_t \text{ for all } t \leq \infty \}$ contains all information up to T.In all our previous formulae (4) - (14) we can substitute T for t and $\mathfrak{F}_T$ for $\mathfrak{F}_t$. The measure theoretical problems arising there are discussed in detail in the author's dissertation (Rhiel,1981).
What do we gain by this "sequential procedure"?
Let us imagine that we are at time t_o. Why shouldn't we use all our information up to time t_o? If we stop observation randomly at time $T \leq t_o$, then we do not use all our information and our decision will be worse,of course. But on the other hand we could perhaps save some money. Let us say that observation up to time t costs $c(t)$, where $c : \mathbb{R}^+ \longrightarrow \mathbb{R}^+$ is an increasing and continuous cost function.
Then we have to add a term $E_\tau c(T) = \int_\Omega c(T(\omega)) \, dP_\tau(\omega)$ to the Bayes risk $r_\tau(T,d_T)$ in equations (6). Taking into account (7) we then have to find a stopping time T^o minimizing

$E_\tau \chi_T$, where

$$\chi_t(\omega) = c(t) + \inf_{a\in\mathcal{A}} \int_\Theta L(\theta,a)\, Q_t(\omega,d\theta) \tag{15}$$

This is the problem of "optimal stopping". The process χ is called "stochastic decision process" (cf.Rhiel 1981).
The term $c(t)$ is increasing, whereas the right-hand term $\inf_{a\in\mathcal{A}} \int_\Theta L(\theta,a)\, Q_t(\omega,d\theta)$ is a P_τ-supermartingale on which the decomposition of Doob-Meyer can be applied.
The existence of an optimal stopping time T^o can be proved in quite general cases, whereas the explicit construction of T^o is very difficult in general (cf. Rhiel 1981).

In example 1.2.1. in case of 0-1 error loss we have

$$\chi_t(\omega) = c(t) + 1 - \max_{i=1..k} Q_t(\omega,\theta_i) \tag{16}$$

Having introduced weights c_{il} we get

$$\chi_t(\omega) = c(t) + \min_{i=1..k} \sum_{l=1}^{k} c_{il}\, Q_t(\omega,\theta_l) \tag{17}$$

In example 1.2.2. we obtain

$$\chi_t(\omega) = c(t) + \int_\Theta \|W(\theta)^{1/2}h(\theta)\|^2 Q_t(\omega,d\theta) - \|(\int_\Theta W(\theta)Q_t(\omega,d\theta))^{-1/2}(\int_\Theta W(\theta)h(\theta)Q_t(\omega,d\theta))\|^2 \tag{18}$$

If $W(\theta) = W$ this reduces to

$$\chi_t(\omega) = c(t) + \int_\Theta \|W^{1/2}h(\theta)\|^2 Q_t(\omega,d\theta) - \|W^{1/2} (\int_\Theta h(\theta)\, Q_t(\omega,d\theta))\|^2 \tag{19}$$

In both examples an optimal stopping time T^o exists.
The example 1.2.2. will be continued at the end of this paper in chapter 2.6..

If a "sequential procedure" does not seem appropriate, then the problem of "optimal stopping" does not arise. Then we are done with the construction of the optimal decision $d_t^o(\omega)$ in (7) . This construction involves the computation of the posterior distribution $Q_t(\omega,\cdot)$.

1.4. Computation of the posterior distribution $Q_t(\omega,\cdot)$

Let us suppose that all probability measures P_θ are mutually absolutely continuous on all σ-algebras $\mathcal{F}_t$, $t<\infty$, i.e.

$$P_\theta \cong P_{\theta_o} \quad \text{on all } \mathcal{F}_t\,, \quad t<\infty \quad , \theta_o \text{ fixed} . \tag{20}$$

This is the only statistically interesting case, because on the orthogonal parts of P_θ and P_{θ_o} we always have good decisions. In most cases the hypotheses P_{θ_o}, P_θ will be mutually orthogonal on the big σ-algebra $\mathfrak{F}_\infty$. Theoretical problems arising in the application of martingale theory are also discussed by Rhiel(1981); let us define

$$g_\theta(\omega,t) = \left.\frac{dP_\theta}{dP_{\theta_o}}\right|_{\mathfrak{F}_t} \tag{21}$$

One easily shows that $P_\theta \ll P_\tau$ on all $\mathfrak{F}_t$, $t<\infty$, hence

$$f_\theta(\omega,t) = \left.\frac{dP_\theta}{dP_\tau}\right|_{\mathfrak{F}_t} \quad \text{exists.} \tag{22}$$

g_θ and f_θ can be chosen right continuous with left-hand limits a.e.. A simple calculation yields

$$f_\theta(\omega,t) = \frac{g_\theta(\omega,t)}{\int_\Theta g_\rho(\omega,t)\, d\tau(\rho)} \quad \text{a.e.} \tag{23}$$

and
$$Q_t(\omega,B) = \int_B f_\theta(\omega,t)d\tau(\theta) \quad \text{for all } B \quad P_\tau \text{a.e.} \tag{24}$$

The last equations are also valid for finite stopping times T . Then the problem of computing $Q_t(\omega,B)$ is reduced to the problem of computing $g_\theta(\omega,t)$. The formula which we are interested in is the following (due to Jacod,1975)

$$g_\theta(\omega,t) = \exp\left(-\int_o^t \int_E (\Lambda_\theta(\omega,s,x) - 1)\nu_{\theta_o}(\omega,ds,dx)\right) \times \prod_{\substack{u\le t\\ \Delta S_u \ne 0}} \Lambda_\theta(\omega,u,\Delta S_u(\omega)) \tag{25}$$

This is the most general formula in the context of multivariate point processes used here. The ingredients of this formula , $\Lambda_\theta(\omega,s,x)$ and $\nu_{\theta_o}(\omega,ds,dx)$, will be explained in the next chapter.

2. MULTIVARIATE POINT PROCESSES IN THE CONTEXT OF MARTINGALE THEORY

2.1. The compensating measure of a multivariate point process

We deal with the process S given in chapter 1.1. This process can be described in two ways: The first way is to consider the times T_n when claims occur and the respective claim amounts X_n.

Then
$$S_t = \sum_{T_n \le t} X_n \, . \tag{26}$$

Remember that the variables X_n are not supposed i.i.d.!
More appropriate for the application of martingale theory is the second way:
We use the counting processes $N(t,B)$ defined by

$$N(t,B) = \sum_{u \leq t} 1_{B-\{0\}}(\Delta S_u) \qquad \text{for } B \epsilon \mathcal{E}. \tag{27}$$

$N(t,B)$ is the number of claims with claim amount in B up to t . Especially we denote the total number of claims up to time t by

$$N_t = N(t,E) \tag{28}$$

The collection of all $N(t,B)$ determines the process (S_t) . For convenience we will make the

$$\text{assumption:} \quad E_\theta N_t < \infty \quad \text{for all } \theta \epsilon \Theta \quad \text{and} \quad t<\infty \quad . \tag{29}$$

If S is under P_θ a process with independent increments, so are all $N(t,B)$. Then $N(t,B) - m_\theta([0,t]\times B)$ is a P_θ-martingale, where $m_\theta([0,t]\times B) = E_\theta N(t,B)$ is a measure on $(R^+\times E, \mathcal{L}^+ \otimes \mathcal{E})$. m_θ is called "Levy measure". By the Levy-Chintchine formula it determines the probability law of the process S w.r.t. P_θ completely. If the application $t \longmapsto m_\theta([0,t]\times B)$ is continuous then $N(t,B)$ is Poisson distributed. If $m_\theta([0,t]\times B) = \lambda_\theta \cdot t \cdot h_\theta(B)$, then S is a stationary compound Poisson process (with independent increments).

If S has no independent increments, then $N(t,B) - E_\theta N(t,B)$ will no longer be a martingale. But by martingale theory we can find a random measure $\nu_\theta(\omega,[0,t]\times B)$ on $(R^+\times E, \mathcal{L}^+ \otimes \mathcal{E})$ such that

$$N(t,B) - \nu_\theta(\omega,[0,t]\times B) \quad \text{is a } P_\theta\text{-martingale.} \tag{30}$$

The application

$$(\omega,t) \longmapsto \nu_\theta(\omega,[0,t]\times B) \qquad \text{is "predictable"} \tag{31}$$

("prévisible"), i.e. measurable w.r.t. the σ-algebra on $\Omega\times R^+$ generated by all continuous real valued stochastic processes ξ for which ξ_t is $\mathcal{F}_t$ -measurable. The application we talk about need not be continuous itself; but is increasing and càdlàg.
ν_θ is called the "generalized Levy measure" or the "compensating measure" or the "dual predictable projection" of the process S w.r.t. P_θ. By Jacod (1979) ν_θ determines the law of S under P_θ completely, esp. S admits independent increments under P_θ , iff ν_θ is non-random (cf. Wang 1981 and Jacod 1979).

ν_θ can also be represented in terms of the common distribution of the sequence $(T_1,X_1,T_2,X_2,T_3,X_3,\ldots.)$. Jacod (1975,1979) derived the following nice formula

$$\nu_\theta(\omega,[0,t]\times B) = \sum_{n\geq 0}\int_{t\wedge T_n}^{t\wedge T_{n+1}} \int_B \frac{1}{P_\theta(T_{n+1}\geq s\mid \mathcal{F}_{T_n})(\omega)} \times \tag{32}$$

$$\times\, dP_{\theta(T_{n+1},X_{n+1}\mid \mathcal{F}_{T_n})}(\omega,ds,dx) ,$$

where $\mathcal{F}_{T_n} = \sigma(T_1,X_1,\dots,T_n,X_n)$.
This formula can be used e.g. for multivariate semi-Markov processes defined by Janssen.
Let us consider the special case $B = E$: Define

$$\rho_\theta(\omega,t) = \nu_\theta(\omega,[0,t]\times E) \quad . \text{ Then}$$

$$N_t - \rho_\theta(\omega,t) \qquad \text{is a } P_\theta\text{-martingale} \quad . \tag{33}$$

$\rho_\theta(\omega,t)$ is the cumulated intensity of (N_t) w.r.t. P_θ.
By the theorem on the existence of conditional distributions (and some martingale techniques) we get the following desintegration formula, which is very helpful in the interpretation and construction of $\nu_\theta.$:

$$\nu_\theta(\omega,[0,t]\times B) = \int_0^t h_\theta(\omega,s,B)\,\rho_\theta(\omega,ds) \quad , \text{ for short} \tag{34}$$

$$\nu_\theta(\omega,ds,dx) = h_\theta(\omega,s,dx)\,\rho_\theta(\omega,ds) \qquad \text{(Wang 1981).} \tag{35}$$

$B \longmapsto h_\theta(\omega,s,B)$ is the conditional distribution of the claim size under condition that a claim occurs at time s.

2.2. Multivariate counting processes

Interesting examples of multivariate point processes are the multivariate counting processes defined by Aalen and Hoem (1978).
Let E be euclidian n-space and S^i the i.th component of S .
We assume that no two of the processes S^i are allowed to jump at the same time. In the proper mathematical sense this is no restriction of generality, but indeed in practical sense.
The jump heights of S are then the canonical vectors $e_i = (0,\dots,1,\dots,0)'$. $\nu_\theta(\omega,[0,t]\times B)$ is concentrated on the set $B^i = \{e_1,\dots,e_n\}$.
$\nu^i_\theta(\omega,t) := \nu_\theta(\omega,[0,t]\times\{e_i\})$ is the cumulated intensity of the component S^i. (36)

2.3. Representation of likelihood ratios

The following theorem is due to Jacod (1975,1979):

Theorem:
Let P_θ be dominated by P_{θ_0} on all $\mathfrak{F}_t$, $t<\infty$.

1. There exists a predictable function

$$\Lambda_\theta : \Omega\times[0,\infty)\times E \longrightarrow R^+ \quad \text{such that}$$

$$\nu_\theta(\omega,[0,t]\times B) = \int_0^t \int_B \Lambda_\theta(\omega,s,x)\ \nu_{\theta_0}(\omega,ds,dx) \ . \tag{37}$$

2. If $\rho_{\theta_0}(\omega,t)$ is continuous in t , so is $\rho_\theta(\omega,t)$ and we have

$$g_\theta(\omega,t) = \exp(\rho_{\theta_0}(\omega,t) - \rho_\theta(\omega,t)) \prod_{\substack{u\leq t\\ \Delta\bar{S}_u\neq 0}} \Lambda_\theta(\omega,u,\Delta S_u(\omega)) \tag{38}$$

Formula (38) is the same formula as (25).
If $\rho_{\theta_0}(\omega,t)$ is not continuous in t , then we have to add a further term in the representation of the density $g_\theta(\omega,t)$ (cf. Jacod 1975). But we will only deal with the "continuous" case. In the theorem we may substitute finite stopping times T for t. We will write for short:

$$\Lambda_\theta(\omega,s,x) = \frac{\nu_\theta\ (\omega,ds,dx)}{\nu_{\theta_0}(\omega,ds,dx)} = \frac{h_\theta(\omega,s,dx)}{h_{\theta_0}(\omega,s,dx)}\cdot\frac{\rho_\theta(\omega,ds)}{\rho_{\theta_0}(\omega,ds)} \tag{39}$$

where
$$\frac{\rho_\theta(\omega,ds)}{\rho_{\theta_0}(\omega,ds)} = \int_E \Lambda_\theta(\omega,s,x)\ h_{\theta_0}(\omega,s,dx) \tag{40}$$

Assumption: In the sequel let $\rho_{\theta_0}(\omega,t)$ be continuous in t . (41)

2.4. Compensators and intensities w.r.t. P_τ

In the chapters 2.1. and 2.3. we indicated that the law of the process S w.r.t. P_θ is determined by the measure ν_θ and hence by the intensity $\Lambda_\theta = d\nu_\theta/d\nu_{\theta_0}$. Of great importance in actuarial mathematics (as well as in Bayesian statistics) is the measure P_τ. Let ν_τ be the Levy measure and $\Lambda_\tau = d\nu_\tau/d\nu_{\theta_0}$ the intensity of S w.r.t. P_τ. Moreover set $\rho_\tau(\omega,t) = \nu_\tau(\omega,[0,t]\times E)$.
The law of the process S w.r.t. P_τ is uniquely determined by ν_τ (or Λ_τ). It is well known that the law of S w.r.t. P_τ is in general much more difficult than w.r.t. all P_θ . Especially if S has independent increments w.r.t. all P_θ , this is not true w.r.t. P_τ. How can we construct ν_τ and Λ_τ? By an application of C. Doléans-Dade's stochastic exponential formula we have the following theorem due to Rhiel.(The lengthy proof is omitted.)

Theorem:

1. $$\Lambda_\tau(\omega,s,x) = \int_\Theta \Lambda_\theta(\omega,s,x)\ f_\theta(\omega,s-)\ d\tau(\theta) \tag{42}$$
2. $$\nu_\tau(\omega,[0,t]\times B) = \int_\Theta\Big(\int_0^t f_\theta(\omega,s-)\nu_\theta(\omega,ds\times B)\Big)\ d\tau(\theta) \tag{43}$$

3. $\rho_\tau(\omega,t) = \int_\Theta (\int_0^t f_\theta(\omega,s-)\, \rho_\theta(\omega,ds))\, d\tau(\theta)$ (44)

For short we can write

4. $\Lambda_\tau(\omega,s,x) = \int_\Theta \Lambda_\theta(\omega,s,x)\, Q_{s-}(\omega,d\theta)$ (42')

5. $\nu_\tau(\omega,[0,t]\times B) = \int_0^t \int_\Theta \nu_\theta(\omega,ds\times B)\, Q_{s-}(\omega,d\theta)$ (43')

6. $\rho_\tau(\omega,t) = \int_0^t \int_\Theta \rho_\theta(\omega,ds)\, Q_{s-}(\omega,d\theta)$ (44')

The preceding theorem clearly shows that even very simple hypotheses P_θ may lead to complicated measures P_τ. If e.g. (N_t) is under P_θ a homogeneous process with independent increments and $\rho_\theta(\omega,t) = \theta\cdot t$, then we have

$$\rho_\tau(\omega,t) = \int_0^t (\int_\Theta \theta\, Q_{s-}(\omega,d\theta))\, ds \; . \qquad (45)$$

Even more difficult is the simple case, where under P_θ the inter-arrival times $T_{n+1} - T_n$, $n\geq 0$, are exponentially distributed with parameter $\lambda_\theta(n)$ and independent and additionally independent of claim amounts. Application of formula (32) yields

$$\rho_\theta(\omega,t) = \int_0^t \lambda_\theta(N_{s-}(\omega))\, ds \; . \qquad (46)$$

2.5. Special examples of likelihood ratios, intensities and compensators

1.Example: Let $\Lambda_\theta(\omega,s,x) = \Lambda_\theta(\omega,s)$ be independent of $x\varepsilon E$. Then $h_\theta(\omega,s,B) = h_{\theta_0}(\omega,s,B)$, i.e. we have the same conditional distribution of claim amounts under P_θ and P_{θ_0} , and $\Lambda_\theta(\omega,s) = \rho_\theta(\omega,ds) \,/\, \rho_{\theta_0}(\omega,ds)$ is the intensity of (N_t) w.r.t. the measure P_θ . Taking into account that $\Delta S_u \neq 0$ iff $\Delta N_u \neq 0$, then we obtain the following density

$$g_\theta(\omega,t) = \exp(\rho_{\theta_0}(\omega,t) - \rho_\theta(\omega,t)) \prod_{\substack{u\leq t \\ N_u\neq 0}} \Lambda_\theta(\omega,u) \qquad (47)$$

only depends on (N_t) , hence (N_t) is sufficient.

2.Example: The general one-dimensional linear model
Let $\Lambda_\theta(\omega,s,x) = \theta\cdot\Lambda(\omega,s,x)$ where $\theta\varepsilon\Theta := (0,\infty)$ and where Λ is a known function. Set

$$\nu(\omega,[0,t]\times B) := \int_0^t \int_B \Lambda(\omega,s,x)\, \nu_{\theta_0}(\omega,ds,dx)$$

$G(\omega,t) := \nu(\omega,[0,t]\times E)$. Then we have

$$\nu_\theta(\omega,[0,t]\times B) = \theta\cdot\nu(\omega,[0,t]\times B) \quad , \qquad (48)$$

$$\nu_\tau(\omega,[0,t]\times B) = \int_0^t (\int_\Theta \theta\, Q_{s-}(\omega,d\theta))\, \nu(\omega,ds\times B) \quad \text{and} \qquad (49)$$

$$g_\theta(\omega,t) = \theta^{N_t(\omega)} \, e^{-\theta\cdot G(\omega,t)} \cdot R(\omega,t) \qquad (50)$$

where the remainder function $R(\omega,t)$ is independent of θ. More general is the following example.

3.Example: The general m-dimensional linear model
Let $B_1,\dots,B_m \in \mathcal{E}$ be pairwise disjoint sets with $0 \notin B_i$.
Let $\Theta := (0,\infty)^m$ and

$$\Lambda_\theta(\omega,s,x) = \sum_{i=1}^{m} \theta_i \cdot 1_{B_i}(x)\cdot\Lambda(\omega,s,x)$$ where the funtion

Λ is known. Then

$$g_\theta(\omega,t) = \prod_{i=1}^{m} \left(\theta_i^{N(t,B_i)(\omega)} \, e^{-\theta_i G^i(\omega,t)} \right) R(\omega,t) \qquad (51)$$

with $G^i(\omega,t) = \int_0^t \int_{B_i} \Lambda(\omega,s,x)\, \nu_{\theta_0}(\omega,ds,dx)$ and the remainder function $R(\omega,t)$ independent of θ.

This example also applies to multivariate counting processes, if we set $B_i = \{e_i\}$; $\Lambda_\theta(\omega,s,e_i) = \theta_i\cdot\Lambda(\omega,s,e_i)$, $i = 1,\dots,m$ the intensity of the i.th component $S_t^i = N(t,B_i)$;

$$G^i(\omega,t) = \int_0^t \Lambda(\omega,s,e_i)\, \nu_{\theta_0}(\omega,ds,\{e_i\}) \quad .$$

Applying formula (51) one can easily compute the Fisher-information matrix.

2.6. Example of a sequential and nonsequential procedure in the linear model 2.5.2.

As an example we will use a one-dimensional gamma distribution with parameters $a>0$ and $b>0$ as a prior τ. The density of τ is given by $\theta \longmapsto b^a\, \Gamma(a)^{-1} \exp(-b\cdot\theta)\, \theta^{a-1}$. τ has mean value $\mu = a/b$ and variance $\sigma^2 = a/b^2$.
Because of formulae (23), (24) and (50) the posterior distribution $Q_t(\omega,d\theta)$ is also a gamma distribution, but with parameters $a + N_t(\omega)$ and $b + G(\omega,t)$.
We will use the loss function $L(\theta,a) = V(\theta)\,(\theta - a)^2$ with $V(\theta) = \exp(-l\theta)\cdot\theta^m$ $(-a<m<\infty$, $-b<l<\infty$).
A straight forward calculation gives the optimal estimator $d_t^\circ(\omega)$ appearing in (7), (13)

$$d_t^\circ(\omega) = \frac{N_t(\omega) + a + m}{G(\omega,t) + b + l} \qquad (52)$$

In the case $al = bm$, which includes $m=l=0$, i.e. $L(\theta,a)=(\theta-a)^2$, a simple calculation yields

$$d_t^{\circ}(\omega) = (1 - Z(\omega,t))\,\frac{a}{b} + Z(\omega,t)\,\frac{N_t(\omega)}{G(\omega,t)} \tag{53}$$

with "predictable"(!) random-credibility factors

$$Z(\omega,t) = \frac{G(\omega,t)}{G(\omega,t) + b + 1} \quad . \tag{54}$$

If $G(\omega,t)$ is non-random, i.e. if (N_t) has independent increments under P_θ for all $\theta\varepsilon\Theta$ (cf. chap. 2.1.), (and if al = bm), then the factors $Z(\omega,t)$ are non-random, too.

In the case m=l=0, i.e. $L(\theta,a) = (\theta-a)^2$, the"stochastic decision process" χ of formula (15) has the form

$$\chi_t(\omega) = c(t) + \frac{N_t(\omega) + a}{(G(\omega,t) + b)^2} \tag{55}$$

This process has to be stopped optimally.
If $G(t) := G(\omega,t)$ is non-random, then we can easily compute

$$E_\tau\,\chi_t = c(t) + \frac{a}{b}\,\frac{1}{G(t) + b} \tag{56}$$

which gives the minimal Bayes risk at time t.

In the case l=0 and m=-1, i.e. $L(\theta,a) = \frac{1}{\theta}\,(\theta-a)^2$, χ has the particular simple form

$$\chi_t(\omega) = c(t) + \frac{1}{G(\omega,t) + b} \tag{57}$$

This process must be stopped optimally in a sequential setting. If $G(\omega,t)$ is non-random, this is a trivial exercise.

In a similar way using m-dimensional gamma distributions as prior τ we can deal with the linear model 2.5.3.

References

Aalen, O.O. and Hoem, J.M. 1978. *Random time changes for multivariate counting processes.* Scand. Actuarial J., pp. 81 - 101.
Delbaen, F. 1984. *Separation of risk parameters.* Lecture at NATO A.S.I. on Insurance Premiums, Leuven.
Dellacherie, C. and Meyer, P.A. 1975. *Probabilités et potentiel.* Hermann, Paris.
Dellacherie, C. and Meyer, P.A. 1980. *Probabilités et potentiel.* Hermann, Paris.
Jacobsen, M. 1982. *Statistical analysis of counting processes.* Lecture notes in statistics 12, Springer, New York, Heidelberg, Berlin.
Jacod, J. 1975. *Multivariate point processes: Predictable projection, Radon-Nikodym derivatives, representation of martingales.*

Z. Wahrscheinlichkeitstheorie verw. Gebiete 31, pp. 235 - 253.
Jacod, J. 1979. *Calcul stochastique et problèmes de martingales.* Lecture notes in mathematics 714, Springer, Berlin, Heidelberg, New York.
Janssen, J. 1982. *Modèles de risque semi-markoviens.* Cahiers du C.E.R.O. 24, pp. 261 - 280.
Liptser, R.S. and Shiryayev, A.N. 1977. *Statistics of random processes I, general theory.* Applications of mathematics 5, Springer, New York, Heidelberg, Berlin.
Liptser, R.S. and Shiryayev, A.N. 1978. *Statistics of random processes II, applications.* Applications of mathematics 6, Springer, New York, Heidelberg, Berlin.
Rhiel, R. 1981. *Die Darstellung optimaler sequentieller Entscheidungsregeln als stochastischer Prozeß bei kontinuierlicher Zeit und beliebig vielen Hypothesen: eine Anwendung der Theorie der Martingale und der stochastischen Integration.* Dissertation, Marburg.
Thompson, M. 1971. *Continuous parameter optimal stopping problems.* Z. Wahrscheinlichkeitstheorie verw. Gebiete 19, pp. 302 - 318.
Wang, J.G. 1981. *Some remarks on processes with independent increments.* Séminaire de Probab. XV à Strasbourg 1979/1980. Lecture notes in mathematics 850, Springer, Berlin, Heidelberg, New York, pp. 627 - 631.

THE ACTUARY IN PRACTICE

D.J. Slee

C.E. Heath Underwriting & Insurance (Aust.) P/L

I have been asked by the Organising Committee to give a short talk on what I do in practice and I propose to concentrate on Workers' Compensation insurance set in a competitive market place, where contracts provide unlimited liability. But before getting down to details, just a few words about the constraints under which the actuary in such circumstances must work.

1. It must not be forgotten that each insurance policy is an individual contract between two parties. It is not possible to break that contract by saying I am sorry I got my sums wrong. As inevitably some of the sums will turn out to be "wrong", a cautious approach is essential.

2. Whatever the actuary suggests must be comprehensible to directors, underwriters, claims people and even brokers. It's no use communicating with greek lettered formulae.

3. Pay attention to what is truely significant. For example in some contracts the significance of an elegant change in premium may be quite insignificant compared to the loading.

4. Remember the underwriter wants an answer now. He can't wait 5 years whilst you collect the data.

5. Know your Broker, and analyse business just as much by source as class.

F. de Vylder et al. (eds.), Premium Calculation in Insurance, 489–492.

Because Workers' Compensation is long tail business we must first of all develop our claim experience, because we need an underwriting answer now. To do this I use the following method:

A. Subdivide the claims data by type of payment and duration since accident. From this data calculate an index of payment per open claim. The reason for this is that inflation does not act evenly over all payment types by the same duration. A subjective future rate of inflation can easily be incorporated.

B. Draw up a simple service table of claims reported and finalised, again by duration.

C. Combine the above to determine an index of total expected to be paid, and an index of payment per reported claim.

D. Using the above calculate development factors at any desired rate of interest.

A sample calculation is shown in TABLES 1 and 2.

Having now calculated what the claims experience really is by the use of development factors, the premium is calculated using one of the formulae as set out by Mark Goovaerts in the first lecture. However, I must point out that credibility theory is not used. Why? Let me just illustrate with 2 situations.

Situation 1: Consider "n" identical insured having a grouped experience. Each of "n" is not big enough to be credible, and the actual experience of each "n" varies. What happens if you apply credibility theory and in effect charge the same premium to each "n"?

Situation 2: Consider 1 very large insurer whose experience is wholly credible. This insurer wishes to buy XOL protection at a level above which it has never had a claim. Is the statistical premium zero?

TABLE 1.

CALCULATION OF INDEX FOR AVERAGE PAYMENT PER OPEN REPORTED CLAIM FOR 1982 ONWARDS

Type of Trans-action	Weekly Payments		Medical		Hospital		Lump Sums		Legal/ Assessors		Other		Index
Devpt Year	f.	p.	f.	p.	f.	p.	f.	p.	f.	p.	f.	p.	
0	100	.15	20	.75	110	.06	110	.01	190	.01	85	.02	100
1	111	.25	25	.60	121	.06	2,500	.01	200	.05	89	.03	212
2	123	.50	30	.20	135	.06	6,000	.04	210	.15	94	.05	852
3	137	.54	34	.16	176	.06	16,000	.03	225	.16	101	.05	1,479
4	152	.57	38	.13	228	.06	20,000	.02	240	.17	108	.05	1,335
5	168	.58	42	.11	296	.06	25,000	.02	257	.18	116	.05	1,627
6	187	.58	47	.10	385	.06	25,000	.02	275	.19	124	.05	1,682
7	208	.58	52	.10	501	.06	30,000	.02	295	.19	133	.05	1,982
8	230	.58	58	.10	651	.06	50,000	.02	315	.19	142	.05	3,015

f. = value of payment standardised to $100 weekly benefit increasing at 11% p.a.
p. = proportion of payments by type of transaction
The Index has been standardised to 100 in development year 0.

TABLE 2.

MODEL CLAIMS SERVICE TABLE FOR 1982 PER 1000 CLAIMS REPORTED IN FIRST YEAR

Devpt Year	Expected Claims Reported in Year	Total Claims to end of Year (Tj)	Expected Claims Finalised in Year	Claims Open During Year	Index of Payment Per Open Claim	Index of Total Expected to be Paid	Index of Payment per Reported Claim (Ij)	Development Factor at 0%	Development Factor at 9%
0	1,000	1,000	528	1,000	100	1,000	100	4.260	3.573
1	102	1,102	436	574	212	2,222	201	1.375	1.191
2	4	1,106	91	142	852	3,427	310	0.535	0.460
3	2	1,108	27	53	1,479	4,210	380	0.250	0.214
4	1	1,109	14	27	1,335	4,570	412	0.151	0.132
5	1	1,110	4	14	1,627	4,798	432	0.097	0.087
6	1	1,111	4	11	1,682	4,983	448	0.056	0.052
7	1	1,112	4	8	1,982	5,141	462	0.024	0.023
8	-	1,112	4	4	3,015	5,262	473	NIL	NIL

$$\text{Estimated Total no. of Claims} = \frac{\text{Actual Claims Reported to end of Development year } j}{T_j} \times 1{,}112$$

$$\text{Estimated Average Cost/Claim} = \frac{\text{Actual Average Paid per Claim at end of development year } j}{I_j} \times 473$$

THE INFLUENCE OF REINSURANCE LIMITS ON INFINITE TIME RUIN PROBABILITIES

M. Van Wouwe F. De Vylder M. Goovaerts
UFSIA U.C.L. K.U. Leuven

Recently a lot of results have been obtained for determining bounds on integrals with integral constraints. As a consequence best bounds on excess of loss and stop loss premiums are determined under various practical constraints. The present contribution shows how these results can be applied to determine the dependence of infinite time ruin probabilities on excess of loss reinsurance limits. Hence the optimal level of excess of loss reinsurance for a portfolio of risks is determined in case the ultimate ruin probability is taken as stability criterion.

1. INTRODUCTION

Assuming the claims have a compound Poisson distribution, the probability of ultimate ruin $\psi(t)$(= prob($T<\infty$) where T denotes the time of ruin) is governed by the following equation [1]

$$\psi(t) = \frac{\lambda}{c}\int_0^t \psi(t-y)(1-F(y))dy + \frac{\lambda}{c}\int_t^\infty (1-F(y))dy \tag{1.1}$$

$$\text{with } \psi(0) = \frac{\lambda p_1}{c}, \quad p_1 = \int_0^\infty (1-F(y))dy \tag{1.2}$$

and F denoting the claim size distribution of one claim.

One could try to examine the dependence on the excess of loss reinsurance limits of the infinite time ruin probability directly by means of (1.1). There is however a more elegant way in studying this dependence by means of the so-called adjustment coefficient R, determined as a non-trivial root of the following equation in r

F. de Vylder et al. (eds.), Premium Calculation in Insurance, 493–504.

$$\lambda + rc = \lambda \int_0^\infty e^{ry}\, dF(y) \tag{1.3}$$

It is a well known fact that [1] the infinite time ruin probability satisfies the inequality

$$\psi(t) \leqslant e^{-Rt} \tag{1.4}$$

valid for all t.

Of course (1.1) and (1.3) can be extended to account for an insurance portfolio consisting of n independent risks, assuming that the claims arising from each risk have a compound Poisson distribution. As in ref [2] we assume that the number of claims arising from the i-th risk is a Poisson process with mean p_i, and the size of each claim has a distribution function F_i. Under the usual independence assumptions the equations (1.1) and (1.3) together with (1.4) can be derived for the special choice of the parameters

$$\lambda = \sum_{i=1}^{n} p_i$$

$$F = \frac{1}{\lambda} \sum_{i=1}^{n} p_i F_i \tag{1.5}$$

If the insurer has arranged an excess of loss reinsurance for the portfolio with a retention limit M_i for the i-th risk then the insurer pays

$$Y_i = \begin{cases} X & \text{if} \quad X < M_i \\ M_i & \text{if} \quad X > M_i \end{cases} \tag{1.6}$$

and the reinsurer pays the excess amount.
Let P denote the total premium received by the insurer and let $P_i(M_i)$ denote the premium for the i-th reinsured risk then equations (1.1) and (1.4) hold with

$$c = P - \sum_{i=1}^{n} P_i(M_i) \tag{1.7}$$

together with (1.5)
and

$$F_i = F_{Y_i} \tag{1.8}$$

Let us still remark that in the sequel the convolution product is denoted by

$$\varphi * H(x) = \int_0^x \varphi(x-y)\, dH(y) \tag{1.9}$$

2. A GENERAL RESULT

Let us consider the case of one insured risk and compare the infinite time ruin probability for two different retention limits $m<M$. From (1.7) we deduce that there are two different premiums $c(m)$ and $c(M)$. Let $F_y(x,m)$, $F_y(x,M)$ denote the distribution of the reinsured risk

$$Y = \begin{cases} X & \text{if } X \leq m \\ m & \text{if } X > m \end{cases} \qquad Y = \begin{cases} X & \text{if } X \leq M \\ M & \text{if } X > M \end{cases} \tag{2.1}$$

We then have to compare the solutions of the equations

$$\psi(t,m) = \frac{\lambda}{c(m)} \int_0^t \psi(t-y, m)\, (1-F_Y(y,m))dy$$

$$+ \frac{\lambda}{c(m)} \int_t^\infty (1-F_Y(y,m))dy \tag{2.2}$$

with

$$\psi(0,m) = \frac{\lambda p_1(m)}{c(m)}$$

$$\text{where } p_1(m) = \int_0^{m+} y\, dF_Y(y,m)$$

$$= \int_0^m y\, dF_Y(y) + m(1-F_Y(m)) \tag{2.3}$$

and the solution of the equation (2.2) with m replaced by M. Use will be made of the following theorem.

<u>Theorem 1</u>. If H is a strictly defective distribution function and if f is bouned, then the renewal equation

$$\xi = f + \xi * H$$

has a unique bounded solution .
If $f \geq 0$, then $\xi \geq 0$. If $f \leq 0$, then $\xi \leq 0$.

Proof
Let $H(\infty) = \alpha < 1$. Then by induction $H^{*n} \leq \alpha^n$ $(n=1, 2, \ldots)$.
Using successively the given equation any solution ξ must satisfy :

$$\xi = f - \xi * H$$

$$\xi = f + (f + \xi * H) * H$$

$$\xi = f * \left(\sum_{i=0}^{n-1} H^{*i} \right) + \xi * H^{*n} \tag{2.4}$$

The last term in (2.4) is bounded by

$$\alpha^n \sup_{x \in \mathbb{R}^+} |\xi(x)|$$

and tends to zero when $n \to +\infty$ if ξ is bounded.
The first term of (2.4) is bounded by

$$\sup_{x \in \mathbb{R}^+} |f(x)| \sum_{i=0}^{n-1} \alpha^n \leq \frac{1}{1-\alpha} \sup_{x \in \mathbb{R}^+} |f(x)| < \infty$$

Clearly the unique bounded solution of the given renewal equation is

$$\xi = f * \left(\sum_{i=0}^{\infty} H^{*i} \right) \tag{2.5}$$

and then this theorem is obvious from this expression.
We then prove the following.

Theorem 2. The function $\psi(t,m)$ is the solution of the equation (with $\lambda = 1$)

$$\psi(m) * (p_1(m) - h_F(m)) = c(m)\, \psi(m) - h_F(m) \tag{2.6}$$

In that equation, or in any equivalent equation replace m by M and let $\psi(m)$ denote the corresponding solution.
Then

$$\psi(m) \leq \psi(M) \quad \text{if} \quad \frac{h_F(m)}{c(m)} \leq \frac{h_F(M)}{c(M)} \tag{2.7}$$

Proof
The equation (2.6) can be transformed into

$$p_1(m) - c(m)\,\psi(m) = (1-\psi(m)) * (p_1(m) - h_F(m)) \tag{2.8}$$

and of course

$$p_1(M) - c(M)\,\psi(M) = (1-\psi(M)) * (p_1(M) - R_F(M)) \tag{2.9}$$

from which we deduce

$$\begin{aligned} -\psi(m) + \psi(M) = {} & \frac{h_F(M)}{C(M)} - \frac{h_F(m)}{c(m)} \\ & + (-\psi(m)+\psi(M)) * \left(\frac{p_1(M) - h_F(M)}{c(M)}\right) \\ & + \psi(m) * \left(\frac{p_1(M) - h_F(M)}{c(M)} - \frac{p_1(m) - h_F(m)}{c(m)}\right) \end{aligned} \tag{2.10}$$

Then the previous theorem 1 can be applied with

$$\begin{aligned} \zeta &= \psi(M) - \psi(m) \\ H &= \frac{p_1(M) - h_F(M)}{c(M)} \\ f &= \left(\frac{h_F(M)}{c(M)} - \frac{h_F(m)}{c(m)}\right) * (1 - \psi(m)) \\ &\quad + \psi(m) * \left(\frac{p_1(M)}{c(M)} - \frac{p_1(m)}{c(m)}\right) \end{aligned} \tag{2.11}$$

with f a bounded function since $h_F(m) - h_F(M)$, $1 - \psi(m)$, $\psi(m)$, $p_1(M)$, $p_1(m)$, are bounded.
Moreover, because of (2.7) $f \geqslant 0$ the function H is a distribution function (it is increasing and $H \geqslant 0$), in fact a strictly defective distribution function because $H(\infty) \leqslant \frac{1}{1+\eta}$

Hence the result of the theorem follows.
We are now in the position to obtain a sufficient condition on $c(m)$ and F_x in order to have $\frac{\partial\psi}{\partial m} \geqslant 0$.
Consequently an optimal choice of the net retention as a value which makes the ruin probability minimum does not exist. This is an alternative result compared to the one in [2] and [3].

Theorem 3.

Let $c(m)/[m(1-F_X(m)) + \int_0^m (1-F_X(y)\,dy]$ be a decreasing function of m (2.12)

then $\frac{\partial\psi(t,m)}{\partial m} \geq 0$ in case F_X exists. (2.13)

Proof.
Of course $h_F(m,t) = \int_t^\infty (1-F_Y(m,y)dy$

$$= \begin{cases} 0 & t \geq m \\ \int_t^m (1-F_X(y))dy + (m-t)(1-F_X(m)) & \end{cases}$$

Consequently application of our theorem 2 results in

$$c(M)\left[\int_t^m (1-F_X(y))dy + (m-t)_+(1-F_X(m))\right]$$

$$\leq c(m)\left[\int_t^M (1-F_X(y))dy + (M-t)_+ \quad (1-F_X(M))\right] \tag{2.14}$$

$$\Rightarrow \psi(t,m) \leq \psi(t,M) \tag{2.15}$$

where of course the explicit dependence of ψ on t is given. Now (2.14) can be transformed as follows

$$\frac{c(M)-c(m)}{c(m)} \leq \frac{\int_m^M (1-F_X(y))dy + (M-t)(1-F_X(M)) - (m-t)(1-F_X(m))}{\int_t^m (1-F_x(y))dy + (m-t)(1-F_x(m))} \tag{2.16}$$

In order to have $\psi(t,m) \leq \psi(t,M)$ this inequality has to be satisfied for $t \in [(0,M)]$. The inequality (2.16) certainly holds in this case

$$\frac{C(M)-c(m)}{c(m)} \leq \frac{\int_m^M (1-F_X(y))dy + M(1-F_X(M)) - m(1-F_X(m))}{\int_0^m (1-F_X(y))dy + m(1-F_X(m))}$$

Dividing (2.17) by M−m in both members and letting m tend to M the following result is obtained

$$\frac{c'(m)}{c(m)} \leqslant \frac{2\,(F_X(m)) - m\,F'_X(m)}{\int_0^m (1-F_X(y))dy + m(1-F_X(m))} \tag{2.18}$$

or

$$\frac{d}{dm}\,\ln \frac{c(m)}{\int_0^m (1-F_X(y))dy + m(1-F_X(m))} \leqslant 0 \tag{2.19}$$

Hence (2.17) implies the fact that

$$\frac{d}{dm}\;\frac{c(m)}{\int_0^m (1-F_X(y))dy + m(1-F_X(m))} \leqslant 0 \tag{2.20}$$

because $c(m) \geqslant m(1-F_X(m)) + \int_0^m (1-F_X(y))dy$

Conversely (2.20) implies the result (2.17). Indeed, integration of (2.19) from m to M gives

$$\ln \frac{c(M)}{c(m)} \leqslant \ln \frac{\int_0^M (1-F_X(y))dy + M(1-F_X(M))}{\int_0^m (1-F_X(y))dy + m(1-F_X(m))}$$

or

$$\frac{c(M)}{c(m)} \leqslant \frac{\int_m^M (1-F_X(y))dy + M(1-F_X(M)) - m(1-F_X(m))}{\int_0^m (1-F_x(y))dy + m(1-F_x(m))}$$

An analogous result can be obtained in case of proportional reinsurance. Indeed, the insurer pays

$$y = ax \qquad \text{with} \quad 0 < a < 1$$

and in this case the reinsurer charges the premium P(a) so that the remaining premium for the insurer is

$c(a) = P - P(a)$

We will show the following.

Theorem 4. In case of proportional reinsurance with proportionality factors a, A ($a \leqslant A$)

$$\psi(t,a) \leqslant \psi(t,A) \text{ in case } \frac{c(A)}{c(a)} \leqslant \frac{A}{a} \frac{\int_{t/A}^{\infty} (1-F_X(y))dy}{\int_{t/a}^{\infty} (1-F_X(y))dy}$$

Proof.

We now have the following situation

$$F_Y(x) = \text{prob } (Y \leqslant x) \qquad \text{prob } (X \leqslant \frac{x}{a}) = F_X\left(\frac{x}{a}\right)$$

Application of theorem 2 results in

$$\psi(a) \leqslant \psi(A) \quad \text{if} \quad \frac{h_F(a)}{c(a)} \leqslant \frac{h_F(A)}{c(A)}$$

The condition on c can be transformed to

$$c(A) \int_t^{\infty} (1-F_X(\frac{x}{a}))dx \leqslant c(a) \int_t^{\infty} (1-F_X(\frac{x}{A}))dx$$

$$\text{or} \quad \frac{c(A) - c(a)}{c(a)} \leqslant \frac{A \int_{t/A}^{\infty} (1-F_X(y))dy - a \int_{t/a}^{\infty} (1-F_X(y))dy}{a \int_{t/a}^{\infty} (1-F_X(y))dy}$$

$$\text{or} \quad \frac{c(A)}{c(a)} \leqslant \frac{A}{a} \frac{\int_{t/A}^{\infty} (1-F_X(y))dy}{\int_{t/a}^{\infty} (1-F_X(y))dy}$$

Theorem 5. In case of a portfolio consisting of n risks with excess of less retention limit M_i for the i-th risk the inequality $\psi(t, m_1, \ldots m_n) \leqslant \psi(t, M_1, \ldots M_n)$ in case all partial derivatives of the first order with respect to $m_1, \ldots m_n$ of

$$P - \sum_{i=1}^{n} P_i(m_i) \Big/$$

$$\sum_i p_i m_i \ (1-F_i(m_i)) + \sum_i p_i \int_0^{m_i} (1-F_i(y))dy$$

are non positive with $M_i \geq m_i$.

3. PRACTICAL RESULTS ON THE ADJUSTMENT COEFFICIENT

In ref [5] it was shown that in case the claim size distributions are ordered by means of the stop-loss ordening

$$X_1 < X_2 \Longrightarrow R_1 > R_2 \quad \text{and hence } \exp(-R_1 x) < \exp(-R_2 x) \tag{3.1}$$

The question that arises is in case the upper bounds (or eventually the lower bounds) on stop-loss premiums are obtained, which are not stop-loss transforms, if these bounds can still be applied to provide for lower bounds on the adjustment coefficient and consequently provide for upper bounds on the corresponding Lundberg bounds for the ultimate probability of ruin. This question can be answered in the affirmative. Indeed, let $v(x)$ be such that $v''(t) \geq 0$, $v'(0) \geq 0$, then making use of the following representation

$$v(x) = \int_0^\infty (x-t)_+ \quad v''(t)dt + x\, v'(x) + v(0) \tag{3.2}$$

we derive the following equality

$$\int_0^\infty v(x) \; dF_X(x) = \int_0^\infty v''(t) \int_0^\infty (x-t) \quad dF_X(x)$$

$$+ v'(0) \int_0^\infty x \; dF_X(x) + v(0) \int_0^\infty dF_X(x) \tag{3.3}$$

In case we obtain an upper bound for h

$$h(t) = \int_t^\infty (x-t)_+ \; dF_X(x) < h_0(t) \tag{3.4}$$

by means of the results in ref [6,7]
the l.h.s. of (3.3) is dominated by

$$\int_0^\infty v(x) \; dF(x) \leq \int_0^\infty v''(t)h_0(t)dt + v'(0)h_0(0) - v(0) \tag{3.5}$$

For the special choice $v(x) = e^{Rx}$ it follows from (1.3) that a lower bound for R can be obtained by solving the equation

$$rc = \lambda r^2 \int_0^\infty e^{rt} h_0(t)dt + \lambda r\, h_0(0)$$

with the root $r=0$ as an explicit solution. The relevant root then is the unique solution of

$$c = \lambda r \int_0^\infty e^{rt} h_0(t)dt + \lambda r\, h_0(0) \tag{3.6}$$

Examples

1. We consider case C_1 of ref [7]
   with assumption : m known
   domain of parameter m : $0 \leqslant m \leqslant M$
   upper bound for stop-loss premium

   $$h_0(t) = (M-t) \cdot \frac{m}{M}$$

   Equation (3.6) can be cast in the form

   $$\frac{m}{M} (e^{Mr} - 1) - rc = 0$$

   From this equation an explicit value for the relevant root R can be obtained by various numerical methods developed to search for the solutions of non linear equations.

2. Case C'_{12} ref [7]

   assumptions : m known, F unimodal
   domain of parameter m : $0 \leqslant m \leqslant M$
   upper bound for stop-loss premium :

   $$h_0(t) = \frac{m}{M^2} (M-t)^2 \qquad \text{if} \quad m \leqslant \frac{M}{2}$$

   $$h_0(t) = \frac{1}{2M} (M-t)^2 + (m - \frac{M}{2}) (1 - \frac{t^2}{M^2}) \quad \text{if} \quad m \geqslant \frac{M}{2}$$

   In case we have the assumption $m \leqslant \frac{M}{2}$ the equation (3.6) is transformed into the following equation

   $$e^{rM} \frac{2m}{M^2} - r^2 c - \frac{2m}{M} r - \frac{2m}{M^2} = 0$$

   The assumption $m \geqslant \frac{M}{2}$ will give raise to the following

   transformation of (3.6)

$$e^{rM}\left(\frac{1}{2Mr} + 2M\left(m - \frac{M}{2}\right) - \frac{2}{r}\left(m - \frac{M}{2}\right)\right)$$

$$+ r\left(- c - M - m - m M^2 + \frac{M^3}{2}\right)$$

$$+ \frac{1}{r}\left(- \frac{1}{2M} + 2m - M\right) - 1 = 0$$

We are also able to prove the following theorem in case we consider two retention levels m, M in case of excess of los reinsurance.

Theorem 6.

If $\frac{h(t,m)}{c(m)} \leqslant \frac{h(t,M)}{c(M)}$ it follows that

$R(m) \geqslant R(M)$ and consequently that

$\exp(-R(m)u) \leqslant \exp(-R(M)u)$

Proof
immediately from the reasoning giving raise to (3.6) (see also (1.7)

Theorem 7. If $c(m) / [m(1-F_X(m)) + \int_0^m (1-F_X(y))dy]$ is a decreasing function of m, then R(m) is a decreasing function of m and hence $\exp(-R(m))$ is an increasing function of m.

Proof
immediately from theorem 6 and an analogeous reasoning as in theorem 3.

Remark.

In case $c(m) / [m(1-F_X(m)) + \int_0^m (1-F_X(y))dy]$ is a decreasing function of m then the optimal excess of loss retention limit is obtained by taking the largest value of m for which

$$\psi(m,u) \leqslant \varepsilon$$

where u denotes the initial reserve and ε the upper bound for the admissible ruin probability on the other hand in order to get more straightforward upper bounds one could choose the largest value of m for which

$$e^{-R(m)u} < \varepsilon$$

Practically this problem is easily solved by solving (3.5) (see e.g. the two examples).

References

1 Gerber, H. "An Introduction to Mathematical Risk Theory" (eds) S.S. Huebner Foundation Monograph Series (1979).

2 Andreakis, M. and Waters, H.R. "The effect of reinsurance on the degree of risk associated with an insurer's portfolio" Astin Bulletin 11 (1980), pp. 119-135.

3 Waters, H.R. "Excess of loss Reinsurance Limits" S.A.J. (1979), pp. 37-43.

4 Waters, H.R. "Some mathematical aspects of reinsurance Insurance : Mathematics and Economics (1983) V2, 1, pp. 17-26.

5 Goovaerts, M. and De Vylder, F. "Upper and lower bounds on infinite time ruin probabilities in case of constraints on claim size distributions" Journal of Econometrics (to be published).

6 De Vylder, F. "Best upper bounds for integrals with respect to measures allowed to very under conical and integral constraints" Insurance Math. Economics 1(2), pp. 109-130.

7 De Vylder, F. and Goovaerts, M. "Analytical best upper bounds on stop loss premiums" Insurance Math. Economics 1,(3), pp. 117-212.

SOME BERRY-ESSEEN THEOREMS FOR RISK PROCESSES

K. JANSEN

Universiteit Antwerpen - U.I.A. - Departement Wiskunde
Universiteitsplein 1, B-2610 Wilrijk - BELGIUM

Abstract. We derive some Berry-Esseen theorems for risk processes in case the underlying counting process is a weighted Poisson process.

F. de Vylder et al. (eds.), Premium Calculation in Insurance, 505–517.

1. INTRODUCTION.

Let $\{N_t : t \in \mathbb{R}_+\}$ be a weighted Poisson process with structure probability distribution S on $\mathbb{R}_+$, i.e.

$$P(N_t=n) = \int \frac{(\lambda t)^n}{n!} e^{-\lambda t} dS(\lambda) . \tag{1.1}$$

Suppose that S has a bounded density function f with

$$\sup_{\lambda \in \mathbb{R}_+} f(\lambda) = m . \tag{1.2}$$

Suppose also that the variance of S exists and is not equal to zero. Let

$$a = \int \lambda dS(\lambda) \quad \text{and} \quad s^2 = \int \lambda^2 dS(\lambda) - a^2, \tag{1.3}$$

and consider a real-valued random variable Y such that $P_Y = S$. Let $\{X_n : n \in \mathbb{N}\}$ be a sequence of i.i.d. random variables, independent of $\{N_t : t \in \mathbb{R}_+\}$, such that the third absolute moment of P_{X_1} exists, and denote

$$p_i = \int x^i dP_{X_1}(x) \qquad \text{for } i = 1,2,3. \tag{1.4}$$

Suppose $X_n > o$, then $p_i > o$ for $i = 1,2,3$.

We define the risk process and the normalised risk process by

$$Z_t = \sum_{i=o}^{N_t} X_i \quad \text{and} \quad Z_t^{\star} = \frac{Z_t - E[Z_t]}{V[Z_t]^{1/2}} . \tag{1.5}$$

It easily follows that

$$E[N_t] = at, \; V[N_t] = at + s^2 t^2, \; E[Z_t] = ap_1 t, \; V[Z_t] = ap_2 t + (sp_1 t)^2 . \tag{1.6}$$

From [2] example 2, it follows that the distribution of $Z_t^{\star}$ converges weakly to a limit distribution Q given by

$$\lim_{t\to\infty} P_{Z_t^{\star}} = Q = P_{\frac{Y-a}{s}} . \tag{1.7}$$

This is in fact the normalised structure distribution of the weighted Poisson process.

The characteristic function of Q is equal to

$$\varphi_Q(u) = \int e^{i\frac{u}{s}\lambda}\, dS(\lambda)\, e^{-iu\frac{a}{s}} . \tag{1.8}$$

If q is the density function of Q then

$$q(y) = sf(sy+a) , \tag{1.9}$$

and so $$\sup_{y\in\mathbb{R}} q(y) = sm \tag{1.10}$$

with m defined in (1.2).

We will need the following lemma which can be found in the book of Feller (1971).

Lemma.

Let P be a probability distribution on $(\mathbb{R},\mathcal{B})$ with distribution function F and characteristic function φ, such that the first absolute moment exists, and Q a probability distribution with bounded density function q ($\sup_{y\in\mathbb{R}} q(y) = M$), distr. function G, char. function χ, such that the first moment also exists. Then the following inequality holds for all $T > o$:

$$\sup_{x\in\mathbb{R}} |F(x)-G(x)| \leq \frac{1}{\pi}\int_{-T}^{T} \left|\frac{\varphi(u)-\chi(u)}{u}\right| du + \frac{24M}{\pi T} . \tag{1.11}$$

Proof.

See [1] p.537-538.

■

Finally we denote F_t for the distribution function of Z_t^* , G for the distribution function of the limit distribution Q, and φ_t for the characteristic function of Z_t^* .

2. THE MEAN RESULTS.

For the next theorem it is sufficient that only the first two moments of P_{X_1} (as defined in (1.4)) exist.

Theorem 1.

Suppose we are in the case as explained in the introduction. Then the following inequality holds:

$$\sup_{x\in\mathbb{R}}|F_t(x)-G(x)| \leq t^{-1/3}[\inf_{\delta>o} \frac{1}{\pi}(\frac{p_2 a}{p_1^2 s^2}(\frac{\delta^2}{2}+\delta t^{-1/3}) + \frac{24ms}{\delta})] \tag{2.1}$$

Proof.

Let ψ be the characteristic function of P_{X_1}, then

$$\begin{aligned}\varphi_t(u) &= E[\exp(iuZ_t^\star)] \\ &= \sum_{n=o}^{\infty} P(N_t=n)\psi^n(\frac{u}{V[Z_t]^{1/2}})\exp(-iu\frac{E[Z_t]}{V[Z_t]^{1/2}}) \\ &= \int \exp(\lambda t(\psi(v)-1) - iatp_1 v)\, dS(\lambda)\end{aligned} \tag{2.2}$$

by (1.1), (1.5) and (1.6), where

$$v = \frac{u}{st(\frac{ap_2}{s^2 t}+p_1^2)^{1/2}} . \tag{2.3}$$

From (1.8) and (2.2) it results that

$$|\varphi_t(u)-\varphi_Q(u)| \leq \int |\exp(\lambda t(\psi(v)-1)-iatp_1 v-i\frac{u}{s}\lambda+i\frac{u}{s}a) - 1|dS(\lambda). \tag{2.4}$$

For every $z \in \mathbb{C}$ with a negative real part ($\Re ez \leq o$) we have

$$|e^z - 1| \leq |z| . \tag{2.5}$$

Since $\Re e[\lambda t(\psi(v)-1) - iatp_1 v - i\frac{u}{s}\lambda + i\frac{u}{s}a] \leq o$, it follows from (2.5), (2.3) and from the inequality

$$|\psi(z) - 1 - ip_1 z| \leq \frac{|z|^2}{2} p_2 \tag{2.6}$$

that

$$|\exp(\lambda t(\psi(v)-1)-iatp_1 v-i\frac{u}{s}\lambda+i\frac{u}{s}a) - 1|$$

$$\leq |\lambda t(\psi(v)-1-ivp_1) + iv(\lambda-a)t(p_1-(\frac{ap_2}{s^2 t}+p_1^2)^{1/2})|$$

$$\leq \lambda\frac{p_2 u^2}{2ts^2 p_1^2} + |u|\frac{|\lambda-a|}{s}\frac{ap_2}{2p_1^2 s^2 t} . \tag{2.7}$$

Combining (2.4) and (2.7), we get

$$|\varphi_t(u)-\varphi_Q(u)| \leq \frac{|u|}{t}\frac{p_2 a}{2p_1^2 s^2}(|u|+1) \tag{2.8}$$

also using (1.3) and the inequality $\int\frac{|\lambda-a|}{s}\,dS(\lambda) \leq 1$.

Now put $T=\delta t^{1/3}$ where $\delta > o$. From the lemma (1.11), (1.10) and (2.8) it finally results that

$$\sup_{x\in\mathbb{R}}|F_t(x)-G(x)| \leq \frac{1}{\pi}\int_{-\delta t^{1/3}}^{\delta t^{1/3}} \frac{1}{t}(\frac{p_2 a}{2p_1^2 s^2}(|u|+1))du + \frac{24ms}{\pi\delta t^{1/3}}$$

$$= \frac{1}{\pi t^{1/3}}\left(\frac{p_2 a}{p_1^2 s^2}(\frac{\delta^2}{2} + \delta t^{-1/3}) + \frac{24ms}{\delta}\right) \tag{2.9}$$

Since (2.9) holds for every $\delta > o$, (2.1) is proven. ∎

Remark.

Since $s\neq o$, the theorem does not give an estimation in case of a Poisson counting process. For such an estimation see [3] theorem (3.6).

In the particular case when the structure probability distribution is a Γ-distribution, we can get a better approximation than in theorem 1 (instead of $t^{-1/3}$ we can obtain t^{-1}).

So we suppose that S is a Γ-distribution with parameters $c>o$ and $p>1$, i.e.

$$f(\lambda) = \frac{c^p}{\Gamma(p)} e^{-c\lambda} \lambda^{p-1} \qquad \lambda>o. \tag{2.10}$$

For (1.1), (1.2), (1.3), (1.6), (1.8), (1.9) we get in this particular case :

$$P(N_t=n) = \binom{n+p-1}{n} \left(\frac{c}{c+t}\right)^p \left(\frac{t}{c+t}\right)^n, \tag{2.11}$$

$$m = \frac{c}{\Gamma(p)} (p-1)^{p-1} e^{-(p-1)}, \tag{2.12}$$

$$a = \frac{p}{c} \quad \text{and} \quad s^2 = \frac{p}{c^2}, \tag{2.13}$$

$$E[N_t] = \frac{p}{c}t, \; V[N_t] = \frac{pt}{c^2}(c+t), \; E[Z_t] = p_1 \frac{pt}{c}, \; V[Z_t] = p_2 \frac{pt}{c} + p_1^2 \frac{pt^2}{c^2}, \tag{2.14}$$

$$\varphi_Q(u) = e^{-i\sqrt{p}u} \left(1 - i\frac{u}{\sqrt{p}}\right)^{-p}, \tag{2.15}$$

$$g(y) = \frac{p^{p/2}}{\Gamma(p)} e^{-\sqrt{p}(y+\sqrt{p})} (y+\sqrt{p})^{p-1} 1_{[-\sqrt{p},\infty)}(y). \tag{2.16}$$

<u>Theorem 2.</u>

In the situation as above, we have the following inequality for $t \geq 4p_2c/5p_1^2$:

$$\sup_{x\in\mathbb{R}} |F_t(x) - G(x)| \leq$$

$$t^{-1}\Big[\frac{I_p}{\pi} \frac{3}{4} \frac{\sqrt{p}p_2 c}{p_1^2} \left(\frac{\sqrt{2}}{\sqrt{3}} \frac{p_2}{\sqrt{p_1 p_3}} + 1\right) \exp\left(\frac{3}{8} \frac{pp_2^2}{p_1 p_3} + \frac{1}{t} \frac{pp_2 c}{p_1^2}\left(\frac{\sqrt{6}}{4} \frac{p_2}{\sqrt{p_1 p_3}} + \frac{3}{2}\right) \right)$$

$$+ \frac{24\sqrt{2}}{\pi\sqrt{3}} \frac{c}{\Gamma(p)} (p-1)^{p-1} e^{-(p-1)} \left(\frac{p_3}{p_1^3}\right)^{1/2} \Big] \tag{2.17}$$

where $$I_p = \int_{-\infty}^{+\infty} (1+ \frac{u^2}{p})^{-p/2} du = \sqrt{p}\, \frac{\Gamma(\frac{p-1}{2})\Gamma(\frac{1}{2})}{\Gamma(\frac{p}{2})} . \tag{2.18}$$

Proof.

Let ψ be the characteristic function of P_{X_1}, then

$$\varphi_t(u) = (\sum_{n=o}^{\infty} P(N_t=n)\psi^n(\frac{u}{V[Z_t]^{1/2}})) \exp(-iu\frac{E[Z_t]}{V[Z_t]^{1/2}})$$

$$= (1 + \frac{t}{c}(1-\psi(v)))^{-p} \exp(-ipp_1\frac{tv}{c}) , \tag{2.19}$$

using (2.11) and the equality $\sum_{n=o}^{\infty} \binom{n+p-1}{n} z^n = (1-z)^{-p}$ for $|z| < 1$,

and putting $v = \frac{c}{t}\frac{u}{\sqrt{p}}(\frac{p_2 c}{t} + p_1^2)^{-1/2}$. (2.20)

From (2.15) and (2.19) it follows that

$$|\varphi_t(u)-\varphi_Q(u)|$$

$$= (1+\frac{u^2}{p})^{-p/2} |\exp(-p(\log(\frac{1+\frac{t}{c}(1-\psi(v))}{1-iu/\sqrt{p}})+i(p_1\frac{tv}{c} - \frac{u}{\sqrt{p}}))) - 1| \tag{2.21}$$

where log is the principal value of the logarithm defined on $\mathbb{C}\setminus(-\infty,o]$ by

$$\log z = \int_{[1,z]} \frac{1}{\zeta}\, d\zeta = \int_o^1 \frac{z-1}{1+(z-1)t}\, dt . \tag{2.22}$$

Hence we have for $z \in \mathbb{C}$ with $\Re z > -1$:

$$|\log(1+z)-z| = |\int_o^1 \frac{z}{1+zt}dt - \int_o^1 z dt|$$

$$\leq |z|^2 \int_o^1 \frac{t}{|1+zt|}dt \leq \frac{|z|^2}{2} \quad \text{if } \Re z \geq o$$

$$\leq \frac{|z|^2}{2(1+\Re z)} \quad \text{if } \Re z \leq o. \tag{2.23}$$

Suppose now that $t \geq \frac{4}{5} \frac{p_2 c}{p_1^2}$ (2.24)

and put $T = (\frac{3}{2})^{1/2} (\frac{p_1^3}{p_3})^{1/2} (\frac{\sqrt{p}}{c} t)$. (2.25)

Also assume that $|u| \leq T$. (2.26)

Then we have using (2.20) and the inequality $|\sin x - x| \leq \frac{|x|^3}{3!}$.

$$|\frac{t}{c} \frac{u}{\sqrt{p}} \int \sin(vy)\, dP_X(y) - \frac{u^2}{p}|$$

$$= |\frac{t}{c} \frac{u}{\sqrt{p}} [\int (\sin(vy) - vy)\, dP_X(y) + \int vy (1 - (\frac{p_2 c}{p_1^2 t} + 1)^{1/2})\, dP_X(y)]|$$

$$\leq \frac{t}{c} \frac{|u|}{\sqrt{p}} [\frac{p_3 |v|^3}{3!} + |v| p_1 ((\frac{p_2 c}{p_1^2 t} + 1)^{1/2} - 1)]$$

$$\leq \frac{u^2}{p} [\frac{p_1}{4} (\frac{p_2 c}{t} + p_1^2)^{-1/2} + 1 - p_1 (\frac{p_2 c}{t} + p_1^2)^{-1/2}] \quad \text{by (2.26)}$$

$$\leq \frac{u^2}{2p} \quad \text{by (2.24)} . \tag{2.27}$$

From (2.27) it results that

$$\mathcal{R}e (\frac{i \frac{u}{\sqrt{p}} + \frac{t}{c}(1 - \psi(v))}{(1 - iu/\sqrt{p})}) \geq \frac{1}{1 + u^2/p} (\frac{t}{c} \frac{u}{\sqrt{p}} \int \sin(vy)\, dP_X(y) - \frac{u^2}{p}) > -\frac{1}{2} . \tag{2.28}$$

Combining (2.23) and (2.28) we get

$$|\log(1 + \frac{i \frac{u}{\sqrt{p}} + \frac{t}{c}(1 - \psi(v))}{(1 - iu/\sqrt{p})}) - \frac{i \frac{u}{\sqrt{p}} + \frac{t}{c}(1 - \psi(v))}{(1 - iu/\sqrt{p})}|$$

$$\leq \frac{t^2}{c^2} |i \frac{u}{\sqrt{p}} \frac{c}{t} + 1 - \psi(v)|^2 (1 + \frac{u^2}{p})^{-1}$$

$$\leq \frac{t^2}{c^2} (|ivp_1 + 1 - \psi(v)| + |ivp_1 - i \frac{u}{\sqrt{p}} \frac{c}{t}|)^2 (1 + \frac{u^2}{p})^{-1}$$

$$\leq \frac{u^2}{p} (1 + \frac{u^2}{p})^{-1} (\frac{p_2 v}{2} + ((\frac{p_2 c}{t} + p_1^2)^{1/2} - p_1))^2 (\frac{p_2 c}{t} + p_1^2)^{-1} \tag{2.29}$$

by (2.6) and (2.20).

Now we are going to estimate the exponent of the power of e in equation (2.21).

$$|p(\log(\frac{1+\frac{t}{c}(1-\psi(v))}{1-iu/\sqrt{p}}) + i(p_1\frac{tv}{c} - \frac{u}{\sqrt{p}}))|$$

$$\leq \frac{u^2}{(1+u^2/p)}\left(\frac{u^2c^2p_2^2}{4t^2pp_1^4} + \frac{u}{2}\frac{c^2}{t^2}\frac{p_2^2}{\sqrt{p}p_1^4} + \frac{3}{2}\frac{p_2c}{p_1^2t}\right) \tag{2.30}$$

(by (2.29), (2.6), (2.20) and the inequalities

$$((\frac{p_2c}{t} + p_1^2)^{1/2} - p_1) \leq \frac{p_2c}{2p_1t} \text{ and } \frac{p_2c}{t} + p_1^2 \geq p_1^2)$$

$$\leq \frac{|u|}{t}\frac{\sqrt{p}p_2c}{p_1^2}\left(\frac{\sqrt{6}}{4}\frac{p_2}{\sqrt{p_1p_3}} + \frac{3}{4}\right) \tag{2.31}$$

(since $|u| \leq T$, $\frac{u^2}{1+u^2/p} \leq p$ and $\frac{|u|}{1+u^2/p} \leq \frac{\sqrt{p}}{2}$) .

(2.30) is also dominated by :

$$p\left(\frac{3}{8}\frac{p_2^2}{p_1p_3} + \frac{\sqrt{6}}{4}\frac{p_2^2}{p_1^{5/2}p_3^{1/2}}\frac{c}{t} + \frac{3}{2}\frac{p_2c}{p_1^2t}\right) . \tag{2.32}$$

For $y \in \mathbb{C}$ the following inequality holds :

$$|e^y - 1| \leq |y|e^{|y|} . \tag{2.33}$$

From (2.21), (2.31), (2.32) and (2.33) we can deduce, for

$|u| \leq T$ and $t \geq \frac{4}{5}\frac{p_2c}{p_1^2}$ with T defined in (2.25), that

$$|\varphi_t(u)-\varphi_Q(u)|$$

$$\leq (1+\frac{u^2}{p})^{-p/2}\frac{|u|}{t}\frac{\sqrt{p}p_2c}{p_1^2}\left(\frac{\sqrt{6}}{4}\frac{p_2}{\sqrt{p_1p_3}} + \frac{3}{4}\right)\exp\left(p\left(\frac{3}{8}\frac{p_2^2}{p_1p_3} + \frac{\sqrt{6}}{4}\frac{p_2^2c}{p_1^{5/2}p_3^{1/2}t} + \frac{3p_2c}{2p_1^2t}\right)\right) \tag{2.34}$$

Finally the theorem follows from (2.34), (2.12) and the lemma.

■

3. A COMPARISON BETWEEN THE TWO ESTIMATIONS.

Suppose that the structure probability distribution in theorem 1 is a Γ-distribution as in theorem 2. Using (2.12) and (2.13) we get for the result of theorem 1 :

$$\sup_{x\in\mathbb{R}}|F_t(x)-G(x)| \leqslant t^{-1/3}\left[\inf_{\delta>0}\frac{1}{\pi}\left(\frac{p_2c}{p_1^2}\left(\frac{\delta^2}{2}+\delta t^{-1/3}\right)+\frac{24\sqrt{p}}{\delta\Gamma(p)}e^{-(p-1)}(p-1)^{p-1}\right]\right. \tag{3.1}$$

(2.17) and (3.1) are estimations of the same distribution functions. It is clear that the estimation (2.17) is much better than (3.1) for large values of t. Call the coefficients in (2.17) and (3.1) resp. A and B, then

$$\sup_{x\in\mathbb{R}}|F_t(x)-G(x)| \leqslant At^{-1} \tag{3.2}$$

$$\leqslant Bt^{-1/3} \tag{3.3}$$

We shall now compare A and B for some values of the parameters. Taking $K^{1/3}$ as an approximation for the value of δ at which B reaches its minimum, where

$$K = \frac{24ms^3p_1^2}{p_2a} = 24\frac{p_1^2}{p_2}\frac{\sqrt{p}}{c\Gamma(p)}e^{-(p-1)}(p-1)^{p-1}, \tag{3.4}$$

gives

$$B \leqslant D = \frac{24^{2/3}c^{1/3}}{\pi}\frac{e^{-\frac{2}{3}(p-1)}(p-1)^{\frac{2}{3}(p-1)}}{\Gamma(p)^{2/3}}\frac{p_2^{1/3}p^{1/3}}{p_1^{2/3}}$$

$$\times\left(\frac{3}{2}+t^{-1/3}\left(\frac{p_2c\Gamma(p)}{24p_1^2\sqrt{p}}e^{(p-1)}(p-1)^{-(p-1)}\right)^{1/3}\right) \tag{3.5}$$

In order to get a smaller number of parameters, we also make the supplementary assumption that P_{X_1} is an exponential distribution with parameter $\theta > 0$. Then

$$p_1 = 1/\theta,\ p_2 = 2/\theta^2,\ p_3 = 6/\theta^3,$$

so A and B do not depend on θ.

Some numerical values.

a) The coefficient D.

t	p\c	0,1	0,5	1	2	3	5	10	100
t=1	1,5	1,90	3,57	4,76	6,39	7,64	9,62	13,3	42,8
	2	1,76	3,31	4,41	5,94	7,11	8,96	12,4	40,3
	3	1,65	3,11	4,15	5,60	6,71	8,48	11,8	38,4
	5	1,58	2,98	3,99	5,39	6,46	8,16	11,3	37,2
	10	1,53	2,90	3,88	5,25	6,29	7,96	11,1	36,4
	∞	1,49	2,83	3,79	5,12	6,14	7,78	10,8	35,7
t=10	1,5	1,76	3,16	4,10	5,35	6,28	7,70	10,3	28,6
	2	1,62	2,91	3,78	4,94	5,80	7,13	9,50	26,8
	3	1,52	2,73	3,55	4,64	5,45	6,70	8,95	25,3
	5	1,45	2,61	3,40	4,45	5,22	6,43	8,59	24,4
	10	1,40	2,53	3,30	4,32	5,08	6,25	8,36	23,9
	∞	1,37	2,47	3,21	4,21	4,95	6,09	8,16	23,3
t=∞	1,5	1,64	2,80	3,53	4,45	5,09	6,04	7,61	16,4
	2	1,50	2,57	3,24	4,08	4,67	5,54	6,98	15,0
	3	1,40	2,40	3,02	3,81	4,36	5,17	6,51	14,0
	5	1,34	2,29	2,88	3,63	4,16	4,93	6,21	13,4
	10	1,30	2,22	2,79	3,52	4,03	4,77	6,02	13,0
	∞	1,26	2,15	2,71	3,42	3,91	4,64	5,84	12,6

b) The coefficient A.

($t \geqslant 8c/5$)

t=1

p\c	0,1	0,5
2	2,09	190
3	3,20	170x10
5	11,7	169×10^3
10	151	671×10^7

t=10

p\c	0,1	0,5	1	2	3
2	1,30	7,92	20,9	79,1	243
3	1,36	9,72	32,0	193	935
5	2,18	22,4	117	169x10	186×10^2
10	4,33	103	151x10	165×10^3	135×10^5

t=∞

p\c	0,1	0,5	1	2	3	5	10	100
2	1,25	6,24	12,5	24,9	37,4	62,4	125	125x10
3	1,25	6,27	12,5	25,1	37,6	62,7	125	125x10
5	1,84	9,19	18,4	36,7	55,1	91,9	184	184x10
10	2,97	14,9	29,7	59,4	89,1	148	297	297x10

Remark.

D is convergent for $p \to 1$ and $p \to \infty$, while A diverges in both cases. The tables show clearly that A is a rather bad beheaving constant in comparison with the constant D. This does not mean that the global estimation is worse, as is shown in the following example. Put p=5, c=1, t=1000, then (we can use the table for t=10, since A and D decrease with increasing t) $A \leq 117$ and $D \leq 3{,}40$, so $At^{-1} \leq 0{,}117$ but $Bt^{-1/3} \leq 0{,}340$.

4. REFERENCES.

[1] Feller W. (1971). An Introduction to Probability Theory and its Applications, vol.II, sec. ed.. John Wiley & Sons, Inc. New York.

[2] Jansen K., Haezendonck J., Delbaen F. (1983). Limit Distributions for Risk Processes in case of Claim Amounts of Finite Expectation. To appear in: Insurance Mathematics and Economics.

[3] Von Chossy R., Rappl G. (1983). Some approximation Methods for the Distribution of Random Sums. To appear in: Insurance Mathematics and Economics.

SOME NOTES ON THE METHODS OF CALCULATION OF LIFE ASSURANCE PREMIUMS IN THE UNITED KINGDOM

A.D. Wilkie, M.A., F.F.A., F.I.A.

Standard Life Assurance Company and Heriot-Watt University, Edinburgh.

Abstract

The way in which life assurance premium rates in the United Kingdom are determined is discussed. The first part of the paper discusses two methods currently in use, both based on deterministic assumptions. The method of the "Cautious Actuary" is to calculate the premium so as to equate the expected present value of income to the expected present value of outgo on the policy, but with everything calculated to be somewhat on the safe side. Implicit margins are thus taken throughout. The second method currently used is the "Profit Testing" approach, which explicitly takes into account the insurance company's capital requirements, both to finance initial expenses and to finance a sufficient policy reserve. It is also assumed that the shareholders wish to earn a satisfactory commercial return on their capital. The premiums are set in such a way as to provide the required return. An alternative approach is then described, based on stochastic methods, and using as an example a new type of policy, an index-linked annuity with minimum money guarantees of various kinds. The cautious actuary would now choose a premium such that the probability that it would prove insufficient is quite small. The profit tester would assume that very strong reserves need to be set up, financed partly by shareholders, and providing them ultimately with profitable return; the remainder of the reserve would come from the premium.

F. de Vylder et al. (eds.), Premium Calculation in Insurance, 519–532.

Introduction

1. The premiums that life offices in the United Kingdom charge for life assurance or annuity contracts are not controlled by government supervision nor by agreement between life offices. Each office is free to charge what it thinks fit. In a competitive market of course each office is constrained by competition not to overcharge. And since each office is required to remain solvent, and required to maintain adequate technical reserves or mathematical reserves (the precise definitions of and difference between which are not well understood in Britain), which are to be determined and certified by the "Appointed Actuary" of the office, the office cannot undercharge by too much without danger of insolvency. Such a structure, however, does not prevent an office running a "loss leader", that is, accepting business at premiums that may be strictly unprofitable when the business is taken on its own, but which are thought to contribute to the overall profitability of the office.

2. Such a "laissez-faire" approach means that life offices and the actuaries who advise them or work for them, are free to use a number of different principles for calculating premiums, as well as being free to use any number of different "bases", by which in Britain is meant the tables of mortality, and the levels of interest and expense loading incorporated in the premium. But because each office "does its own thing" the outside observer cannot tell what method or basis has been used. He can only see the final premium rates.

3. This note can therefore do no more than describe some of the commonly used methods, but it will then go on to discuss some new developments in stochastic methods of premium determination which I have been working on.

The Traditional Method

4. The traditional method of premium calculation, which I shall call the method of the "cautious actuary", is the one that has appeared in actuarial textbooks for more than a century. Bases of mortality, interest and expenses are chosen, and the premium is calculated to equate the expected present value of income on the policy to the expected present value of outgo. The bases are usually chosen to be appropriate to the particular office and to the particular type of policy, but somewhat on the safe side. Thus for assurances one would base the rates of mortality on a table appropriate to assured lives (such as is published by the joint Continuous Mortality Investigation Committee of the Faculty and Institute of Actuaries, basing their tables on the experience of a

substantial number of life offices). Since mortality has generally been improving, to use a recently constructed table has generally been sufficiently safe. However, since the mortality of annuitants is expected to improve, and is in any case much lighter than that of assured lives, it has been usual to base annuity rates on an annuitant's table with an appropriate allowance for forecast improvements in mortality. It is customary to use "select" tables, but not to use "generation" tables directly.

5. In the traditional formulae the interest rate would be taken as uniform over the whole duration of the policy, and for annual premium policies a moderately cautious estimate of the long-term rate of interest would be used. For single premium policies and immediate annuities the rates used would take into account those currently obtainable on suitable fixed interest stocks. However, over the past twenty years interest rates have risen substantially, and have always seemed to be unusually "high". Various methods have been used, therefore, to allow for the high returns that may be obtained on the investment in the early years of the policy, on the assumption that subsequent premiums and future interest income may be reinvested at lower interest rates.

6. The expenses allowed for have usually been of the form: initial expenses as a percentage of the sum assured or a percentage of the first premium (depending in part on the customary commission rates in force at the time) plus a fixed charge; renewal expenses as a percentage of the premium plus a fixed charge. It has long been customary to adjust the premium rate for the size of the sum assured. The older practice was to use a "sliding scale", whereby the premium rate per cent was reduced by small fixed amounts as the sum assured exceeded certain fixed limits. More recent practice has tended towards charging a "policy fee", that is, a fixed monthly or annual addition to the basic premium which is taken as proportionate to the sum assured.

7. Usually no explicit loading for risk or uncertainty is included in the premiums. It is felt that the uncertainty about the true basis factors (mortality, interest, etc) for the future outweighs any stochastic fluctuations because of variations in the lifetimes of individual policyholders. To use a sufficiently cautious basis is therefore sufficient, without explicit security loadings.

8. A substantial amount of life assurance business is written in Britain on a without profits basis, including, for example, almost all term assurance and immediate annuity business. However, a very much larger proportion of business is with profits. The usual method of distributing profits to policies in Britain has been by means of "reversionary bonus", that is, by an addition to the sum assured payable on death or maturity, the premium remaining unchanged. The bonus additions are most usually calculated on a uniform compound system, with the bonus each year being a level percentage of the sum assured and previous bonuses. However, both simple bonus systems (the bonus being based only on the sum assured) and super-compound systems (the bonus being declared at a higher rate on existing bonus) are also known. A terminal bonus payable on death or maturity is now quite common. It is customary to charge in the premiums a "bonus loading" assuming a bonus of say half the presently declared rate.

9. In practice many offices keep their with profits rates almost unchanged for very many years, though they do calculate from time to time the effective bonus loading that is inherent in the rates. Without profits premium rates are changed much more frequently, and the rates for immediate annuities may change very frequently along with the prices of Government securities. When premium scales for new policies are changed, premiums for existing policies remain unchanged.

10. Within such a framework of premium rate calculation, which is suitable for a mutual life office, or a well established proprietary life office (i.e. one with shareholders) which has a large proportion of with profits business, and where a large part of the profits are returned to the policyholders, the actuary's main concern is decide what are appropriate bases to use at any time, and how to be equitable between different generations of with profit policyholders in the distribution of surplus.

Profit-testing

11. However, the formation in the last two decades of a number of new proprietary life offices selling almost entirely without profits business, in particular "unit linked" business (which will be described below), has encouraged some actuaries to think in terms of the profitability to the shareholders of particular products. The "profit testing" approach has been described in papers by Benjamin (1976) and Smart (1977).

12. The problem of the new life office is that it must lay out substantial capital in formation expenses, advertising and selling, both in aggregate and in respect of each policy sold. Even if it uses Zillmerized reserves it may not treat any policy as an asset, so it requires initial capital to finance initial expenses which it hopes to recoup from future renewal premiums. In addition, the system of taxation of British life offices provides the new office with an intangible asset in the form of unrelieved expenses; in effect, tax that would be due in future on interest income can be reduced by a carry forward of initial expenses, but this amount carried forward is only of value if future interest income is in fact received; such an asset cannot practicably enter a balance sheet.

13. Shareholders of such offices are naturally concerned to make profits on their investment, and will wish to know how much that profit might be and how quickly it will emerge. The profit testing approach to premiums therefore assumes explicitly that the shareholders finance initial expenses for each policy out of capital, and wish to earn a satisfactory commercial return on that capital, at a rate of interest suitably higher than that assumed to be earned on the investments. In addition, a specific profit per policy may be added, to cover the formation expenses and setting up of necessary contingency reserves. The specific cash flow position of a group of policies, or of the whole office, is considered, taking account of the incidence of expenses and of taxation. By "trial and error", or rather by successive approximation, a premium can be determined that provides an adequate profit margin. It may seem surprising that premiums calculated in this way are competitive with those charged by established life offices, but it is often said in Britain that "life assurance is sold, not bought", and it appears to be the case that a life office, by using skilful though expensive selling techniques can sell business at premium rates that adequately cover these expenses, at least within reasonable limits.

Linked Life Assurance

14. In the past two decades in Britain there has been a very substantial expansion in "unit linked" life assurance. With such policies a specified large part of each premium is deemed to be invested in units of a "fund", i.e. a specified set of assets allocated to these policies by the life office, and run very like a unit trust or mutual fund. In early policies of this type the investment portion of the premium was in fact invested in units of a unit trust, and many such policies still continue, though

the internal fund is now more common. The remainder of the premium is available to the life office for commission and expenses, and to cover the mortality risk, i.e. the amount (if any) by which the specified sum assured exceeds the value of the units at death. In addition, the policy usually prescribes that an annual management charge can be made, equal to a small fraction (0.5 to 1.0%) of the value of the units. The income from the investments is first used to pay this management charge, and is then reinvested to increase the value of units.

15. It can be seen that the policyholder in principle receives all the investment profits, subject to the life office making an equitable charge for taxation. The policyholder also suffers almost all the investment losses (except perhaps on death). The life office's risks are limited to mortality and expenses, and there are no large margins which it can take a small share of. The life office's income includes the management charge on the assets, which will depend on the value of the assets; and its outgo includes the sum it has to make up on death, which is also affected by the value of the units. The "sterling reserve", i.e. the reserve that must be held by the office in addition to those units credited to the policyholder, may easily turn out to be negative, either in the initial years or later during the policy term (because of the incidence of mortality) and any such negative reserve implies that shareholders' capital is required. It has been particularly for this type of policy that the profit testing approach has been developed, though it has also been found useful for conventional policies.

16. The usual approach for such policies has been to assume a given rate of return on the units, as in the classical actuarial approach to interest rates. But such a policy clearly invites the application of a stochastic model for the assets. Such an approach has been used in the assessment of contingency reserves to cover "performance guarantees" which have been included in many such policies, although not in those issued nowadays. By a performance guarantee is meant that the sum payable on maturity (or perhaps on death, or on surrender at any chosen date), will not be less than some fixed money amount, such as the total of the premiums paid. Many such policies were invested wholly in ordinary shares (common stock) and the experience of the 1974 Stock Market crash in Britain demonstrated that there was no guarantee that the prices of ordinary shares would move perpetually upwards, even when reinvested dividend income was taken into account. The Maturity Guarantees Working Party in its Report (1980) used a

stochastic model for ordinary share dividends and yields to evaluate the contingency reserves required to cover such guarantees with a given small probability of ruin. It is possible that offices now use a stochastic asset model in their profit testing, but I do not know the details of how such methods are used.

Stochastic models

17. I want now to describe two approaches to premium determination, both based on stochastic models, but using as an example a completely new type of policy for which traditional methods provide limited assistance.

Index Linked Stocks

18. In March 1981 the British Government issued its first index linked stock, 2% Index Linked Treasury 1996. For each £100 subscribed, a notional interest payment of £1 per half-year is made until 1996, when a notional £100 is repaid. However, each interest payment and the repayment of capital is to be "index linked", that is, the notional amount (£1 or £100) is to be multiplied by the value of the Retail Price Index (RPI) applicable to the date of payment divided by the value of the RPI applicable to the date of issue. (In practice the "applicable" RPI is that for a date eight months before the date of payment or date of issue; this has practical consequences, but does not affect the principle). Since that date several more such stocks have been issued, and at the time of writing (June 1983) some £6,000 million has been raised in this way).

Index Linked Annuities

19. The existence of such index linked stock makes possible the issue by life offices of index linked annuities. It has always been difficult to devise annuities with a suitable with profits element, or ones suitably linked to the price of real assets such as ordinary shares or property. Yet in times of inflation a fixed money annuity is very unsatisfactory for the annuitant. Index linked annuities therefore provide a product which some policyholders may find attractive, in particular those making their own individual arrangements for a pension (through tax-efficient self-employed retirement annuities) or the trustees or employers of pension schemes wishing to provide pensions that are suitably protected against inflation.
Of course the starting yield for an index linked annuity is considerably lower than that for a level money annuity, and many policyholders are not willing to accept this lower starting yield, so as yet the amount of business of this type that has been sold is comparatively small.

20. The basic form of such an index linked annuity is straightforward. Each payment is equal to the initial rate of payment multiplied by the RPI at the time of payment and divided by the RPI at the issue date. The method of calculating the price is also straightforward. Mortality and expense assumptions are conventional (except that one might perhaps assume that those who choose such annuities are backing their longevity, so may experience particularly light mortality). The "real" rate of interest to use is also straightforward. The index linked Government stocks have been issued with "coupons" (i.e. nominal interest rates" of either 2 or 2.5%), and they have usually traded to give a real yield of between 2.5 and 3%. Since this is the sort of rate of interest that was obtained during the nineteenth century on first class Government stock, at a time when inflation was assumed not to exist, one can expect these stocks to continue to trade at these yield levels. In any case, for immediate annuities it is appropriate to use the current yield on a suitably dated stock.

Miniumum guarantees

21. Such annuities can be made more attractive by the inclusion of a minimum money guarantee. This could be of three forms: the amount of annuity paid will never fall below the initial value; the amount will never fall below the value in any previous year; the amount will be increased each year by the rate of inflation but will remain unchanged if prices fall. We can express this formally. Let B(t) be the amount of annuity payment at time t. Let Q(t) be the Retail Price Index applicable to time t. Let the initial amount of annuity at time t = 0, B(0), be 1. Then the four types of annuity are as follows:

1. No guarantee:

$$B(t) = Q(t)/Q(0)$$

2. Not less than initial amount:

$$B(t) = \text{Max}\left\{1,\ Q(t)/Q(0)\right\}$$

3. Not less than previous amount:

$$B(t) = \text{Max}\left\{B(t-1),\ Q(t)/Q(0)\right\}$$

4. Increased by rate of inflation but unchanged if prices fall:

$$B(t) = B(t-1) \times \text{Max} \left\{1, Q(t)/Q(t-1)\right\}$$

22. If prices rise continuously, then each of these annuities pays the same amount each year, and has the same present value. But there can be no certainty that prices will always rise; indeed many times in the past they have fallen; and there would be no point in including such guarantees if there were not some chance that prices would fall. It is therefore necessary to make a higher charge for annuities with such a guarantee, and the charge should be higher the stronger the level of guarantee. Classical methods of premium calculation seem to me to give no assistance in deciding on the appropriate amount of charge. One must use some technique that explicitly allows for the probability of variation in inflation rates.

23. The approach I have adopted, which has been described with results more fully in a paper submitted to the 22nd International Congress of Actuaries in Australia in 1984 (Wilkie (forthcoming)) uses a stochastic time-series model for inflation. This is based on investigations described in Wilkie (1981).

24. The model is a simple one: the force of inflation each year, $\nabla \ln Q(t)$, is assumed to follow a first order autoregressive process so that:

$$\nabla \ln Q(t) = \mu + \alpha (\nabla \ln Q(t-1) - \mu) + \sigma \epsilon (t),$$

where ∇ is the backwards difference operator, so that:

$$\nabla x(t) = x(t) - x(t-1),$$

$\epsilon(t)$ is a unit normal random variable, not correlated with $\epsilon(u)$, for $t \neq u$.

The three numerical parameters in the formula are:

μ, the mean force of inflation,

α, the autoregressive parameter,

and σ, the standard deviation of residuals.

These three numbers form a "basis", and it is of interest to investigate the effect of using different bases in the calculations. A reasonably plausible basis is:

$$\mu = 0.05$$
$$\alpha = 0.6$$
$$\sigma = 0.05,$$

but the details of the basis are not of importance here.

25. A simulation technique is used to determine the premium rates. Computer-generated pseudo-random unit normal variables are produced for as many years ahead as an annuitant might live. These are used to construct values of the Retail Price Index for a single possible future experience. Knowing the values of the RPI one can determine exactly the money amounts of annuity that will be paid in each future year. Their "real" values can also be calculated, discounting for inflation. We can use the traditional assumptions about mortality, and also assume a fixed real rate of interest for discounting. If we wish we can add expenses on an appropriate basis, linking them, if we wish, to the level of price index in any future year. This process gives us a discounted present value for the annuity for this particular possible future experience.

26. By repeating the simulation technique many times, to generate many different possible futures, we can construct an empirical distribution for the discounted present value of the annuity, which is therefore the single premium that should be charged. However, this single premium is a random variable, and we cannot ask the policyholder to pay a series of varying amounts with specified probabilities.

27. The traditional actuarial approach might be first to calculate the expected value of the premium, or at least the mean of the empirical distribution, and to consider charging this amount. It will be found that this would give an extremely small addition to the basic premium for the non-guaranteed annuity, which of course is a fixed amount, with no variation. However, to charge the expected value implicitly assumes that variations about that expected value are of no importance, or are at least sufficiently small to be ignored. This is sufficiently true for fluctuations because of mortality, at least when sufficiently many policies are written. But with this type of annuity there is no such averaging over

simultaneous policies. All policies that start at the same date will experience the same values for the RPI, and will therefore have the same structure of benefits. The fact that annuitants are of different ages and die at different times does not reduce this variability at all; the fact that annuities may be taken out over a number of years and therefore have different starting dates would usually reduce the variability, though it would not in fact do so for the strongest guarantee of type 4. If the guarantees are invoked they will be invoked for all policies simultaneously. The office will lose on its whole portfolio, and, except in the very long run, has no possibility of corresponding gain. It must therefore charge more than the expected value premium, and the user of the traditional approach would probably add an arbitrary loading to the premium.

28. Another method of determining the premium to charge I shall describe as "the risk of ruin" approach. The office chooses some sufficiently low probability of ruin, and charges a premium that will be sufficient to cover the liabilities in all but these extreme cases. For example, it might choose a risk of ruin (on this portfolio) of 1 in 200. An estimate of the appropriate quantile of the distribution can be obtained by taking say the fifth highest premium value out of 1,000 simulations. More precise estimates could be obtained by using the appropriate order statistic out of a larger number of simulations. This figure is then the chargeable premium.

29. In practice an office would expect to offer two types of such annuity: one with no guarantee, and one with a guarantee, probably of the third type described above, i.e. the benefit never reduces. It is convenient to find a simple relationship between the chargeable premiums on these two bases, which will apply to annuitants of both sexes and a variety of ages, and will be much the same whatever the real rate of interest used. Whether such a relationship can be found depends on the particular assumptions used, the range of ages, interest rates, etc. the office wishes to consider, and the extent to which the office is prepared to tolerate approximations. However, the investigations I carried out suggested that a uniform percentage loading on the basic premium was of the right shape to cover the guarantee. The amount of this percentage loading would depend on the bases used, but 10% was a typical figure. If the charge for the guarantee is too high, as was the case with certain other products investigated, then the office would presumably not be willing to quote at all.

30. However, it can be pointed out that except for 1 case in 200, when the office, or at least this portfolio, is "ruined", policyholders are being overcharged. In many circumstances, therefore, the office will make profits that appear to be superfluous, and do not appear to "belong" to anybody. One approach is to retain the excess premiums as additional reserves for the portfolio, allowing premiums for future annuities to be reduced, or alternatively, allow the risk of ruin to be reduced. But it is a rather curious idea to ask present policyholders to pay more, so that future policyholders may pay rather less.

31. An alternative approach is to use the profit-testing model. We start by assuming that the life office wishes to set up reserves for these contracts on a very safe basis, perhaps with a risk of ruin of 1 in 1,000. Let us assume that we have found that this can be done by setting up a reserve which we can determine. It might, for example, be 20% higher than the reserve for the non-guaranteed annuity. One would need to devise a reserve basis for this annuity in each future year. Strictly this should depend on whether or not the guarantee is being invoked, but it would probably be sufficient to decide on a reserving basis that depended only on the then current annuity at any time.

32. We now postulate that the office must set up initial reserves on this relatively strong basis. Some of the reserves will come from the initial premium paid by the annuitant. The remainder will be provided by the shareholders of the company. It is assumed that this total amount is invested in suitable index-linked stocks, which will yield a constant real rate of interest. The problem is again approached through simulation. Within one simulation the values of the Retail Price Index for each future year will be determined. An accounting procedure is then carried out for each future year. These assets at the start of the year are assumed to earn interest at the appropriate real rate, the appropriate annuity is paid out, and reserves are required at the end of the year on the same strong basis, for an annuitant a year older (strictly for the survivors among a large group of annuitants). Unless the guarantee is being invoked, the amount of assets will be somewhat greater than the required reserves, and will thus fall into "surplus" for that year. If the guarantee is required, this surplus for that year may be reduced, and may even become negative. In this case we assume that the shareholders

have to supply additional capital. But we wish to count the number of such cases, since, if no shareholders' capital is available, the office is "ruined". We wish to check that the reserving basis chosen is in fact strong enough for the risk of such ruin to be sufficiently small.

33. Because the shareholders are putting up risk capital initially, they will expect a higher return on it than can be obtained on secure Government Index-Linked Stock. For example, when such index-linked stock was yielding about 2.75%, the average of ordinary shares in the U.K. was yielding about 5%. There is no guarantee that dividends will rise pari passu with inflation, though on average they may be expected to do so. The excess return, of some 2.25%, can be treated as the risk premium required by investors for securities with the average level of risk of ordinary shares. It will be necessary to decide what the appropriate higher return should be for shareholders with this type of contract. This is a decision that must be taken by the management of the office. However, having decided on a total real return required by the shareholders, we can discount the amounts of surplus that emerge each year, using this higher rate of return, and obtain a present value of shareholders' profit.

34. This discounted amount of shareholders' profit will itself depend on the details of each simulation, and is thus a random variable. We can build up an empirical distribution for this shareholders' profit. In all the many cases when the guarantee has not been invoked, the amount of profit will be the same. In other cases it will be less, and if the portfolio has been sufficiently much "ruined" during the simulation, the profit might even be negative. But the use of a higher discount rate already takes care of the variability. It is therefore appropriate to estimate the expected value of this profit by taking the mean of its empirical distribution. This gives the amount which shareholders should be prepared to provide initially. The rest should be provided by the policyholder, and forms the premium.

35. My experimental calculations have not gone sufficiently far at this stage to establish whether the premium calculated in this latter way is greater or less than that calculated according to the risk of ruin approach discussed earlier. This would depend on how strong the reserving basis was, what rate of return the shareholders are expecting, and on the stochastic basis used for simulating inflation. However, the second basis is more coherent and more justifiable than the first. The first

appears to give the shareholders something for nothing; the second gives them something for something. It also provides a coherent reserving basis, in the sense that the reserves can be maintained from year to year, without the injection of further capital except in the most extreme circumstances. This is considered a desirable feature by some British actuaries.

Summary

36. This note has described two principles of premium calculation that are widely used in Britain for determining life assurance premiums; both of these work on deterministic methods of calculation. It has then described two stochastic methods, which to some extent parallel the first two and can be thought of as the stochastic equivalents of the cautious actuary and the profit-testing approaches. They have been applied in a situation where deterministic methods would fail. But their application is much wider than this. With a satisfactory stochastic model for inflation and investment, and readily available computer techniques for simulation, the stochastic approach could be applied to conventional policies quite easily. Whether such a development will in fact occur depends in part on the discussion of ideas at seminars such as this, and on their subsequent diffussion.

Acknowledgement

37. I should like to thank in particular my colleague, I.C. Lumsden, for his helpful and constructive suggestions on this note.

References

BENJAMIN, S. (1976) "Profit and other Financial Concepts in Insurance", Journal of the Institute of Actuaries, 103, 233.

MATURITY GUARANTEES WORKING PARTY (1980), Report, Journal of the Institute of Actuaries, 107, 101

SMART, I.C. (1977) "Pricing and Profitability in a Life Office", Journal of the Institute of Actuaries, 104, 125.

WILKIE, A.D. (1981) "Indexing Long Term Financial Contracts". Journal of the Institute of Actuaries, 108, 299 and Transactions of the Faculty of Actuaries, 38, 55.

WILKIE, A.D. (forthcoming) "The Cost of Minimum Money Guarantees on Index-Linked Annuities", submitted to 22nd International Congress of Actuaries, 1984.

A STOCHASTIC MODEL FOR INVESTMENT VARIABLES IN THE UNITED KINGDOM

A.D. Wilkie, M.A., F.F.A., F.I.A.

Standard Life Assurance Company and Heriot-Watt University, Edinburgh.

ABSTRACT:
The note describes briefly a stochastic model for investment variables in the United Kingdom that has been developed on the basis of data for the last sixty years, and which can be used for long-term simulations of the assets of financial institutions such as insurance companies and pension funds.

1. This note describes a stochastic model that has been developed using data from 1919 to 1982, at annual intervals, for the series described below, which relate to the United Kingdom.

   Q(t) The Retail Prices Index, or its predecessors (similar to a Consumer Price Index).

   D(t) An index of dividends on a succession of share indices, the last being the Financial Times-Actuaries All-Share Index.

   Y(t) The dividend yield on the same share indices as the dividend index, that is, the dividend index at the specified date divided by the price index at that date.

   C(t) The yield on 2.5% Consols, at the end of June each year; Consols is an irredeemable Government stock, paying a fixed coupon in perpetuity; it is taken as a measure of the general level of fixed interest yields in the market.

2. The data were analysed using essentially Box-Jenkins methods (see Box & Jenkins, 1970). However, the model

F. de Vylder et al. (eds.), Premium Calculation in Insurance, 533–538.

chosen was not the one that gave statistically the best fit, rather it was one that made intuitive economic and investment sense, without being statistically unreasonable. The model chosen was a "cascade" one, which can be shown diagrammatically below, where the arrows indicate the direction of influence.

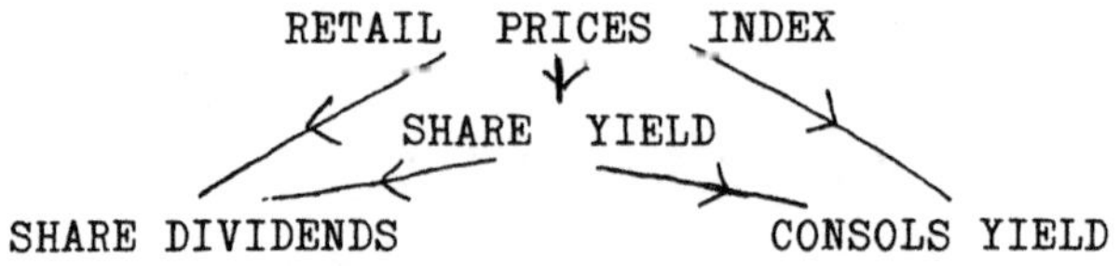

3. The Retail Prices Index series, $Q(t)$ is described entirely in terms of itself and a random white noise series. The model is

$$\nabla \ln Q(t) = QMU + QA(\nabla \ln Q(t-1) - QMU) + QSD.QZ(t),$$

where the backwards difference operator ∇ is defined by

$$\nabla X(t) = X(t) - X(t-1).$$

and $QZ(t)$ is a sequence of independent identically distributed unit normal variates.

4. This model says that the annual rate of inflation follows a first order autoregressive process, with a fixed mean QMU, and a parameter QA such that the expected rate of inflation each year is equal to the mean plus QA times last year's deviation from the mean. Appropriate values for the parameters are:
$QMU = 0.05$, $QA = 0.6$, $QSD = 0.05$.

There is fairly little uncertainty about the appropriate values for QA and QSD, but considerable uncertainty about the value to use for QMU, where anything between 0.04 and 0.10 might be justifiable, depending on the period of observation.

5. The share yield depends both on the current level of inflation and on previous values of itself and on a white noise series. The model is:

$$\ln Y(t) = YW.\nabla \ln Q(t) + YN(t),$$

where $YN(t) = \ln YMU + YA(YN(t-1) - \ln YMU) + YE(t),$

$$YE(t) = YSD.YZ(t),$$

and $YZ(t)$ is a sequence of independent identically distributed unit normal variates.

6. This model says that the natural logarithm of the yield consists of two parts: the first is directly dependent on the current rate of inflation (a high rate of inflation implying a high share yield and vice versa), and the second, $YN(t)$, follows a first order autoregressive model, similar to that of the rate of inflation itself. Appropriate values for the parameters are:

   $YMU = 0.04$, $YA = 0.6$, $YW = 1.35$, $YSD = 0.175$.

7. The index of share dividends is made to depend on inflation, both with an exponentially lagged effect and an additional direct effect, and on the residual, $YE(t)$, from the yield model, plus a white noise series, which has both a simultaneous and a lagged effect. The parameters are such that a given percentage increase in the Retail Prices Index ultimately results in the same percentage increase in the dividend index, so the model is said to have unit gain. The model is:

$$\ln D(t) = DW \left(\frac{DD}{1-(1-DD)B} \right) \nabla \ln Q(t) + DX \quad \nabla \ln Q(t)$$

$$+ DMU + DY.YE(t-1) + DE(t) + DB.DE(t-1),$$

where the backwards step operator B is defined by

$$BX(t) = X(t-1),$$

$$DE(t) = DSD.DZ(t),$$

and $DZ(t)$ is a sequence of independent identically distributed unit normal variates. The term in parenthesis involving DD represents an infinite series of lag effects, with exponentially declining coefficients:

DD,
$DD\ (1-DD)$,
$DD\ (1-DD)^2$,
etc.

8. Appropriate values for the parameters are:

   $DW = 0.8$, $DD = 0.2$, $DX = 0.2$, $DMU = 0.0$, $DY = -0.2$, $DB = 0.375$, $DSD = 0.075$.

9. The model makes the dividend index appear to depend on the residual of the share yield. In fact share prices to some extent correctly anticipate changes in dividends. For example, an unusual rise in dividends may be correctly forecast by investment analysts, so that share prices take

account of this and so rise. The yield is calculated on the previous year's dividend, and so falls. Although this is the causal sequence, it is convenient in the model to reflect the temporal sequence, so that an unexpected fall in yields results in an upwards change in the dividend index in the following period.

10. Although the parameter DMU is set to zero, it is retained in the model, since one may wish to investigate the results of assuming a small positive or negative value for it, implying a positive or negative long-term change in real dividends.

11. The Consols yield is assumed to consist of a real part, CN(t), plus an allowance for expected future inflation. The latter is based on the actual values of present and past inflation. The real part is defined by a third order autoregressive model, together with an influence from the residual of the yield series, YE(t), and a residual white noise series. The model is:

$$C(t) = CW \left(\frac{CD}{1-(1-CD)B} \right) \nabla \ln Q(t) + CN(t),$$

where $\ln CN(t) = \ln CMU + (CA1.B + CA2.B^2 + CA3.B^3)(\ln CN(t) - CMU$

$$+ CY.YE(t) + CSD.CZ(t),$$

where CZ(t) is a sequence of independent identically distributed unit normal variates.

12. Appropriate values for the parameters are:
CW = 1.0, CD = 0.045, CMU = 0.035, CA1 = 1.20,
CA2 = -0.48, CA3 = 0.20, CY = 0.06, CSD = 0.14.

The value of CW is 1.0, and it might appear that this term could be omitted; however, it is of interest to investigate variations in this parameter.

13. A simplified version of the model, which produces rather similar long-term forecasts, is given by setting three of the parameters equal to zero, and changing the values of three of the others, viz:
CW = 1.0, CD = 0.05, CMU = 0.035, CA1 = 0.91,
CA2 = 0.0, CA3 = 0.0, CY = 0.0, CSD = 0.165.

This form of the model says that the influence of inflation on the Consols yield is reflected by using as expected inflation an exponentially weighted moving average of past inflation, with a parameter of 0.05. The real rate of return has a mean of 3.5%, and follows a first order autoregressive series with a parameter of 0.91, so that it tends back towards its mean rather slowly.

14. It will be seen that the complete model is wholly self-contained. The only inputs are the four separate white noise series, and no exogenous variables are included. It is felt that, whatever may be the case for short-term forecasting, such a self-contained model is better for long-term simulations. The level of inflation, the level of company dividends, the level of interest rates, and the prices at which shares trade may well depend on such extraneous factors as government policy, business conditions and the world political, military, economic and climatic situations. Wars, famines and natural disasters may or may not occur. But they are not forecastable in the long run and their influence is subsumed in the white noise series.

15. Any number of particular realisations of the model described above can be simulated on a computer, or with some labour by hand calculation. A few results can be obtained analytically. But in general one would also wish to take into account the level of share prices, given by $D(t)/Y(t)$ or the price of fixed interest stock, given by $1/C(t)$. From these one can derive the accumulated amount of assets which result from some particular investment policy.

16. The potential applications of the model range very widely. It can be used to investigate the results of different investment strategies. In life assurance it can be used to estimate appropriate premium rates or policy reserves, by estimating the probability that a particular premium rate will prove insufficient, or that a particular level of reserves may prove insufficient. In order to do this one may need to postulate a dynamic investment strategy, where the choice of investments depends on the actual experience in the simulation so far; if with profits business is being investigated one may need to postulate also a dynamic bonus strategy, where the level of bonus distributed (either in cash or as an addition to the sum assured) may depend on the investment performance in the simulation so far.

17. The approach has been used, with a simpler model, by the Maturity Guarantees Working Party in their report (1980) to investigate the contingency reserves needed for unit linked life assurance contracts with a minimum money guarantee at maturity. The investment model has also

been used by Wilkie (forthcoming) to investigate the premiums needed for index-linked annuities with minimum money guarantees, which is also discussed in another paper in this volume ("Some notes on the methods of calculation of life assurance premiums in the United Kingdom").

References

Box G.E.P. and Jenkins G.M. (1970), Time Series Analysis, Forecasting and Control. Holden-Day.

Maturity Guarantees Working Party (1980), Report, Journal of the Institute of Actuaries, 107, 101.

Wilkie, A.D. (forthcoming) "The Cost of Minimum Money Guarantees on Index-Linked Annuities", submitted to 22nd International Congress of Actuaries, 1984.

INFLATIONARY EFFECTS ON PENSION PLANS: WAGE AND BENEFIT PATTERNS***

Jacob Antler* & Yehuda Kahane**

*Manager of the Actuarial Department, The Union of Israeli Workers (Histradrut).
**Academic Director, Erhard Insurance Center, Tel Aviv University, Israel.

ABSTRACT

Most studies on pension funds under inflationary circumstances emphasize the protection of the purchasing power of the investments as the major problem. The high inflation rates in Israel makes it possible to isolate and examine additional important effects which are related to the relationships among the wage patterns, the insured wage and the pension schemes under inflation. It emphasizes the impact of inflation on the contributions to the funds, and on the benefit patterns, under various methods of indexation. The study, which is based on the experience with approximately one half of the employees in Israel, may relate to general problems which are experienced in other countries having lower rates of inflation.

1. INTRODUCTION

The experience of countries suffering from high inflation rates could be very instrumental for countries which have lower inflation rates, simply because the researcher in an inflationary environment can observe patterns which are strongly magnified and therefore tend to standout. The researcher is practically working in a laboratory without needing a microscope. Due to this reason we felt it might be of some interest to explore some of the problems experienced in the Israeli environment. Although the analysis may

***An earlier version of this paper has been presented also at the 9th Seminar of the European Group of Risk and Insurance Economists, in Geneva, Switzerland, September 22-24, 1982.

F. de Vylder et al. (eds.), Premium Calculation in Insurance, 539–560.

emphasize some specific effects which relate only to the Israeli experience it is believed that the findings may relate to general problems which are experienced in other countries as well.

The major problem of pension funds in time of inflation is the protection of the purchasing power of the pensions. This problem is strongly related to the possibility of guaranteeing the real value of the investment portfolio. Pension funds in Israel are relieved of this problem, since a major part of their portfolios consist of fully indexed governmental bonds. These bonds bear a relatively high real interest rate and are fully indexed (both the principal and interest) to the Consumer Price Index (CPI). We shall not relate to the indexation methods and to the investment problem in this paper.

Being relieved from most investment risks enables the researchers to concentrate on other important inflationary effects. This study will focus on the relationships among the wage patterns, the insured wage, and the pension scheme under inflationary circumstances.

This study explores the inflationary effects on the actuarial balance of the Federation of Labor's (Histadrut) pension funds. These funds serve approximately one half of all the employees in Israel.

Inflation affects the contributions to the pension schemes through its effects on the insured wages; Inflation erodes the real wages despite the existence of a cost of living allowance (which is added to every employee's wage or salary in Israel). This is so since the allowance covers only a certain part of the CPI changes and has an absolute upper limit. Active employees often manage to protect their real wages through adjustments made within general work contracts (like, automatic promotion in ranks, introduction of additional wage components, and periodical adjustments of wages). Unfortunately, the insured income is not always protected by these adjustments, since most of the adjustments relate to the wage components which are uninsured. This creates a rapid erosion in the real insured income, which is transmitted to the pension funds. In other words, inflation creates similar effects to those observed in non-life insurance — i.e., the employees are "under insured" in pension protection and therefore may have to face the "coinsurance penalty". The inflationary burden is fully shifted to the shoulders of the insured. Although under some inflationary scenarios, pension funds may still achieve a short-term improvement of their actuarial balance, it almost always deteriorates in the long run.

Inflation affects also the pension benefits. Recent reform in the Israeli pension funds schemes changed the existing method of calculation of pension benefits based on the "final wage" base to a more complex method, which takes into consideration the relative wages throughout all the years of employment. The reform affected also the indexing of pensions: cost of living indexation is made rather than linkage to the "rank" in a wage scale. The effects of this reform depend on the wage patterns. When the real insured wage deteriorates over a relatively long period, the reform improves the pensioners' positions and disturbs the actuarial balance of the fund.

The findings may be extremely relevant for pension schemes in other countries and especially for the U.S. Social Security System. The American Social Security has only recently introduced a new system, where benefits are based on the income profile throughout the years of employment. The Israeli pension schemes system is quite similar and, therefore, both systems may be exposed to the same problems with inflation. The analysis, although crude because of the lack of detailed relevant statistics, points out very important inflationary effects on the insured wage which should not be ignored by the pension funds and the social security system.

Section 2 of the article is devoted to a survey of some of the specific patterns of wages in Israel and to the study of the uninsured wage patterns under inflation. Section 3 will examine the implications for funded pension plans in which pensions are based on the salary in the latest years of employment. Section 4 explains the newly introduced plans in which pensions are determined according to the income profile throughout the entire period of employment. The vulnerability of such plans to inflation is examined later while pointing out some interesting factors which should be considered when similar pension schemes revisions are made. The paper is summarized in Section 5 with some concluding remarks.

2. THE INSURED WAGE STRUCTURE IN ISRAEL

It is necessary to understand the main features of the employees' remuneration system in Israel in order to be able to examine pension schemes and their problems; The contributions to pension schemes are typically based on a certain proportion of salaries and wages. Benefits to pensioners are also often linked to salaries and wages and are often indexed by either a certain general price index (usually the CPI) or by the individual's pattern of earnings prior to retirement (indexed by job classification).

2.1. Wage Scales

Wages and salaries in Israel exhibit a few common features: most employees are paid according to general work contracts signed from time to time (usually annually or bi-annually) by the major employees' unions. Special work contracts usually follow the main lines of the general contracts of the leading groups.

In each sector, salaries and wages are determined according to a "scale" which is typical for that sector. Each salary scale represents a set of employment "ranks". Most employees are, thus, divided into job categories with specific payments designed to each category. The most common scales are those of government officials, teachers, academicians in social sciences, industrial workers, and construction workers. Other scales are directly linked to the terms of these scales or at least follow their general terms. It could, therefore, be said that the above mentioned scales determine the wages of most Israeli employees.

Historical developments created a complex remuneration system under each of these "scales". Each contract defines who will be eligible for certain benefits, how much he will get, and under what conditions. The work contracts define which elements of the salary will be taken into consideration in calculating the contributions to pension funds and in calculating the benefits.

2.2 The Cost of Living

All employees in Israel are eligible for a cost of living allowance, which is set in a contract signed by employers, employees, the Union, and the Government (as an employer). All Israelis are eligible for this allowance. The rate of increase is determined periodically (at present four times a year) according to the changes in the CPI. Note, that the rate is determined according to the CPI changes which occurred in the past quarter. The time gap between the actual price changes and the effective date of adjusting the wages (and pensions) could be highly significant in times of rapid inflation.

The cost of living allowance is often set as a certain percentage of the actual CPI change (at present only 80% of the actual CPI change). Moreover, this allowance is calculated in a way whereby only certain elements of the salary are indexed, while others remain unchanged. The remaining elements may be adjusted independently, often at a different rate and at different times. The cost of living allowance is given on the "basic salary" (which is a stated part of the wage for each rank in the scale).

The cost of living allowance is granted only up to a certain maximum salary; i.e., there is an upper limit in absolute terms. It is, therefore, clear that employees cannot preserve the purchasing power of their income, unless they get additional compensation, which usually takes the form of additional fringe benefits.

2.3. Inflation and the Insured Wage

Contributions to the pension funds are made from a certain part of the wage (the insured wage). Certain components, such as premiums, overtime payments, and various remunerations, are excluded from the insured wage. The development of the insured wage under inflationary circumstances is the key parameter in the evaluation of inflationary effects on the pension funds and their actuarial balance.

This problem is brought to light in Table 1, which presents the development of the insured wage of employees in two of the largest salary scales, during the period 1964-1981. Although the average real wage in Israel continuously increased during the last years, despite the high inflation rates, the insured's real wage has deteriorated drastically. The erosion has not been the same in all ranks and tended to be more drastic at higher wage levels, due to the upper absolute limit on the cost of living allowance. The wage scales were adjusted from time to time, but the real effects of such corrections last for only a short period. The continuing inflationary erosion sooner or later wiped out all the prior achievements.

The same trend is exhibited graphically in Figure 1. This Figure exhibits on a semi-log scale, the developments in the insured wage in relation to the CPI. The figure shows the insured wage for the month of January in each of the years 1971-1981, for a certain rank which has been chosen for demonstration purposes only (Rank 9 of the academician scale). A more detailed analysis of the monthly development of the insured wage could also be of much interest (see Figure 2). This Figure shows the behavior of the insured salary at this rank. The step function form is interesting since it clarifies the need for more frequent and higher adjustments as the inflation has accelerated.

2.4. The Timing of Wage Adjustments

In times of rapid inflation, short term trends can also be quite meaningful. Inflation is a continuous process while the development of the wages (and insured wages) is not. Only a few wage components are automatically adjusted to inflation, the rest are adjusted periodically. The retroactive adjustments once every three months (and twice a year in the past) leave a very noticeable gap between the adjusted wage and the real value of the initial wage. This gap increased further due to the fact that the cost of living adjustment is only partial. Moreover, the cost of living allowance, and other agreed wage adjustments (which are granted from time to time) are actually paid only after additional time lags (between the signing of the agreement and the actual payment). This contributes further to the erosion of the purchasing power of the adjusted wages.

At higher inflation rates it is important to adjust the wages more frequently. This has been done in Israel where due to the recently high inflation rates, wages were adjusted after shorter and shorter periods, even monthly. (See Figure 2.)

Israeli wage contracts often grant retroactive wage adjustments. This is not shown in Figure 2, which represents the changes at the date due, and not at the date of actual payment. This time lag could be very significant in times of inflation and may affect the actuarial balance of the pension funds. However, in the Israeli case, this has little effect on the funds, since their

Table 1: The Real Insured Wage[1], For Selected Ranks
1964 - 1981 (January 1964 = 100)

Wage Group & Rank	1964	1973	1974	1975	1976	1977	1978	1979	1980	1981
Rank in General Scale[2] ("Derug Ahid")										
"Low"	1.00	.91	.83	.75	.68	.72	.63	.68	.65	.62
"Middle"	1.00	.88	.85	.70	.64	.60	.62	.66	.64	.63
"High"	1.00	.85	.90	.77	.70	.65	.65	.74	.76	.78
Rank in Social Sciences & Humanities[3] ("M.H.R.")										
"Low"	1.00	1.28	1.19	1.08	1.02	0.95	0.88	0.92	0.87	0.83
"Middle"	1.00	1.12	1.04	0.85	0.80	0.75	0.70	0.86	0.78	0.75
"High"	1.00	1.14	1.04	0.82	0.78	0.72	0.67	1.06	0.88	0.90

1 For each year, January insured salary divided by the Consumer Index and brought to common base (1964) used.

2 In the General Scale ("derug Ahid"), "Low", "Middle", and "High" correspond to the following ranks: "Yud bet", "Yud het", and "Caf", respectively.

3 In Social Sciences and Humanities ("M.H.R."), "Low", "Middle", and "High" correspond to ranks 2, 6 and 10, respectively. The scale is divided into ten ranks, ranging from 1 to 10.

Table 2: The Insured Wage in Relation to the Average Monthly Wage in Israel, Selected Ranks, 1964 - 1981.

Wage Group	1964	1973	1974	1975	1976	1977	1978	1979	1980	1981
General Scale ("Derug Ahid")[1]										
"Low" Rank	1.98	1.22	1.11	1.03	.90	.93	.79	.78	.75	
"Medium" Rank	3.08	1.84	1.75	1.50	1.31	1.20	1.22	1.16	1.16	
"High" Rank	3.65	2.08	2.20	1.95	1.69	1.53	1.50	1.54	1.62	1.58
Social Sciences & Humanities (M.H.R.)2										
"Low" Rank	1.52	1.33	1.22	1.14	1.02	.94	.85	.80	.77	.71
"Medium" Rank	2.59	1.96	1.80	1.54	1.38	1.26	1.14	1.27	1.18	1.10
"High" Rank	3.18	2.42	2.20	1.81	1.62	1.48	1.34	1.92	1.62	1.58

1 In the General Scale ("Derug Ahid"), "Low, "Middle", and "High" correspond to "Yud bet", "Yud het", and "Caf", respectively.

2. In the Social Sciences and Humanities ("M.H.R."), "Low", "Middle", and "High" correspond to ranks 2, 6 and 10 respectively. The scale is divided into ten ranks, ranging from 1 to 10.

FIGURE 1
Nominal Insured Salary vs. C.P.I.
(January)
Rank 9 Academicians Scale

(January 1978 = 1.0)

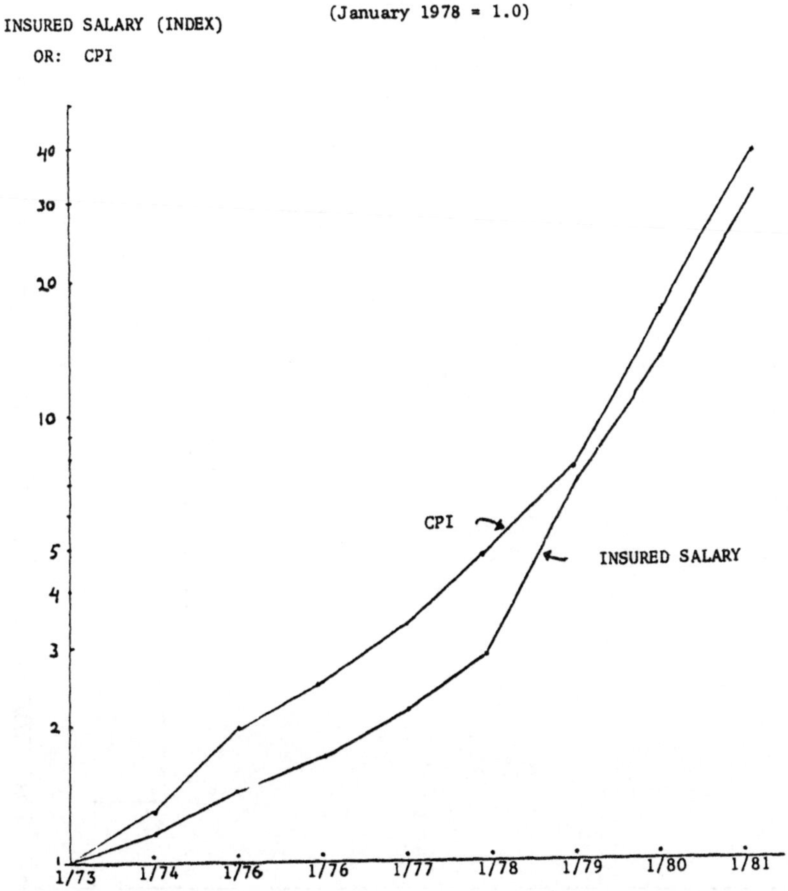

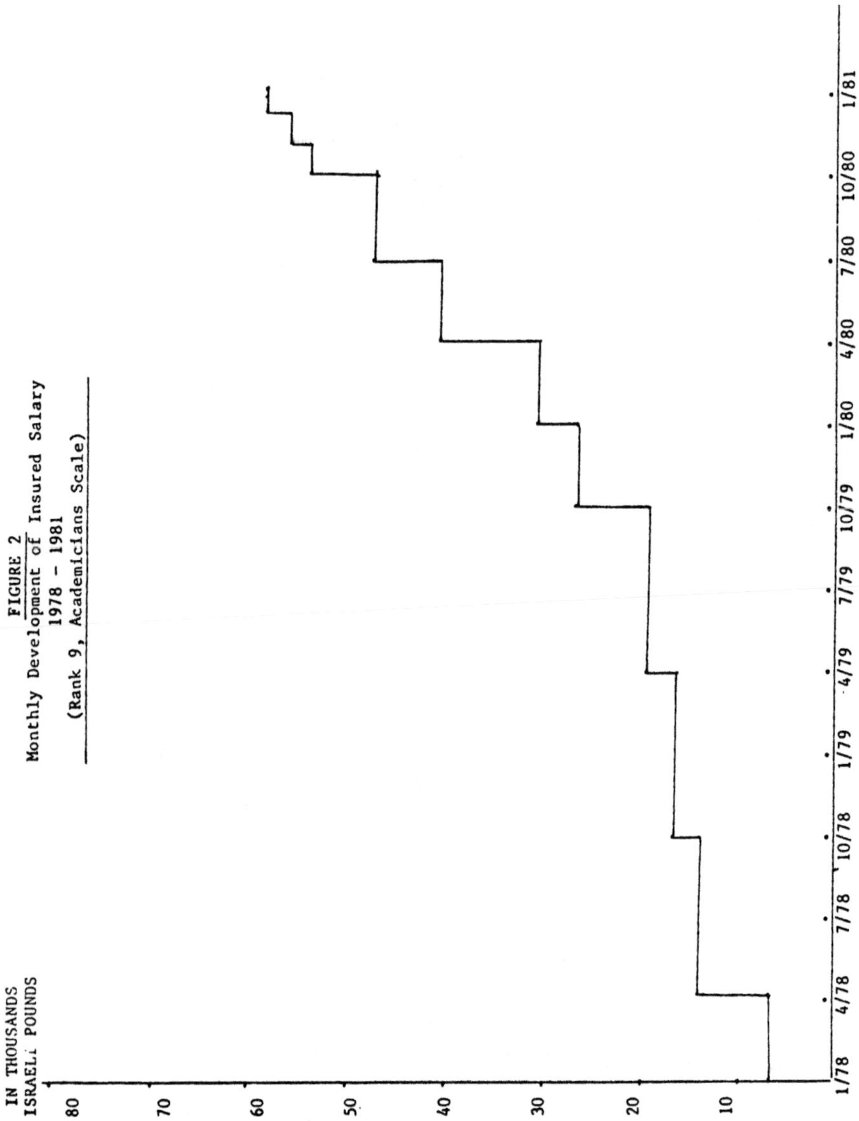

FIGURE 2
Monthly Development of Insured Salary
1978 - 1981
(Rank 9, Academicians Scale)

liabilities are based on the wages as paid, and the pensions are adjusted according to the wage contracts, only after they are published and the contributions are paid accordingly.

2.5. Uninsured Wage Components

The main thrust of wage negotiations in times of inflation is placed towards maintaining the purchasing power of wages and closing the gap between the automatic cost of living allowance and the full price changes. A typical way of reaching such a goal is by "inventing" new wage components (under various and often peculiar names). Evidently, there is a preference for fringe benefits which have additional tax advantages (increased payments for the use of private cars for business use, premiums for improved productivity, employer's participation in employees' expenses, etc.). The names of these components are selected in such a way to give the impression that they are granted to a specific sector, or specific jobs (to prevent a chain of claims in other sectors for similar adjustments).* In 1980, some 20-40% of the basic wage of a typical employee were given in the form of such uninsured components, depending on the scale to which he belongs. In some exceptional cases, the uninsured elements are approximately equal to the basic wage.

The introduction and updating of such wage components helps to maintain the real income of employees despite inflation. However, the new wage components usually create a problem for pensioners and for those about to retire. The main reason is that most components are often uninsured in pension programs and therefore the basic wage for pension calculations is computed on only a fraction of the actual wage obtained prior to retirement.** No data is available in Israel to show the development of the insured and the uninsured wage components. The available statistics simply average all wage components, with no distinction between the insured and uninsured elements. Only an indirect evidence about the erosion of the insured part (due to the insufficient cost of living allowance and the uninsured components), may be obtained. For that purpose the ratio between the insured income and the average wage will be examined.

Table 2 reveals the development of the ratio between the insured wage and the average monthly income of employees in Israel. The average monthly income series reports all payments, including the uninsured wage components. Therefore, the ratios in Table 2

*It is however, noteworthy that this method has not disappeared after the 1975 Ben Shahar tax reform, despite the fact that the reform has abolished the tax advantages for which such wage components has been eligible.

**Until recently, only the basic salary (Sechar Yesod) seniority allowance (Tosefet Vetek) and Cost of Living Allowance (Tosefet Yoker) were insured.

give a better idea of the combined effects of the declining purchasing power and the inclusion of uninsured wage components on the basic pensions. Note, for example, that a pensioner in one of the M.H.R. ranks, who had not been promoted would retire in 1981 with a final wage of some 50-65 percent of his real wage eight years earlier.

It is noteworthy that inflation affects the real wages in a non-uniform way. This happens since the cost of living allowance is only partial and is constrained by an upper limit. Therefore, the erosion of the wages of higher ranks has been more severe. The introduction of other wage components creates additional distortions. Table 2 shows that the ratio between the real wage in various ranks of the same scale are not preserved. Moreover, inflation distorts the ratio between the real wages of employees who received their wages according to different scales.

2.6. Wage Scale Shifts and Their Effects on Pensioners

The introduction of new wage components is only one of the possible ways to prevent the erosion of real wages. Another method which has become popular in Israel is the automatic promotion of employees to higher salary ranks, without a real change in their job or responsibilities. Such promotions are sometimes made across the board, according to the work contracts with the Union. The effect of such shifts as shown in Figure 3, which presents the distribution of governmental employees employed under the terms of the "general scale" (Derug Ahid). The entire distribution shifts by approximately one rank per year. As a result, new ranks are added at the top, while the ranks at the bottom of the scale practically disappear.

The automatic promotion affects the basic pension at retirement, as shown in Table 2. Since the basic pension has been determined by the wage during the last thirty six months of employment, erosion of the basic pension may occur even given automatic promotions.

3. INFLATION AND PENSION PLANS BASED ON THE FINAL WAGE

3.1. "Final Wage" Plans Under Inflationary Circumstances

Pension schemes generally tie the pension level to the insured wage during the years of employment. The level of pension is typically based on the "final wage", i.e., the wage level at retirement, or during a short period of years immediately prior to retirement.

In order to avoid moral hazard, it is clear that the basic pension should bear a reasonable relationship to the wages during a relatively long period prior to retirement. For example, the pension may be based on the arithmetic average of the earnings

FIGURE 3: CHANGES IN DISTRIBUTION OF EMPLOYEES BY RANK (GENERAL SCALE)

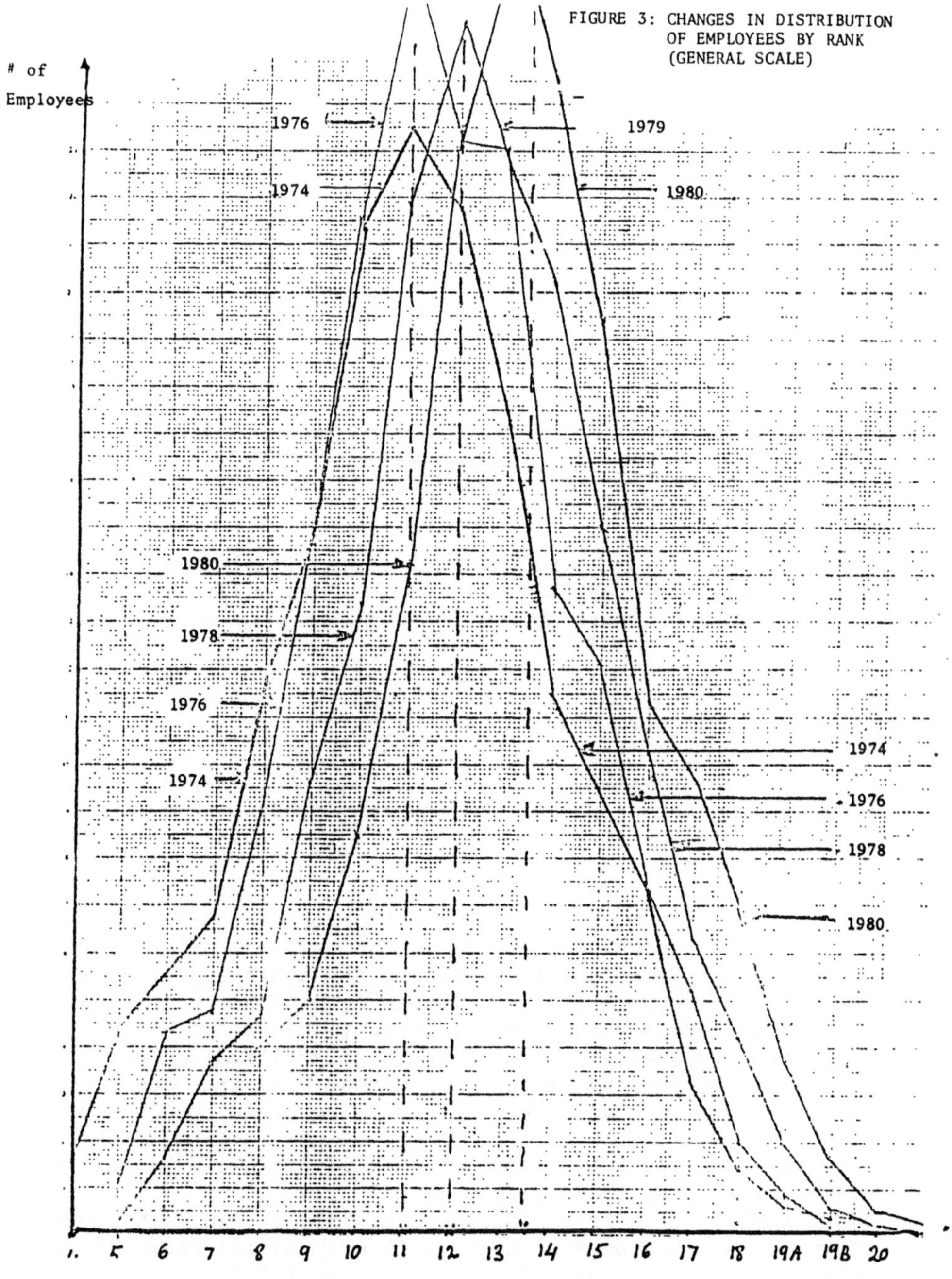

during the last 36 months of employment. During an inflationary period, the use of such a method could drastically lower the pension base even when the current salaries are exactly adjusted to preserve their purchasing power. For example, at a constant monthly inflation rate of 2%, $1 of the monthly salary in real terms would be presented by a series of $1.00, $1.02, $1.04... for the first, second, and third month respectively.* The final wage for the 36th month would be $1.70. The arithmetic average would be approximately $1.35. If the pension benefits are based on this later figure, the basic pension would be some 79% of the real value that the pensioner would obtain at zero inflation ($1.35/$1.70 = 0.79). Thus, even at relatively low inflation rates, the affect can be severe.

The bias could be even larger in practice, since wages do not follow the price changes exactly, but often lag behind. (For example, assume that the monthly inflation rate is 2% but the initial wage of $1.00 is adjusted at the monthly rate of only 1%. Under such conditions, the average nominal wage during the 36 consecutive months is approximately $1.18 and its real value, after adjustments to the 70% price increase, is only $0.69 ($1.10/1.70=$0.69).)

In order to avoid some of the inflationary biases and in order to reduce the possible moral hazards of a final wage system, the Israeli unions' pension schemes were based on a modified final wage system. Until 1980, the level of pensions was based on the average rank held by the employee during the 36 months prior to retirement.** The average rank was translated into financial terms according to the appropriate official wage scale at the retirement date.***

The Israeli system is applicable only where wages are paid according to official job classifications scales. Elsewhere, some sort of indexing must be used prior to the calculation of the average of the latest salaries in order to bring the salaries to a common basis at the date of retirement.

The Israeli system corrects for some of the inflationary biases since the pension is based on the official salary scale at the date of retirement, rather than on the nominal average salary. Thus, it prevents part of the erosion of the pension base during inflation. It does not, however, correct the erosion which may result from temporal changes in the job classification nor does it correct for the erosion of purchasing power of wages due to inadequate cost of living adjustments and the exclusion of certain

*For the sake of simplicity the examples are based on additive rather than compounded growth.

**This system has been replaced in the 1980 reform. The current formula is based on the earnings throughout all years of employment. This reform will be discussed later.

***Rank averaging reduces the effects of moral hazards (which may cause rapid promotion of employees soon prior to retirement).

wage components from the insured income.

Even the average rank system does not fully prevent the possible erosion of pensions in inflationary periods. The above discussion of the Israeli experience reveals certain mechanisms which led to erosion of the purchasing power of the insured wages. Since the average final rank method is based on the insured wage, the level of basic pensions will be subject to the same inflationary biases as the insured wage.

The wages of the active employees were protected against inflationary erosion, but pensioners remained unprotected, since the insured wage has been subject to inflationary erosion and biases. The basic sources of such biases related to the cost of living allowance arrangements, the uninsured wages components, and the habit of "automatic rank promotion".

The cost of living allowance does not fully compensate for price changes. Therefore, the real value of the average wage for the terminal period prior to retirement is often understated (even beyond the above mentioned biases). This may drastically lower the basic real benefits obtained by pensioners.

Table 2 showed that although the average real wage continuously increased during recent years, despite the high inflation rate, the insured part has deteriorated drastically in real terms and this leads directly to a further reduction of the purchasing power of the pensioners. It has also been shown that the erosion has not been the same in all ranks and has tended to be more drastic at higher wage levels.

The development of the insured wages, described above, creates a situation in which those already retired are not eligible for benefits which active employees receive. The insured wage has deteriorated, due to the inclusion of uninsured components; the pensioners are therefore eligible for only a certain part of the real wage of an active employee holding the same rank.

3.2. Automatic Promotion of Active Employees

A pensioner who retired at the highest rank may find a few years later that his pension is based on a much lower rank than that of the person currently holding his job. It may thus be concluded that the retired do not benefit from the techniques used to overcome the erosion of wages of the active employees. They suffer from a quick erosion of the pensions.* This means that inflation tends to create situations of "under-insurance" also in pension programs and that pensioners are, in fact, suffering from

*Note that in certain cases the wage at the "final rank" exceeded the upper limit of the cost of living allowance, while the pension itself (a certain percentage of the wage in the relevant rank) did not reach that maximum. In the union pension funds, the pensions were adjusted by the cost of living allowance, and the pensioners were thus eligible to higher pensions than those who were paid only according to a constant proportion of the adjusted wage.

FIGURE 4a
The ratio between the Insured Salary
and the Average Salary in Israel
(General Scale "Derug Ahid")

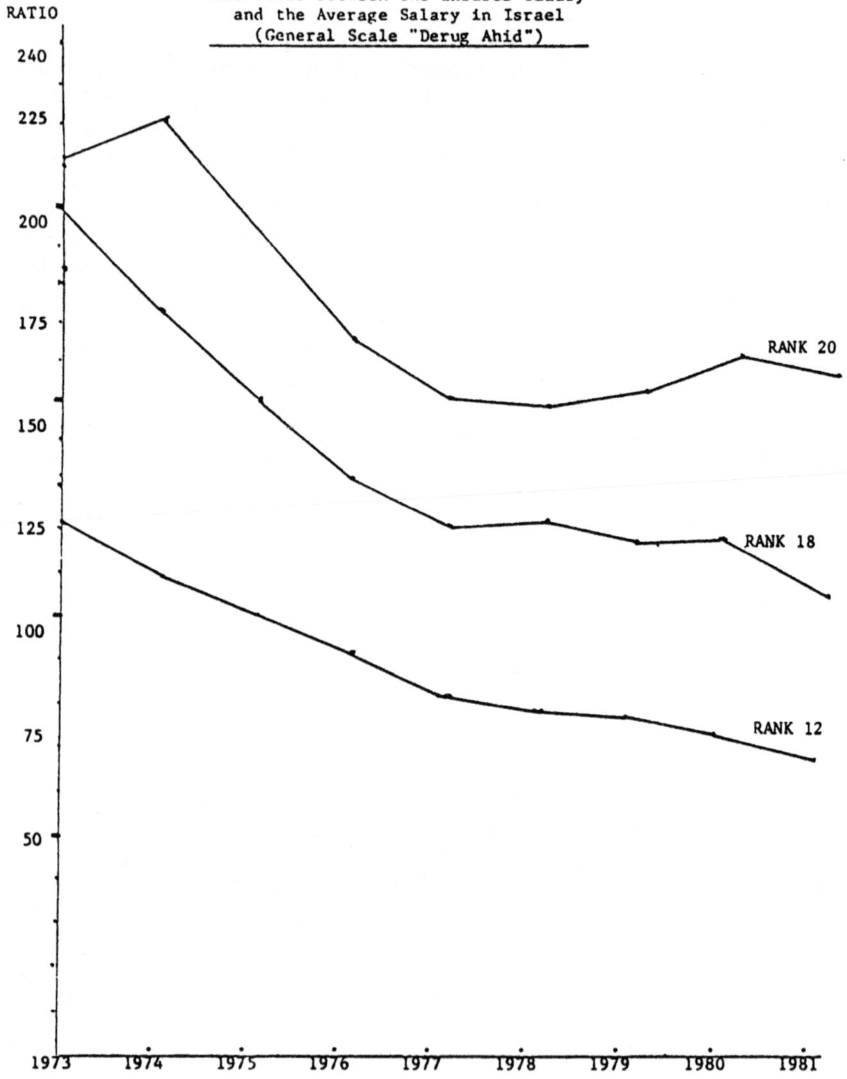

FIGURE 4 b
The ratio between the Insured Salary
and the Average Salary in Israel
(Academicians Scale "M.H.R.")

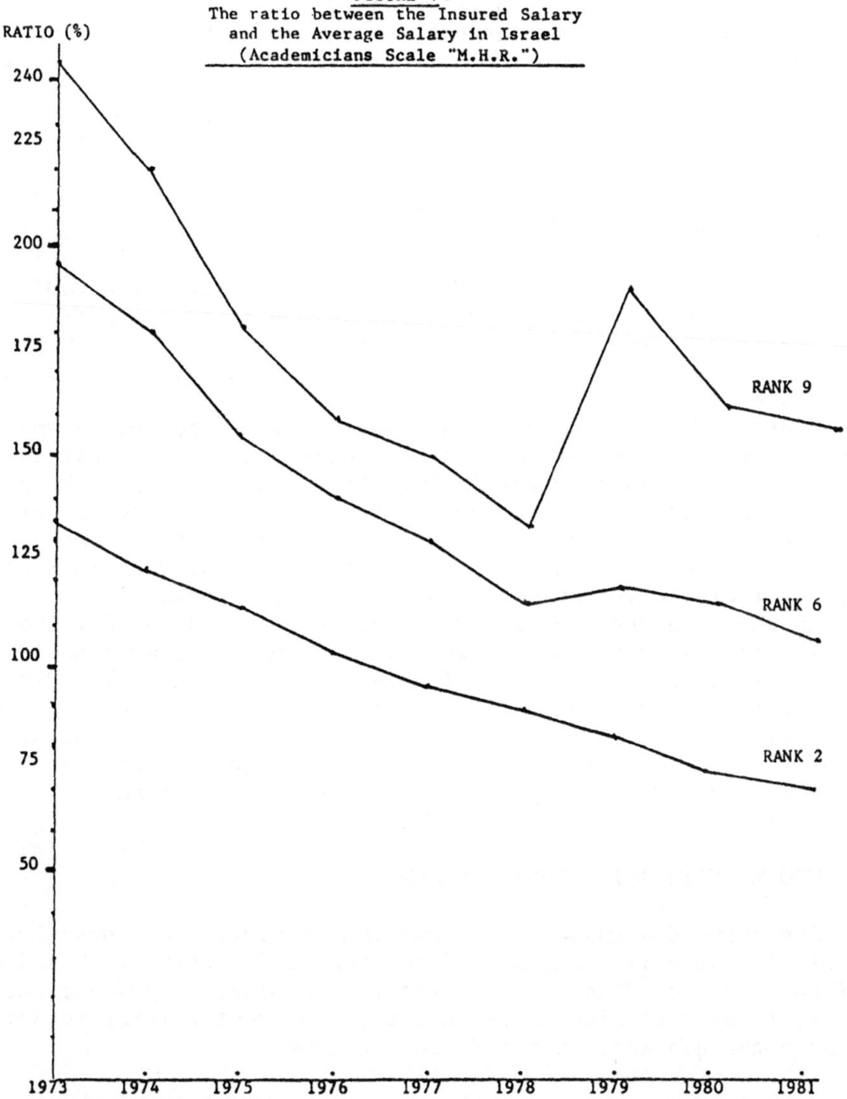

"non-insurance penalty".

3.3. The Effects of Pension Funds

The wage developments described above have a significant impact on pension funds. Assume that at a certain initial point in time the fund is actuarially balanced. The wage developments which were described above affect the actuarial balance of pension funds. In the short run, the erosion of pensions improves the actuarial balance of the pension funds since its real obligation to the active members decreases, as the past contributions to the fund were based on higher real wages. The same holds true with respect to the liabilities towards those already retired and who are now eligible to reduced real pensions. In other words, the inflationary burden is fully shifted to the shoulders of the insured, similarly to other insurance contracts where inflation creates a situation of "under-insurance". The inflation could be advantageous to the pension fund and these effects may offset, at least partially, the possible losses resulting from moral hazard problems (which could be quite significant in a system based on "last rank" or "final wage").

Under a final rank or a final wage system, the active employees have less incentive to take countermeasures against the erosion of the insured wages in the short run. However, the pensioners press the unions to rectify the situation. One way of doing so would be to introduce ad-hoc adjustments to the pensions paid to existing pensioners. Such corrections could be introduced on account of the inflationary earnings of the funds.

In the long run, the insured wages must be adjusted (otherwise it will rapidly approach zero). Such an adjustment could drastically affect the actuarial balance of the fund. The direction and intensity of the effect will depend on the demographical characteristics of the fund. In general, the higher actuarial age of the insured population, and the longer period for which they were insured, the higher the bias against the fund.

4. REFORM IN ISRAELI PENSION SCHEMES

The above discussion has shown that inflation may drastically distort the actuarial balance of pension funds which are based on a "final rank" or "final wage" system. In order to prevent such biases, Israeli pension funds recently underwent a major reform. The main changes were in the following areas:

(a) initial pension based on the wages throughout the working period;
(b) Pension indexation to the CPI, rather than the final rank, and
(c) broader definition of the insured wages and the inclusion of additional components.

The initial pension is a product of the "basic wage" times the accumulated pension rights (rate of pension). The pension rate is based on seniority in the program (number of years of employment) multiplied by 2% per working year, up to a maximum of no more than 70% in total (until the reform, the pension plans used various scales, which were non-linear with seniority).

The "basic wage" computation takes into consideration the wages during all years of employment, rather than the last 36 months only. For each year, the ratio between the insured wage and the average wage in Israel is calculated. The average of these periodical ratios during the years of employment is multiplied by the final average wage in Israel to get the "basic wage". For those employed for more than 35 years, the average ratio is determined according to the ratios of the "best" 25 years.

Such a system eliminated the moral hazard which created biases against the funds in the old system (where the employees were promoted just prior to their retirement, thus increasing the final wage).

The reform introduced additional changes. Pension rights are now accumulated at a uniform rate (2% of the basic wage for each year of employment). In the past, the rate had been uneven; a faster accumulation during the first years of employment and lower rates for later years. This change has little impact on the pension plan when the employees retire at the maximum pension rates. In Israel, however, the reform was needed because of the entrance of large groups of employees, including a relatively large number of older people into the pension programs.

The reform has been accompanied by a change in the definition of insured income. In order to establish a tight relationship between the contributions to the pension fund and the benefits, most wage components which were uninsured up to that point were included in the insured wage (only certain elements, such as employers participation in expenses remain uninsured).

4.1. The Effects of the Reform on the Actuarial Balance

The analysis of the actuarial balance of the unions' pension funds is impossible at this point because of insufficient data. Therefore, only major expected implications of the reform can be analysed.

The switch from the "final wage" system to the "lifetime income" approach may be quite significant. The intensity of the effect depends on various factors, and especially the wage pattern through life. The typical wage pattern shows an increase in the absolute (and relative) wage during the first period of employment, which is the period of learning and gaining experience. The high growth rate is explained by the ability of the employee to work over-time and to earn higher premiums. However, towards retirement, the ratio between the individual's wage to the average wage in the country declines. Such a pattern is presented in

FIGURE 5

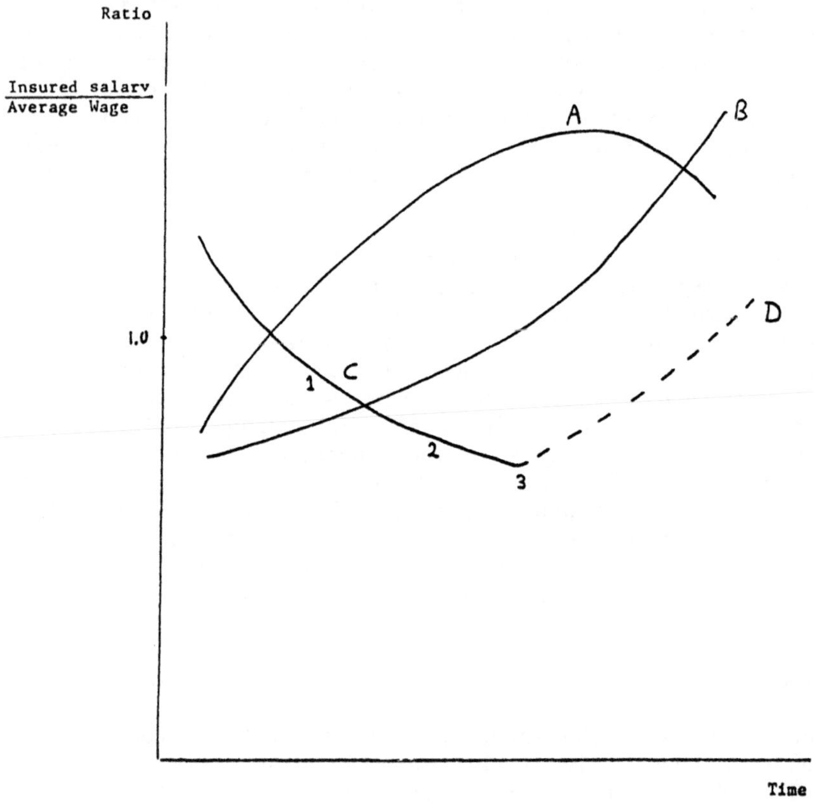

Figure 5, curve A. Under such a pattern, employees would gain from the reform, since the "final wage base" tends to be higher than the "average wage base". In such a case the pension funds would enjoy a significant improvement of their actuarial balance, at the introduction of such a reform.

The analysis in the previous sections showed, however, that the wage pattern could have been quite different from these expected patterns. Although the analysis is partial, it shows that for a large segment of Israeli employees, the insured wage has drastically declined in real terms, independent of age, experience in work, and other parameters. The same picture is obtained when the ratios of the insured wage to the average wage are examined. This means that the "average ratio" system would generally yield higher pensions than the final wage method. Pensioners and those about to retire would thus benefit from the reform in the short run, but this would threaten the actuarial balance of the fund.

The inclusion in the insured wage of most wage components, which were uninsured until the reform, stops the erosion of the insured wages and improves the actuarial balance of funds in the long run. However, in the short run this generates a large deficit in the funds, since they must pay larger benefits to pensioners still based on the "final rank" system and still linked to the current wage in the appropriate rank including all newly added components. The option which has been granted to pensioners, to select the preferred scheme ("average ratio base" or "final rank") would in turn prove to be quite expensive from the pension funds' point of view.

The inclusion of newly insured wage components thus creates an actuarial problem to the funds. Its intensity depends on the timing of the changes. When new wage components are introduced into the insured income, there is a significant jump in the insured income. Since large new groups of employees joined the programs after the reform, there might be a strong effect on the actuarial balance of the funds, but its direction could be determined only after a careful analysis of the age and wage distributions in newly joining groups.

4.2. The Effects of the Timing of the Reform

The "average wage ratios" (career benefit formula) method reflects the life time earnings pattern and is regarded as a better system from a social point of view. The shift from one method (say the final wage base) to the average wage ratios method could affect the actuarial balance of the pension funds, since their obligation to the insured may change.

The change in the actuarial balance depends on the specific wage pattern. It is commonly assumed that the typical pattern follows a trend similar to that described by curve A in Figure 5, which reflects increasing wages during the first years of employment and decreasing wages during the later years (reflecting slower promotion rate in relation to younger workers and reduced premiums and extra time benefits). The exact relationship between the average and the final depends on the slope and duration of growth and the decline of the relative wage. However, under the above wage pattern, the reform would tend to improve the basic pension at retirement, since the average wage will typically be higher than the final wage. This negatively affects the actuarial balance of the pension fund.

The analysis in this paper has shown that insured wages in Israel have actually declined for most employees due to the effects of inflation (i.e., a trend like curve C in Figure 5). Under such conditions the final wage method leads to a much lower pension than the average wage method. Thus, the introduction of the average wage method will lead to deterioration of the actuarial balance of the fund. The funds, however, may be forced into a situation in which every action causes the actuarial balance to deteriorate. The inflationary erosion of the insured wages could not continue indefinitely and at a certain point the decision will be made to include the uninsured wage components in the insured wage. This would cause the wage pattern to experience either a sudden "jump" to a higher level or a gradual increase (see curve D in Figure 5). Such a change would immediately increase the liabilities towards the already retired population (whose pensions are based on the final rank method), as well as towards the active employees, whose pensions will be set at a higher level (while their contributions would be higher only at a certain part of their lifetime). If the fund has been balanced up to that point (meaning that part of the inflationary investment income is used to finance the pensions), it will become imbalanced.

The turning point of the wage curve (point 3 in Figure 5) could be very important for determining the timing of the reform. The funds would clearly prefer an early introduction of the reform prior to this point. In that way, the newly included wage components which will be introduced after the reform would have only a gradual effect on the pensions. If the new wage components will be included in the insured wage prior to the reform, the planners of such a reform would face very severe difficulties in introducing the new method of pension calculation. At such a stage, there will be a resistance of pensioners and of those about to retire against the reform, since they will be unwilling to replace the final wage pensions (which include the added components) by the average wage method.*

*Only one Israeli pension fund was successful in introducing a gradual change. The insured wage has been adjusted gradually to include the previously uninsured components and the pensions were adjusted gradually to that change (the pensions were adjusted by 10% of the newly included components each year and a full adjustment after 10 years). The effects on the actuarial balance depend crucially on the population age structure.

5. CONCLUDING REMARKS

Pension plans must be designed in such a way as to remain balanced even when prices and wages are subject to inflationary conditions. The major problem of funded schemes (and often unfunded schemes as well), is to maintain an adequate level of economic security to pensioners while preserving the actuarial balance of the program. The literature often tends to emphasize only one aspect of the problem, i.e., how to maintain the adequacy of benefits, or how to protect the actuarial balance. These two problems are, in fact, two sides of the same coin and should therefore be handled simultaneously.

A major problem of pension funds, even when balanced at a zero inflation rate, is the protection of the value of their investments in times of inflation. This problem has been partially solved in Israel through the availability of indexed government bonds, which serve as the main investment media for pension schemes. This gives the researcher the opportunity to examine the relationships among the wage pattern, the insured wage and the pension benefits under inflation, while disregarding the investment aspects. The Israeli experience shows that there are many problems which are not yet solved by indexing. Indexing by the CPI helps to maintain the purchasing power of pensions, but pensioners lag behind the increasing standard of living in the economy (due to increases of the real wage).*

The data show the drastic inflationary erosion of the insured wage due to the partial compensation offered by the cost of living allowance, due to exclusion of certain wage components from the insured income, and due to the shift in wage scales.

In the Israeli environment there has been an attempt to handle the problem through indexation of the pension benefits by the "final rank" method. The analysis in this paper has shown that such a method has often failed even to maintain the real value of the initial pensions (and did not succeed in guaranteeing a growth of real pensions). In times of rapid inflation and especially under the specific Israeli wage structure, a switch from the existing "final wage" base method to a more complex pension base which takes into consideration the relative wage through all the years of employment has been an inevitable move. The reform in the Israeli pension funds introduced such changes and replaced indexation by rank prior to retirement by indexation to the cost of living.

The timing of the reform, too, is of great importance. The reform has to precede the inevitable adjustments of the wage structure. Introducing the reform at a later stage could generate

*The opposite situation, i.e., where real wage decrease over a long period, is less common. Under such circumstances pensioners enjoying indexed benefits will be better off in relation to the working population.

strong reaction against the reform.

Although this analysis has focused on the specific Israeli experience, the situation in other countries suffering from high inflation rates would be similar. Under inflation there is a strong motivation to compensate employees by alternative means other than wages. This causes a decline of insured pensionable income. Such a decline must eventually be stopped. While reforms outside of Israel might take different forms, these effects would eventually be similar.

REFERENCES

1. George Heubeck, "Theoretical and Practical Aspects of the Adjustment of Pensions to Changes in Prices & Wages", International Social Security Association, Acapulco, 8-11, May (1979).

2. Kahane Yehuda, Life Insurance Pension Funds and Retirement Saving Programs - A Handbook for Business and Financial Planning (Hebrew), Ateret Publishing House, P.O. Box 21589, Tel Aviv, Israel, June (1983).

3. Myers Robert J., Indexation of Pensions and Other Benefits, Richard D. Irwin Inc., Homewood, Illinois (1978).

4. International Social Security Association, "Problems of Social Security Under Economic Recession and Inflation", Studies & Research No. 10., Geneva (1979).

5. Howard E. Winklevoss, Pension Mathematics, Richard D. Irwin Inc., Homewood, Illinois (1977).

INDEX

GPSR Compliance

The European Union's (EU) General Product Safety Regulation (GPSR) is a set of rules that requires consumer products to be safe and our obligations to ensure this.

If you have any concerns about our products, you can contact us on ProductSafety@springernature.com

In case Publisher is established outside the EU, the EU authorized representative is:

Springer Nature Customer Service Center GmbH
Europaplatz 3
69115 Heidelberg, Germany

Batch number: 10816188

Printed by Printforce, the Netherlands